土壤污染修复丛书

丛书主编 朱永官

农业土壤-植物系统中硒的行为与调控

朱永官 孙国新 等 著

科学出版社

北京

内容简介

本书以土壤-植物系统为对象，系统阐述了土壤环境中，特别是我国土壤中硒的分布规律及其成因，土壤中硒含量、形态及其有效性，微生物、土壤动物对硒的生物转化及其机制，植物对硒的吸收和同化的分子机制，硒的生物营养强化技术与应用，植物对硒的超积累及其机制，硒与其他元素的交互作用等内容。

本书可供高等院校和科研机构土壤学、环境科学、生态学相关领域的研究生和科研人员参考。

图书在版编目（CIP）数据

农业土壤-植物系统中硒的行为与调控/朱永官等著.—北京：科学出版社，2023.5

（土壤污染修复丛书/朱永官主编）

ISBN 978-7-03-075037-2

Ⅰ.①农… Ⅱ.①朱… Ⅲ.①硒-土壤成分-关系-植物-研究 Ⅳ.①S153.6 ②Q948.12

中国国家版本馆 CIP 数据核字（2023）第 038995 号

责任编辑：王海光 刘 晶/责任校对：郑金红
责任印制：吴兆东/封面设计：东方人华

科学出版社出版
北京东黄城根北街 16 号
邮政编码：100717
http://www.sciencep.com

北京建宏印刷有限公司印刷
科学出版社发行 各地新华书店经销

*

2023 年 5 月第 一 版 开本：787×1092 1/16
2025 年 1 月第二次印刷 印张：14 1/2
字数：344 000

定价：198.00 元
（如有印装质量问题，我社负责调换）

"土壤污染修复丛书"编委会

主　编　朱永官

副主编　冯新斌　潘　波　魏世强

编　委（按姓氏汉语拼音排序）

陈玉成　代群威　段兴武　李楠楠　李廷轩

林超文　刘承帅　刘　东　刘鸿雁　孟　博

秦鱼生　宋　波　孙国新　王　萍　吴　攀

吴永贵　徐　恒　徐小逊　杨金燕　余　江

张乃明　郑子成

特邀审稿专家　林玉锁　刘杏梅

丛 书 序

　　土壤是地球的皮肤，是地球表层生态系统的重要组成部分。除了支撑植物生长，土壤在水质净化和储存、物质循环、污染物消纳、生物多样性保护等方面也具有不可替代的作用。此外，土壤微生物代谢能产生大量具有活性的次生代谢物，这些代谢产物可以用于开发抗菌和抗癌药物。总之，土壤对维持地球生态系统功能和保障人类健康至关重要。

　　长期以来，工业发展、城市化和农业集约化快速发展导致土壤受到不同程度的污染。与大气和水体相比，土壤污染具有隐蔽性、不可逆性和严重的滞后性。土壤污染物主要包括：重金属、放射性物质、工农业生产活动中使用或产生的各类污染物（如农药、多环芳烃和卤化物等）、塑料、人兽药物、个人护理品等。除了种类繁多的化学污染物，具有抗生素耐药性的病原微生物及其携带的致病毒力因子等生物污染物也已成为颇受关注的一类新污染物，土壤则是这类污染物的重要储库。土壤污染通过影响作物产量、食品安全、水体质量等途径影响人类健康，成为各级政府和公众普遍关注的生态环境问题。

　　我国开展土壤污染研究已有五十多年。20世纪60年代初期进行了土壤放射性水平调查，探讨放射性同位素在土壤-植物系统中的行为与污染防治。1967年开始，中国科学院相关研究所进行了除草剂等化学农药对土壤的污染及其解毒研究。60年代后期、70年代初期，陆续开展了以土壤污染物分析方法、土壤元素背景值、污水灌溉调查等为中心的研究工作。随着经济的快速发展，土壤污染问题逐渐为人们所重视。80年代起，许多科研机构和大专院校建立了与土壤环境保护有关的专业，积极开展相关研究，为"六五""七五"期间土壤环境背景值和环境容量等科技攻关任务的顺利开展打下了良好基础。

　　习近平总书记在党的二十大报告中明确指出：中国式现代化是人与自然和谐共生的现代化。必须牢固树立和践行绿水青山就是金山银山的理念，站在人与自然和谐共生的高度谋划发展。

　　土壤环境保护已经成为深入打好污染防治攻坚战的重要内容。为有效遏制土壤污染，保障生态系统和人类健康，我们必须遵循"源头控污-过程减污-末端治污"一体化的土壤污染控制与修复的系统思维。

　　由于全国各地地理、气候等各种生态环境特征不同，土壤污染成因、污染类型、修复技术及方法均具有明显的地域特色，研究成果也颇为丰富，但多年来只是零散地发表在国内外刊物上，尚未进行系统性总结。在这样的背景下，科学出版社组织策划的"土壤污染修复丛书"应运而生。丛书全面、系统地总结了土壤污染修复的研究进展，在前沿性、科学性、实用性等方面都具有突出的优势，可为土壤污染修复领域的后续研究提供可靠、系统、规范的科学数据，也可为进一步的深化研究和产业创新应用提供指引。

　　从内容来看，丛书主要包括土壤污染过程、土壤污染修复、土壤环境风险等多个方面，从土壤污染的基础理论到污染修复材料的制备，再到环境污染的风险控制，乃至未来土壤健康的延伸，读者都能在丛书中获得一些启示。尽管如此，从地域来看，丛书暂时并不涵盖我国大部分区域，而是从西南部的相关研究成果出发，抓住特色，随着丛书相关研究的进展逐渐面向全国。

丛书的编委，以及各分册作者都是在领域内精耕细作多年的资深学者，他们对土壤修复的认识都是深刻、活跃且经过时间沉淀的，其成果具有较强的代表性，相信能为土壤污染修复研究提供有价值的参考。

与当前日新月异、百花齐放的学术研究环境异曲同工，"土壤污染修复丛书"的推进也是动态的、开放的，旨在通过系统、精炼的内容，向读者展示土壤修复领域的重点研究成果，希望这套丛书能为我国打赢污染防治攻坚战、实施生态文明建设战略、实现从科技大国走向科技强国的转变添砖加瓦。

朱永官　中国科学院院士

2023 年 4 月

前　言

硒是细菌、古菌、藻类及动物必需的微量元素，而在真菌和植物中为非必需，可能后者在进化过程中已经遗失了代谢硒的能力。与其他微量元素相比，硒拥有最小的缺乏（<40 μg/d）和毒害（>400 μg/d）浓度范围，二者只相差一个数量级。人体中硒缺乏与土壤中硒含量低的趋势一致，目前全世界超过10亿人患有因硒缺乏导致的各种疾病。

硒虽然为植物非必需元素，但作为有益元素可促进植物生长，提高植物抗氧化活性。不同的植物，吸附和代谢硒的能力存在显著差异，有些偏好富硒土壤，进化为硒超积累植物。植物对硒的积累、代谢和耐受能力差异显著，从而能够影响人体、动物健康。植物中硒浓度和形态与食用人群健康状况显著相关，因为人体直接从植物获得硒或间接从食草动物获得硒。植物体内硒浓度和形态可通过传统育种方法或基因操作的方式提高。硒浓度和形态不仅与植物基因型有关，还与土壤中硒浓度、形态和有效性相关，多种生物与非生物因素影响植物体内硒的积累。植物体内硒浓度也可通过生物强化的方式提高。哪些选择压力会促进或阻碍植物体内硒的积累？植物体内硒如何影响其与食草动物、传粉昆虫、微生物及其他植物的相互关系？土壤硒如何影响植物与生态进化？土壤与植物中的硒含量和形态如何影响食物链、生态系统及全球硒循环？本书作者及其团队成员围绕上述问题进行深入研究，取得了一系列研究成果，经系统整理、归纳、总结形成本书。

本书以土壤-植物系统为对象，系统阐述了土壤环境中，特别是我国土壤中硒的分布规律及其成因，土壤中硒含量、形态及其有效性，微生物、土壤动物对硒的生物转化及其机制，植物对硒的吸收和同化的分子机制，硒的生物强化技术与应用，植物硒的超积累及其机制，硒与其他元素交互作用与相关关系等内容。最后，本书还对农业环境硒研究的前沿做了简要概述，希望与读者一起探讨当前的研究空白和未来需加强研究的领域。

本书各章撰写分工如下。第一章硒元素概述，由孙国新、郑瑞伦、朱永官撰写；第二章土壤中硒的含量、形态及其有效性，由梁东丽、周菲、齐明星、程楠、陈萍、马元哲、樊尧撰写；第三章土壤中硒的生物转化，由李刚、韩瑞霞撰写；第四章植物硒吸收和同化的分子机制，由黄新元、陈杰、赵方杰撰写；第五章硒生物营养强化技术与应用，由宋佳平、刘晓航、王张民、袁林喜、牛珊珊、王晓虎、尹雪斌撰写；第六章植物硒的超积累及其机制，由王子威、涂书新撰写；第七章硒与重金属的交互作用，由万亚男、王琪、李花粉撰写；第八章农业环境中硒研究的前沿与展望，由朱永官、孙国新撰写。在本书成稿过程中，中国科学院南京土壤研究所陈怀满研究员给予了许多具体的指导和帮助，并参与了全书的审阅和修改工作，在此表示衷心感谢。

本书由中科环境科技研究院（嘉兴）有限公司院士工作站资助出版。

由于著者水平有限，书中难免存在不足之处，敬请读者批评指正。

朱永官
2022 年 9 月 15 日

目 录

第一章 硒元素概述 ... 1
第一节 土壤-植物系统中硒的重要性 ... 1
一、硒与土壤质量、土壤健康和土壤安全 ... 1
二、硒与植物健康和农产品安全 ... 3
三、硒与人体健康 ... 4
第二节 硒的基本性质 ... 5
一、硒的物理化学性质 ... 5
二、硒的地球化学特征 ... 8
三、土壤中硒的来源及分布 ... 20
四、硒的生物化学 ... 26
第三节 土壤-植物系统中硒的研究现状与问题 ... 30
一、研究现状 ... 30
二、研究中存在的问题 ... 32
参考文献 ... 33

第二章 土壤中硒的含量、形态及其有效性 ... 41
第一节 土壤中硒的来源及其含量 ... 41
一、土壤中硒的来源 ... 41
二、世界土壤硒含量 ... 42
三、中国土壤硒含量 ... 43
第二节 土壤中硒的形态 ... 44
一、土壤中硒的价态 ... 44
二、土壤中硒的形态分级 ... 45
第三节 土壤中硒的化学转化过程 ... 46
一、溶解-沉淀 ... 47
二、氧化-还原 ... 47
三、配位作用 ... 48
四、吸附-解吸 ... 48
五、挥发作用 ... 50
第四节 土壤硒的有效性及其影响因素 ... 50
一、土壤中硒的有效性及其表征 ... 50
二、影响土壤硒有效性的因素 ... 54

第五节　土壤硒有效性的调控 …… 62
第六节　结论 …… 62
参考文献 …… 63

第三章　土壤中硒的生物转化 …… 77
第一节　微生物对硒的吸收转运 …… 77
第二节　硒的微生物还原 …… 78
一、硒的同化还原 …… 78
二、硒酸盐的异化还原 …… 79
三、亚硒酸盐的异化还原 …… 81
四、硒的生物成矿 …… 86

第三节　硒的微生物氧化 …… 88
第四节　硒的微生物甲基化 …… 88
一、硒的甲基化过程 …… 89
二、土壤中硒的甲基化 …… 90
三、植物根际硒的甲基化 …… 91

第五节　土壤动物对硒的转化 …… 91
一、蚯蚓对硒的转化 …… 92
二、线虫对硒的转化 …… 92

第六节　结论 …… 93
参考文献 …… 93

第四章　植物硒吸收和同化的分子机制 …… 103
第一节　植物根系对硒的吸收 …… 104
一、植物根系对硒酸盐的吸收 …… 104
二、植物根系对亚硒酸盐的吸收 …… 107
三、植物根系对有机态硒的吸收 …… 108

第二节　硒在植物中的迁移与转运 …… 109
一、硒酸盐在植物中的迁移与转运 …… 109
二、亚硒酸盐在植物中的迁移与转运 …… 110
三、有机态硒在植物中的迁移与转运 …… 111

第三节　植物叶片吸收转运硒的过程 …… 112
第四节　植物体内硒的同化与代谢 …… 114
参考文献 …… 117

第五章　硒生物营养强化技术与应用 …… 125
第一节　硒生物营养强化的理论基础 …… 125
一、土壤中的硒 …… 125
二、植物对硒的吸收、转运与代谢 …… 128

第二节　硒生物营养强化的实用技术 132
　　　一、土壤强化 132
　　　二、叶面强化 135
　　　三、天然富硒区自然富硒 136
　　　四、传统生物育种 137
　　　五、基因工程 139
　　第三节　硒生物营养强化案例研究 140
　　　一、芬兰的硒生物营养强化 140
　　　二、中国宁夏硒砂瓜的硒生物营养强化 142
　　　三、天然硒生物营养强化国际合作计划 143
　　第四节　硒生物营养强化发展展望 144
　　参考文献 145

第六章　植物硒的超积累及其机制 151
　　第一节　概述 151
　　第二节　硒超积累植物的发现和分类 152
　　　一、硒超积累植物的发现和特点 152
　　　二、植物超积累硒的一般假说 153
　　　三、硒超积累植物的种类 155
　　第三节　植物硒的超积累及其影响机制 155
　　　一、土壤中硒的含量与形态对植物硒积累的影响 155
　　　二、硒超积累植物的吸收、转化与分配机制 160
　　　三、植物解硒毒与硒的超积累 165
　　　四、微生物与植物硒超积累 166
　　　五、硒代谢与植物硒超积累 169
　　第四节　植物硒超积累的生态效应 173
　　　一、富硒土壤上植物群落组成和植被特征 173
　　　二、硒超积累植物对硒循环的影响 174
　　　三、其他物种对硒超积累植物的反应 174
　　第五节　硒污染土壤的植物修复及案例研究 175
　　　一、硒超积累植物的污染修复应用 175
　　　二、硒污染土壤修复的典型案例研究 178
　　　三、超积累植物硒生物强化与人类健康 179
　　参考文献 181

第七章　硒与重金属的交互作用 191
　　第一节　硒对镉、铅吸收和累积的影响 191
　　　一、硒与镉铅的拮抗作用 191

二、硒与镉的协同作用 ·· 199
第二节　硒对砷铬吸收和累积的影响 ··· 200
第三节　硒对汞吸收和累积的影响 ··· 202
一、硒对水稻吸收和累积汞的影响 ·· 202
二、硒对其他植物吸收和累积汞的影响 ··· 204
第四节　重金属对硒吸收和累积的影响 ·· 205
第五节　硒对重金属胁迫的缓解作用 ··· 206
一、硒的抗氧化作用 ··· 207
二、硒对细胞的修复作用 ·· 208
三、硒与植物螯合肽 ··· 208
四、硒与重金属复合物的形成 ··· 209
参考文献 ··· 209

第八章　农业环境中硒研究的前沿和展望 ··· 217
第一节　土壤化学过程 ·· 217
第二节　土壤生物学过程 ··· 217
第三节　植物对硒的吸收、转运及积累机制 ··· 218
第四节　植物内生菌、叶际微生物对硒积累的作用及机制 ···················· 218

第一章 硒元素概述

第一节 土壤-植物系统中硒的重要性

一、硒与土壤质量、土壤健康和土壤安全

硒（Se）是人和动物所必需的微量元素（FAO and WHO，2001）。硒对人体的重要性最早由 Schwarz 和 Foltz（1957）发现。1973 年，研究证实了硒对谷胱甘肽过氧化物酶的作用（Rotruck et al.，1973）。同年，世界卫生组织（WHO）正式将硒列入人体必需的微量元素，确认其具有清除过氧化物、防止细胞损伤、延缓细胞衰老等作用。

全球大部分地区环境中硒含量处于偏低水平。我国的农业依据 400 mm 等降水量线，总体呈现为"东耕西牧、南稻北麦"，虽然目前已不完全如此，但总趋势仍如此。从东北到西南的低硒带处在东耕地区，即农田区。农产品中硒的含量与土壤硒及其有效性关系密切，一般情况下土壤硒水平偏低或者其有效性低导致植物可食部分硒含量偏低，造成粮食中硒含量缺乏。除了藻类等生长需要硒元素外（Yokota et al.，1988），其他高等植物生长过程是否需要硒目前仍存疑。虽然硒可能不是植物生长的必需元素，但它是一种有益元素，土壤中硒浓度高可保障食品或饲料中有足量的硒，满足人体和动物的需要。在动植物组织中，硒主要与蛋白质结合形成含硒蛋白，因此蛋白质含量高的肉类及海产品中硒含量也比较高。人体中的硒主要来源于肉类、海产品及粮食作物。粮食作物虽然硒含量较低，但消费量大，因此植物硒也是人体摄入硒的主要来源之一（图 1-1）。

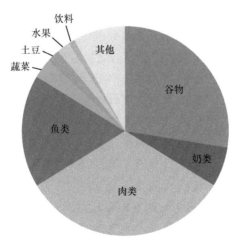

图 1-1 英国成人饮食中硒的主要来源（Michalke，2018）

土地是民生之本。我国人均耕地面积远低于世界平均水平，因此在确保耕地面积的同时，更要关心耕地的质量。长期以来，由于我国重视氮、磷和钾等大量元素而对中、

微量元素的重视程度不足,不平衡施肥现象普遍,造成农田土壤退化,土壤所含中、微量元素的缺乏日趋严重,土壤质量下降明显。包括硒在内的微量元素虽然不是所有植物都必需的,但对某些植物是必需的,或者对大多数植物是有益的或有用的。植物中的硒等微量元素不足,常常引起人畜中、微量元素的缺乏,导致由中、微量元素不足引起的地方性疾病屡有发生,如家禽缺硒造成的白肌病、消瘦症、胸水肿、胰腺萎缩及羽毛发育不良等。

食物是人体中硒的主要来源,而土壤是粮食作物中硒的主要来源。人体中缺乏硒而形成的大骨节病、克山病等地方病,主要原因是农田土壤中硒含量过低或者植物对硒的吸收能力差(Johnson et al., 2000; Fernandez-Martinez and Charlet, 2009)。据报道,世界范围内,每7个人中就有一个人缺乏硒(Combs, 2001),而硒缺乏的主要原因是农田硒浓度低或生物有效性较低。我国中低产田面积占耕地总面积的2/3,其中大部分存在中、微量元素缺乏的问题,这一缺素问题成为影响耕地质量的重要因素。合理施用中、微量元素肥不仅可提高产量,而且可明显提高农产品的品质。针对我国土壤退化、主要农区耕地质量下降、严重制约土地产出能力的状况,"十五"期间,农业部组织实施了以耕地质量建设为重点的"沃土工程",并将其作为一项基本国策和长期任务。而测土配方施肥(国际上通称为"平衡施肥")是"沃土工程"建设的关键技术。各地普遍采用的是五项基础化验,即碱解氮、速效磷、速效钾、有机质和pH。然而,即使是测土工程,目前也还没有涉及硒等中、微量元素,离真正的"平衡施肥"还有很大的距离。我国土壤硒含量严重不均,西北和东南地区较高,而中部地区(从东北到西南地区)是典型的低硒带。植物可食部分硒含量低下造成我国约70%以上地区处于不同程度缺硒状态,其中约有30%为严重缺硒地区(Sun et al., 2017)。继高产农业、绿色农业之后,目前我国农业发展开始进入第三个阶段,即功能农业阶段。在人们越来越关注膳食营养的今天,通过日常饮食来提高硒摄入量被认为是一种安全可行的途径,因而有关如何提高作物硒含量的研究近年来受到更多的关注。

成土母质是影响土壤硒含量的一个重要因素,结晶基岩和第四纪沉积物是地球土壤的主要母质,其中沉积岩占75%,变质岩和火成岩占25%(Systra, 2010)。我国土壤由77.3%的沉积岩风化而成。除个别岩石如页岩含硒较高外,其他岩石硒含量较低,世界土壤平均硒浓度为0.05~0.09 mg/kg,我国土壤平均硒浓度为0.058 mg/kg。我国土壤硒浓度分布和母质分布完全不一致,说明土壤母质不是我国低硒带的主要成因,很可能与影响土壤硒累积和损失的气候及生物等因素有关(Sun et al., 2016)。其中,土壤硒损失主要是硒挥发去除,使土壤中的硒含量减少。但目前农田系统硒挥发通量,以及相关微生物生态、功能基因多样性等仍未见报道。探索农田生态系统硒挥发通量及不同植物和微生物对硒挥发的贡献,揭示土壤理化性质、当地气候条件及地质条件对硒挥发损失的影响,可以为通过调控农业措施减少土壤硒挥发及提高农作物硒含量提供理论基础,对改善我国人口微量元素硒摄入量、提高人体健康、控制由于隐性饥饿造成的地方病都有重要意义。

二、硒与植物健康和农产品安全

植物中硒的功能已有大量研究，一些证据表明藻类生长需要硒，但其他植物中硒的作用仍不明确。硒不是维管植物生长所必需的微量元素，但作为植物的有益元素，适量的硒可以缓解植物的氧化胁迫，促进植物的生长和发育，提高作物产量和品质（Pilon-Smits and LeDuc，2009；White and Brown，2010）。除了个别硒富集植物外，硒的生态学意义目前仍不清晰。硒富集植物，特别是硒超积累植物可在体内积累大量的硒而产生毒性和威慑力（元素防御），从而抵御细胞破坏性昆虫或草食性动物的取食（El Mehdawi et al.，2012）。

农田土壤中适量硒的添加或植物地上部喷施适量浓度的硒都能够显著降低水稻、小麦、蔬菜等多种作物中重金属的积累（陈松灿等，2014）。据报道，硒能够与镉、铅、铬、汞等多种重金属元素产生拮抗效应，包括降低土壤中重金属的生物有效性及其向地上部的转运能力，从而降低植物地上部或可食部分对重金属的累积，缓解重金属对植物的毒害。另外，硒在植物体内主要以有机态形式存在，包括硒代甲硫氨酸、硒代半胱氨酸等，硒代半胱氨酸是谷胱甘肽过氧化物酶的活性中心。谷胱甘肽过氧化物酶具有清除体内自由基、保护细胞膜的功能，因此硒可通过有效调节过氧化物酶等抗氧化酶的形成与活性减少植物体内重金属产生的自由基造成的损伤，从而降低重金属的对植物的毒害作用（陈松灿等，2014）。

我国农业发展可概略地分为三个阶段。

第一阶段，通过土壤改良、化肥和农药的大量施用及良种的研发，提高农作物产量。这是农业发展的第一阶段，即高产农业。高产农业在今后一段时间仍将持续。

第二阶段，即绿色农业阶段。绿色农业以农产品的生态安全为诉求，强调对农产品生产过程的投入严加管理，控制有害物质的引入，包括重（类）金属等。硒可阻控多种重金属在作物中的积累。广义上的绿色农业包含无公害、绿色和有机农业，狭义上指按照绿色标准生产的绿色农业。绿色农业在中国的发展至今已有 20 多年。截至 2013 年，我国很多地区的绿色农业认证面积已超过 50%，总量上已基本能够满足我国人民需求，保障生产过程绿色无污染，但是"质"的问题仍需提高。

第三阶段，即功能农业阶段。功能农业是通过生物营养强化技术或其他生物工程生产出具有健康改善功能的农产品。简单地说，功能农业就是要种植出具有保健功能的农产品，以解决一些营养元素过量和不足的问题。粮食中微量营养素缺乏和失衡，可导致营养不良，并由此引发各种慢性疾病，包括克山病、大骨节病等。

隐性饥饿（hidden hunger）是指机体由于营养不平衡或者缺乏某种维生素及人体必需的矿物质，同时又存在其他营养成分过度摄入，从而产生隐蔽性营养需求的饥饿症状。目前，全球约有 25 亿人在遭受不同程度的隐性饥饿，其中约 10 亿人遭受硒缺乏导致的隐性饥饿。我国隐性饥饿的人口数量约 3 亿（Biesalski，2013），其中硒、锌、铁、钙、碘缺乏造成的隐性饥饿在我国比较普遍。特别是我国有 70% 以上的耕地处于不同程度的缺硒状态，其中约 30% 的耕地严重缺硒（Sun et al.，2017）。功能农业的主要作用和发展使命就是以日常食物来补充这些微量元素的不足，解决隐性饥饿问题，开发出更有效、安全性更高的功能性农产品，助力我国农业高质量发展。

三、硒与人体健康

硒对人体健康具有双重作用,既是人体必需的微量元素(缺乏会产生健康风险),但摄入过量又会导致中毒,只有在某一合理浓度区间才有益(图1-2)。硒缺乏可导致心脏疾病、克山病、大骨节病、甲状腺机能减退及免疫功能低下等多种疾病。硒含量过高会造成人和动物中毒。在合适的含量范围内,硒化合物具有抗癌作用。在一项长期的双盲研究中,补硒与肺癌、结肠癌和前列腺癌发病率的明显下降有关。曾有膳食营养素参考摄入量表明(Fairweather-Tait et al., 2011),硒缺乏和中毒之间的摄入量范围相对较窄,估计摄入量低于 30 μg/d 是不足的,而超过 900 μg/d 是潜在的有害量。

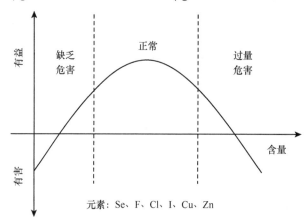

图 1-2 生命健康有益/有害微量元素剂量-效应曲线(王学求等,2021)

目前已有研究证实人体中的硒有多种作用(图1-3),包括:清除自由基,提高机体抗氧化能力;改善免疫功能,提高抗癌能力;抑制肿瘤血管形成,预防肿瘤生长、转移;保护肝脏,阻抗人群中病毒性肝炎的传播;通过抗氧化功能防治心、脑、肾血管疾病;具有拟胰岛素样作用及与胰岛素有协同作用,提高糖尿病的治愈率;改善肺功能,降低哮喘发病率;调节甲状腺激素的代谢平衡,改善甲状腺功能;促进睾酮合成,提高精子活动能力及维持雄性生殖力(Preedy, 2015)。

1973年,世界卫生组织(WHO)宣布硒(Se)是人生命中必需的微量元素,被医药界称为"生命的火种",享有"长寿元素""防癌抗癌之王""心脏守护神""自然解毒剂"等佳誉。

1983～1996年,美国亚利桑那大学亚利桑那癌症中心流行病研究组主任 Larry C. Clark 教授等在美国进行了为期 13 年的补硒双盲干预试验(Clark et al., 1996)。这项试验在 7 家临床中心进行,受试者为 1312 名有非黑色素皮肤癌病史的患者。试验将 1312 名患者随机分为两组,其中一组患者(653名)每天补硒。结果发现,每日补充 200 μg 硒,总癌的发病率和死亡率分别降低了 37% 和 50%,其中发病率下降最为明显的 3 种癌症为前列腺癌(63%)、结直肠癌(58%)、肺癌(46%)。此研究被认为是硒防癌研究的"里程碑",引起国际社会强烈关注硒对健康的重要性。

2003年9月,美国食品药品监督管理局(FDA)通过一项重大决议,确认硒是抑癌剂。FDA 在美国做出以下合法声明:①硒能降低患癌危险,科学证据表明,摄入硒可降

图 1-3 硒对人体健康的有益作用（Preedy，2015）

低患某些癌症的风险率；②硒在人体内具有抗癌作用，一些证据表明摄入硒可在人体内产生抗癌变作用。

硒在土壤健康、食品安全、人体健康等方面有着不可或缺的作用，因而对其基本性质等方面的深刻理解显得十分重要。

第二节 硒的基本性质

一、硒的物理化学性质

硒（selenium），化学符号为 Se，在 1817 年由瑞典化学家琼斯·雅可比·贝采里乌斯（Jons Jakob Berzelius，1779—1848）采用黄铁矿制取硫酸时发现，是典型的稀有分散元素（在地壳中平均含量低，一般为 $10^{-9} \sim 10^{-6}$ 级，而又十分分散的元素）。这种元素起初被认为是碲（tellurium），tellurium 来自拉丁语 tellus，原意为地球（earth）。由于此物质在化学上与碲的性质相似，遂定名为硒（selenium），selenium 来自拉丁语 selene，原意为月亮（moon）（Schilling，2010）。硒是氧族元素之一，位于元素周期表第四周期，第ⅥA族，在硫和碲之间的第 16 组，原子序数为 34，相对原子量为 78.963，离子半径 1.98 Å，电负性 2.55，属两性元素（Patai，1987；刘英俊和曹励明，1987；周令治和陈少纯，2008）（表 1-1）。

表 1-1 硒的物理化学性质（刘英俊和曹励明，1987；周令治和陈少纯，2008）

物理化学性质	值
原子序数	34
原子量	78.963 Da

续表

物理化学性质	值
原子体积	16.5 cm^3/mol
电子排布	$[Ar]3d^{10}4s^24p^4$
电子构型	$4s^24p^2$
地球化学价态	-2, 0, +4, +6
原子半径	103 pm
共价半径	116 pm
范德华半径	190 pm
Se^{2-}有效离子半径	198 pm
Se^{4+}有效离子半径	50 pm
Se^{6+}有效离子半径	42 pm
第一电离能	941.0 kJ/mol
第二电离能	2045.0 kJ/mol
第三电离能	2973.7 kJ/mol
还原电位（$Se \rightarrow Se^{2+}$）	-0.78 V
离子电位（-2）	-1.05 V
电负性	2.55
电子亲和力	-4.21 eV
熔点	494 K，221 mol
沸点	958 K，685 mol
键能 M—M	184.2 kJ/mol
键能 M—H	280.5 kJ/mol
密度	4.81 g/cm^3（20℃）
地壳丰度	0.05 mg/kg

硒具有 6 种稳定的天然同位素：^{74}Se、^{76}Se、^{77}Se、^{78}Se、^{80}Se、^{82}Se，丰度分别为 0.89%、9.36%、7.63%、23.78%、49.61%、8.73%，也有 27 种放射性同位素，如 ^{73}Se、^{75}Se、^{79}Se（Greenwood and Earnshaw，1997）。硒的外层电子构型为 $3d^{10}4s^24p^4$，常见的价态有 +6 价、+4 价、0 价、-2 价，实验室中还观测到 +1 价、+2 价、+3 价、+5 价，因此，硒既具有氧化性又具有还原性。此外，硒还存在很多阴离子（如 Se^-、Se^{2-}、Se^{3-}、Se^{4-}）、多聚阴离子（Se_3^{2-}）、多聚阳离子（Se_4^{2+}、Se_8^{2+}）等（周令治和陈少纯，2008）。单质硒化学性质稳定，只溶于强酸及强碱中，在常温下不与氧作用，但加热能燃烧，产生蓝色的火焰，生成二氧化硒（黄硒矿，SeO_2），散发出一股蒜臭味。干燥硒粉可与格氏试剂、金属有机化合物在一些低极性溶剂中发生剧烈反应，得到相应的金属硒醇盐。硒可与氢、氟、氯、溴及许多金属元素化合，也可以与硝酸和硫酸反应，形成包括硒化物、卤化物、氧卤化物、氧化物、酸类和含氧酸盐等在内的无机态硒化合物，所以硒更多以无机硒化合物的形式存在，形成 170 种固体化合物、3 种液态化合物（Se_2Cl_2、SeF_4、CSe_2）和 2 种气态化合物（H_2Se、SeF_6）（Chizhikov and Shchastlivyi，1968）。硒与金属元素或氢元素形成

化合物时，表现出-2价；与氧元素形成化合物时，则表现为+4价和+6价，属于铜型离子（刘英俊和曹励明，1987）。随着不同 Eh 和 pH 的环境因子的变化，硒元素的价态也随之变化（李家熙等，2000；Tabelin et al.，2018）（详见第二章）。硒与活泼金属生成离子型硒化物，而与其他元素生成共价键型化合物。和硫化物类似，Se^{2-}只存在于强碱性溶液中，中性环境中为 HSe^-，酸性溶液中为 H_2Se。硒与氧化态为+1的金属可生成两种硒化物，即正硒化物（M_2Se）和酸式硒化物（MHSe）。碱金属和碱土金属硒化物的水溶液会使元素硒溶解，生成多硒化合物（M_2Se_n），这与硫可形成多硫化物相似。多硒化物是含 Se_n^{2-} 阴离子的化合物，在水溶液中可以和大的有机阳离子化合物发生复分解反应，得到的大阳离子硒化合物可溶于有机溶剂中：

$$2Na + nSe \longrightarrow Na_2Se_n$$

$$Na_2Se_n + 2\,R_4NCl \longrightarrow (R_4N)_2Se_n + 2\,NaCl$$

除无机硒化合物外，有机硒也是硒元素的主要存在形式。不过硒在有机硒中常以-2价存在，受合成技术限制，到20世纪50年代硒的有机化学才开始快速发展。有机硒包含许多种类，如硒醇、硒醚及其衍生物、硒杂环化合物、含硒氨基酸及其衍生物、硒蛋白、硒核酸、硒多糖及其降解物、硒氨基酸和硒肽、含硒蛋白、含硒甾醇、黄酮类、多酚类等，参与生物体的多种生物化学过程（程水源，2019）。在毒性和稳定性方面，虽然单质硒毒性较小，但硒化物、亚硒酸盐、硒酸盐及氟化硒等毒性较大（王越，2018）。例如，H_2Se 是一种无色、有恶臭气味并且有毒的气体，热稳定性差，溶于水，具有强酸性，且能溶于多种有机溶剂。SeO_2 是白色针状固体，受热易挥发，易溶于水和苯，有较强氧化性，常温下可被空气中的灰尘或痕量有机物还原成单质红硒。H_2SeO_3 是一种无色固体，极易溶于水，弱酸性，除碱金属外，金属亚硒酸盐在水中的溶解度较小。在酸性环境中，SeO_3^{2-} 可被强还原剂或细菌还原成 SeO。SeO_3^{2-} 还能与多种金属氧化物及其氢氧化物形成稳定难溶的配合物。SeO_3^{2-} 是易吸潮的白色固体，热稳定性差，特别是强氧化剂，与有机化合物可发生激烈反应，甚至引起爆炸。SeO_3^{2-} 可与水反应生成 H_2SeO_4。H_2SeO_4 的氧化性远高于 H_2SO_4，能氧化氯化物放出氯气，可与金属化合物反应生成硒酸盐和酸式硒酸盐。

物理性质方面，硒单质是红色或黑色粉末，带灰色金属光泽。硒性脆，温度较高时有塑性。硒是典型的半导体，其导电能力随光照的增强或温度的升高而增强，在光照下其导电性比在黑暗中增加上千倍，这一性质被用于制造复印机和激光打印机中的核心部件"硒鼓"。硒单质有三种同素异形体，存在于三种形式。第一种是无定形硒，也叫"液态"硒，在 40~50℃开始软化，100℃时可以流动，220℃时变成液态，并在 70~210℃时可转化为六方灰硒，无定型硒可分为粉末状硒、玻璃状硒和胶体硒。玻璃状硒为深灰色（接近黑色），硬度与玻璃相同，但较玻璃脆，具有链状结构，微溶于二硫化碳（CS_2），是绝缘体和不良热传导体；粉末状硒为红色；胶体硒颜色由紫到红。第二种是灰色的水晶三角螺旋链形，称为六方晶体硒或灰硒、金属硒，是热力学最稳定的一种形式，不溶于 CS_2，具有良好的导电性、导热性及优良的光导性。第三种是有 Se_8 环的红色水晶单斜晶体，又称红硒，分 α 和 β 两种，其在大于110℃时可转化为六方灰硒（Patai，1987；Hoffman and King，1997；Minaev et al.，2005）。金属硒化物难溶于水，多存在于金属硫化物的矿床中。而硒酸盐溶解度高，多存在于溶液中，仅在极高的氧化还原条件下，固态硒酸盐才大量存在。

二、硒的地球化学特征

（一）硒矿物学

1. 硒矿床类型

硒在地壳中的赋存状态主要有三种：以独立矿物形式存在；以类质同象形式存在于硫化物或硫盐矿物中；以微粒形式存在于非金属矿物中，如与有机物共存硅岩或板岩中（刘英俊和曹励明，1987；温汉捷和肖化云，1998；刘家军等，2020；肖鹏等，2020）。根据硒的工业利用情况，可将硒矿床划分为独立硒矿床和伴生硒矿床两大类；根据成矿作用的不同，可以把硒矿分为内生矿床和外生矿床两大类，内生矿床又分为岩浆岩型矿床、火山沉积型矿床和热液型矿床。根据硒的主要来源又可以将伴生硒矿床划分为以下几种工业类型：斑岩型矿床（铜矿床、铜钼矿床、铅锌矿床、锡石硫化物矿床、硒和锑的金银矿床、含硒沥青铀矿）、岩浆岩型矿床（铜镍硫化物矿床）、矽卡岩型矿床（铜矿床、铁矿床）、海底喷流型矿床（硫化物矿床）、火山沉积型矿床（黄铁矿床）、热液型矿床（铀-汞-钼-钒多金属矿床、铜矿床、铅锌矿床、金矿床）、沉积型独立硒矿床（黑色页岩、炭质硅质岩、煤、磷块岩等矿床）（Sindeeva，1964；Simon and Essene，1996；Simon et al.，1997；刘家军等，2001；Saunders and Brueseke，2012；John and Taylor，2016）。在这些硒矿类型中，岩浆岩型、斑岩型、热液型和沉积型矿床最重要，硒含量约占其总储量的90%（冯彩霞等，2002；彭大明，1997）。

地壳中硒的含量为 0.05~0.09 mg/kg（Taylor and Mclennan，1995；Wedepohl，1995；Rudnick and Gao，2003），因含量很低，主要呈分散状态存在，其作为典型的稀散元素，传统上认为很难形成独立的硒矿。已有研究表明，在岩浆结晶和较高温热液阶段（>300℃），硒能扩大硫的类质同象置换范围，从而更容易进入硫化物晶格内；而在中低温热液阶段，且硫含量较低时，便可形成硒的独立矿物（刘英俊和曹励明，1987）。虽能形成某些独立矿物，但硒与硫、碲在结晶及地球化学方面性质相似，大部分硒与硫元素类质同象分布于方铅矿、黄铜矿、黄铁矿、闪锌矿及辉钼矿等硫化矿床中（周令冶和陈少纯，2008；翟秀静和周亚光，2009）。所以硒极少形成具有工业价值的独立硒矿床，其主要来源于综合性含硒矿床，大部分作为其他矿床的副产物进行回收（彭大明，1997；George，1998）。我国位于湖北恩施的双河渔塘坝硒矿床是独立硒矿床，属于典型的沉积型硒矿床（宋成祖，1989）。西秦岭南亚带的拉尔玛-邛莫金-硒矿床，硒矿化富集的成因为热水沉积-有机质固定；玻利维亚的帕卡哈卡矿床属于典型的热液硒矿床。

从目前已探明的硒矿储量来看，硒矿主要以伴生矿种产出，大多数硒伴生在铜、镍等矿中。斑岩铜矿床和铜钼矿床是含硒综合性矿床的主要成矿类型，次要类型包括含硒多金属矿床、汞硒矿床、含硒砂岩矿床、硒银金矿床、硒钒铀矿床等（彭大明，1997）。以铜矿石的储量计算，全世界硒储量约为12万t。另外，煤可能作为硒的另一个潜在来源，因为世界范围内，煤矿中硒含量相当于铜矿中硒含量的80倍（涂光炽等，2004）。

2. 硒与有机质的关系

硒在地质演化过程中具有明显的、与有机质结合的趋势。全球陆相沉积普遍贫硒，海相沉积黑色页岩硒含量最高,达到 1~277 mg/kg(彭大明,1996)。海洋中有机亲和指数 K_{OA}(海生植物/海水)大于 5000 的元素为生物制约元素,硒的有机亲和指数达到 8900,属于典型的生物制约元素(温汉捷,1999)。硒在某些黑色岩系、煤炭、石油中强烈富集(肖鹏等,2020)。硒含量按含碳质硅质岩型、含硅质碳质页岩型、半暗腐泥煤型递增,其中进入有机碳的硒达到 65.8%,进入硒铁铜矿和含硒黄铁矿的占 33.9%,剩余不到 0.3% 的分布在脉石等中(Weiss et al.,1971;彭大明,1996;周令治和陈少纯,2008;Blazina et al.,2014;Sun et al.,2016)。硒含量与有机碳含量也有显著的正相关关系。例如,在我国兴山白果园银钒矿床中发现的硒银矿、辉硒银矿和富硒硫锗银矿,有机碳含量为 2.8%,硒含量达到 3340 mg/kg,与原岩相比,平均富集系数达到 10.5;西秦岭拉尔玛-邛莫金-硒矿床中发现的硒硫锑矿、硒硫锑铜矿、灰硒铅矿和硒镍矿等,有机碳含量在 0.12%~8.5%,平均高达 2.3%,硒富集系数达到 50.3~64.0,平均为 55.7,矿床中硒的有机结合态占 75%,无机结合态占 25%,不同成矿阶段的有机结合态和无机结合态比例略有不同。湖北恩施渔塘坝硒矿床中发现的硒铜蓝、蓝硒铜矿和方硒铜矿等,有机碳含量高达 19.7%。国外某些高硒的黑色页岩中有机碳含量也很高,例如,西伯利亚黑色页岩中有机碳含量为 10.3%;芬兰 Talvivaara 含碳质硅质岩,有机碳含量为 9.2%。世界煤的硒含量约为 0.2~10 mg/kg。我国煤的平均硒含量高于美国和世界平均值(Ren et al.,1999),主要是由于我国部分石煤中硒含量较高。高硒石煤中硒含量最高可达 1150 mg/kg,为世界罕见。世界各地煤的硒含量如表 1-2 所示。我国煤中硒含量以东北地区内蒙古较低,西南地区较高,山西煤中硒含量也较高。从成煤时代来看,煤层由老到新,煤中硒含量逐渐降低,古生代煤中硒含量明显高于中生代、新生代(表 1-3)。从煤级来看,硒含量有随煤级增高而增加的趋势,无烟煤中硒含量普遍较高(Ren et al.,1999)。

表 1-2　世界各地煤的硒含量(张军营等,1999)

地区	范围/(mg/kg)	算术平均值/(mg/kg)	几何平均值/(mg/kg)	样品数
世界	0.2~10	3.0		
美国	≤150	2.8	1.8	7563
中国	0.12~56.7	6.22	3.64	118
中国山西	<0.8~12.6	4.938		69
中国贵州	0.35~4.38	2.41	2.11	32
中国东北地区、内蒙古东部	0.062~3.506	0.671	0.333	203
中国四川东部燃煤型氟病区		4.477		29
高硒石煤分布区	32~1150			28

表 1-3　中国不同时代煤中硒的含量(张军营等,1999)

时代	平均含量/(mg/kg)	样品数
晚第三系(N)	0.80	5
早第三系(E)	1.63	15
晚侏罗世-白垩纪(J3-k)	0.47	4
早侏罗世-中侏罗世(J1-2)	0.40	15

续表

时代	平均含量/（mg/kg）	样品数
晚三叠世（T3）	1.15	5
晚二叠世（P2）	7.63	47
早二叠世（P1）	4.44	22
晚石炭世（C3）	4.54	23

3. 中国的硒矿

我国是世界主要硒资源国之一，硒蕴藏量占世界硒储量的1/3以上，保有工业储量居世界第四位，仅次于加拿大、美国和比利时（Zhu et al.，2009）。2018年和2019年全球硒产量分别2810 t和2800 t。2019年，中国、日本、德国、比利时、俄罗斯五个国家硒产量总和占全球总产量的83.9%，是全球硒产量的主要来源国家。2019年，中国硒产量为930 t，占全球总产量的33.2%，占比最大；日本硒产量为770 t，占全球总产量的27.5%；德国硒产量为300 t，占全球总产量的10.7%；比利时硒产量为200 t，占全球总产量的7.14%；俄罗斯硒产量为150 t，占全球总产量的5.4%。我国的硒矿产地现已探明的有十几处，在已探明的硒矿中，岩浆岩型铜镍硫矿化物矿床占硒总储量的50%以上，大多是伴生在铜、镍等矿床中，分布上主要集中在我国的西北部和长江中下游。陈炳翰等（2020）在前人工作的基础上进行研究，结果表明我国硒矿资源主要集中在甘肃、江西、新疆，并将我国硒矿类型划分为斑岩型、岩浆型、热液型、矽卡岩型、沉积型硒矿五大类型，其中以热液型、矽卡岩型硒矿最为重要（表1-4）；将中国硒矿划分为24个成矿带，主要分布在华北板块边缘、扬子板块边缘、喀喇昆仑-三江成矿带等地（表1-5）；构建了中国硒矿的成矿谱系（图1-4），划分出39个与硒矿成矿作用有关的成矿系列（表1-6）。由成矿谱系图（图1-4）可以看出，空间上，北方地区以华北和东北地区的金矿为主，中部地区集中在华北板块南缘的小秦岭-熊耳山地区的金矿，南方地区集中在长江中下游和滇川黔地区的矽卡岩型铜矿。湖北恩施渔塘坝硒矿是目前我国发现的唯一的小型独立硒矿床，硒的质量分数高达$8.59×10^{-3}$，且易探、易采，属于沉积再造型层控矿床。下二叠统的茅口组灰岩顶部含炭硅质岩是渔塘坝硒矿床的主要富硒层位。硒矿化层与碳酸盐台地海盆浅部沼泽环境的黑色碳质硅质岩相带密切相关（冯彩霞等，2002）。

在成矿时间方面，硒矿赋存地层的时代和硒矿伴生的主矿种的形成时间决定了硒矿的成矿时代。硒矿的成矿时代大致可以划分成元古代、早古生代、晚古生代、中生代、新生代（表1-7，图1-4），大部分矿床形成时代集中在古生代和中生代（陈炳翰等，2020）。其中，早古生代（主要在寒武纪）形成的硒矿矿床类型有沉积型和热液型，分布在扬子板块边缘和内陆、阿尔金-祁连造山带、伊犁板块内陆；晚古生代成矿期形成的硒矿矿床类型有岩浆型、沉积型和热液型，广泛分布在伊犁地块、扬子板块、阿尔金-祁连造山带、华北陆块、巴颜喀拉-松潘造山带、大兴安岭成矿省、准噶尔成矿省；中生代成矿期的矿床类型有矽卡岩型和热液型，分布在秦岭-大别造山带西缘、华北陆块边缘、华南板块边缘、喀喇昆仑-三江造山带和扬子板块边缘。元古代和新生代成矿期形成的硒矿分布较为集中。元古代成矿期形成的硒矿分布在华北陆块和扬子成矿省，矿床类型有岩浆型、热液型和沉积型。新生代成矿期则集中在喜马拉雅期，矿床类型有热液型和岩浆

表1-4 中国富硒矿床预测类型划分表

硒矿预测类型	类型	典型矿床					
		西北区	华北区	东北区	西南区	中南区	华东区
与岩浆分异铜镍硫化物有关	岩浆型	华阳川、黑山、金川	小南山	赤柏松	白马寨		平水
区别于典型矽卡岩型、斑岩型的其他热液型矿床	热液型	金堆城、拉尔玛、达坂城、德尔尼、九个泉、花沟、祁连山-折腰山、可乃克、火焰山、东流水	高尔旗		翁子、盆兰、里伍、小明山、咚寨	铜山岭、方家屯、巷子口、徐家山、下高跃	银山、张十八、寺门口、金山、木瓜岭
主矿种类型为矽卡岩型矿床	矽卡岩型	铜峪沟			烂头山	鸡冠咀、李家湾、康家湾、七宝山、大宝山	城门山、狮子山
主要为沉积型矿床	沉积型					双河鱼塘坝、杆子坪-汪家寨、大坪、晓坪、郑家湾、白果园	富家坞
斑岩矿床	斑岩型		乌鲁格吐山、铜矿峪	多宝山	马拉松多		

表 1-5 中国硒矿成矿带划分表

成矿带编号	成矿带名称	所属III级区带编号	主要成矿类型	主要成矿时代	代表性矿床
Se1	准噶尔	III-6	热液型	中二叠世	达坂城
Se2	伊犁	III-11-2	热液型	奥陶系、早泥盆世	可可乃克、彩花沟
Se3	海拉尔	III-47-3	斑岩型	燕山早期（177 Ma）	乌奴格吐山
Se4	东乌珠穆沁旗-嫩江	III-48-1、III-48-2	斑岩型、热液型	奥陶纪（475.1±5.1 Ma）、早二叠世	多宝山、高尔旗
Se5	华北陆块北缘西段	III-58-1	岩浆型	中二叠世（272.7±2.9 Ma）	小南山
Se6	辽东	III-56-2	岩浆型	晚古生代和早中生代	赤柏松
Se7	南祁连	III-23	岩浆型	晚泥盆世（374.6 Ma±5.2 Ma）	黑山
Se8	阿拉善	III-18	岩浆型	中-新元古代	金川
Se9	北祁连	III-21	热液型	中、下奥陶统	九个泉、祁连山、火焰山-折腰山
Se10	西秦岭	III-28-2	热液型、斑岩型	燕山期（127 Ma）、古元古代（2117±13 Ma）	拉尔玛、铜岭沟
Se11	巴颜喀拉	III-29	热液型	石炭纪	德尔尼
Se12	五台-太行	III-61-3	热液型	古元古代	东流水
Se13	小秦岭-熊耳山	III-63	热液型、矽卡岩型、斑岩型	下古炭统、燕山期、印支期	桃园、金堆城、大石沟、华阳川、银水寺、铜矿岭
Se14	昌都	III-36-1	斑岩型	喜马拉雅期（36.9 Ma）	马拉松多
Se15	义敦-香格里拉	III-32-2	热液型	燕山期	里伍
Se16	上扬子中东部	III-77-1	沉积型、斑岩型	寒武纪、雪峰期	杆子坪-汪家寨、白果园、双河、雪花村、猫子坪、凉风垭、猫子坪、下高跃、客寨
Se17	江南古陆西段	III-78	沉积型、热液型	寒武纪	大坪-晓坪、郑家湾、方家屯
Se18	长江中下游	III-69-1、III-69-4、III-69-5	矽卡岩型、热液型	侏罗纪、早白垩世、加里东期、燕山期	李家湾、巷子口、鸡冠咀、徐家山、老鸦岭、铁山头、仙人冲、狮子山、寺门口、张十八、穿山洞、城门山
Se19	江南隆起东段	III-70	矽卡岩型	燕山期	七宝山、康家湾、平水、银山、富家坞
Se20	武功山-杭州湾	III-71	热液型、矽卡岩型、斑岩型	中晚罗世、燕山期、寒武纪	金山、木瓜岭
Se21	昌宁-澜沧	III-38	热液型	早喜山期	翁子
Se22	哀牢山	III-75-3	岩浆型	喜马拉雅期	白马寨
Se23	滇黔桂	III-88	热液型	印支期	益兰、小明山
Se24	南岭	III-83-2、III-83-3	矽卡岩型、热液型	燕山期	大宝山、铜山岭、烂头山

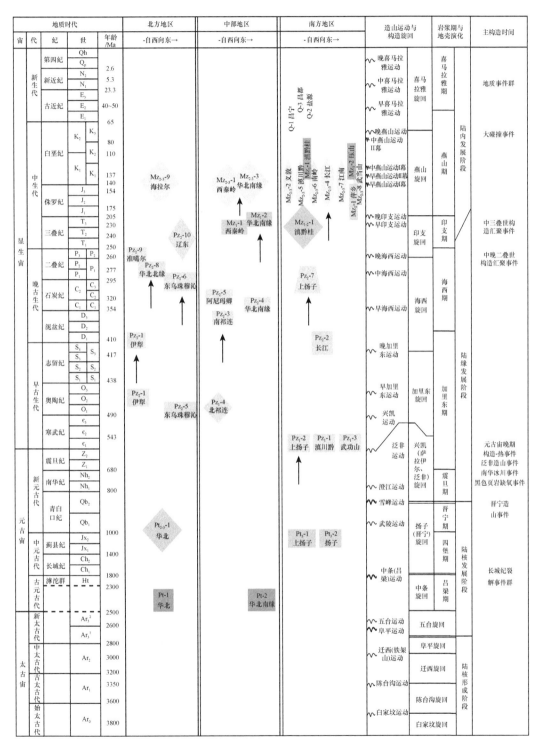

图 1-4 中国硒矿成矿谱系图（陈炳翰等，2020）

表 1-6　与中国硒矿有关的成矿系列简表

序号	时代	编号	成矿系列（组）名称	代表矿床
1	前寒武纪	Pt-1	华北地台元古代与陆核形成阶段有关的铁、铜矿床成矿系列	东流水
2		Pt-2	华北南缘古元古代与变斑岩有关的铜、钴、钼、金矿床成矿系列	铜矿峪
3		Pt_{2-3}-1	华北地台中、新元古代与镁铁-超镁铁质岩浆作用有关的铜、镍、铂、铁矿床成矿系列	金川
4		Pt_3-1	上扬子中东部与新元古代海相化学沉积作用有关的磷、锰、铁矿床成矿系列	下高跃、白果园
5		Pt_3-2	扬子板块新元古代与海相细碧角斑质火山-侵入活动有关的铜、锌矿床成矿系列	平水
6	古生代	Pz_1-1	滇东、川南、黔西与早古生代碳酸盐建造有关的铅、锌、锰矿床成矿系列	郑家湾、大坪-晓坪
7		Pz_1-2	上扬子中东部早古生代与海相化学沉积作用有关的磷、锰、铁矿床成矿系列	杆子坪-汪家寨
8		Pz_1-3	武功山、杭州湾与寒武纪含碳硅泥质岩类有关的铜、铁、铌、钽矿床成矿系列	木瓜岭
9		Pz_1-4	北祁连早古生代与海底火山喷发沉积作用有关的铜、铅、锌、铁矿床成矿系列	祁连山
10		Pz_1-5	东乌珠穆沁-嫩江奥陶纪与海相基性-中酸性岩浆活动有关的铁、铜矿床成矿系列	多宝山
11		Pz_1-6	伊犁微板块南缘与加里东期中基性岩浆作用有关的铁、锰、铅、锌矿床成矿系列	可可乃克
12		Pz_2-1	伊犁微板块南缘早泥盆与中酸性岩浆作用有关的铜、镍、金、铁、锰矿床成矿系列矿带	彩花沟
13		Pz_2-2	长江中下游与加里东旋回陆缘发展阶段有关的铜、金、铁、锌矿床成矿系列	徐家山
14		Pz_2-3	南祁连与海西旋回造山作用陆缘发展阶段有关的金、铜、镍矿床成矿系列	黑山
15		Pz_2-4	华北陆块南缘与海西旋回陆缘发展阶段有关的铁、铜、金矿床成矿系列	桃园
16		Pz_2-5	阿尼玛卿与海西期蛇绿岩建造有关的铜、锌、钴矿床成矿系列	德尔尼
17		Pz_2-6	东乌珠穆沁旗-嫩江海西旋回与基性岩浆侵入作用有关的铜、铅、锌矿床成矿系列	高尔旗
18		Pz_2-7	上扬子中东部与海西旋回海陆过渡相碳酸盐、碎屑岩建造有关的铅、锌、铁、锰矿床成矿系列	猫子山、雪花村、横石溪、杨柳坪、凉风垭、双河
19		Pz_2-8	华北陆块北缘西段与海西旋回镁铁质-超镁铁质岩浆作用有关的金、银、铁、铜、镍、铂矿床成矿系列	小南山
20		Pz_2-9	准噶尔南缘与晚古生代沉积作用有关的铜、锰、磷、钒矿床成矿系列	达坂城
21		Pz_2-10	辽东与海西旋回基性-超基性岩类有关的铜、镍、钴、铂矿床成矿系列	赤柏松

续表

序号	时代	编号	成矿系列（组）名称	代表矿床
22	中生代	Mz_1-1	西秦岭与印支旋回火山喷发-沉积作用有关的铜、钴、锌矿床成矿系列	铜峪沟
23		Mz_1-2	华北陆块南缘与印支旋回陆内发展阶段有关的金、钼、钨、铁矿床成矿系列	大石沟
24		Mz_{1-2}-1	滇黔桂北部与印支旋回、燕山运动陆内发展阶段有关的金、锑、汞、银矿床成矿区	小明山
25		Mz_2-1	萍乡与燕山期中酸性-酸性岩浆侵入-喷发活动有关的铜、钼、钨、锡、铅、锌、银矿床成矿系列	金山
26		Mz_{2-3}-1	西秦岭与燕山期深源岩浆侵入活动有关的金矿矿床成矿系列	拉尔玛
27		Mz_{2-3}-2	义敦、香格里拉与燕山期岩浆侵入活动有关的铅、锌、银、金、铜矿床成矿系列	里伍
28		Mz_{2-3}-3	华北陆块南缘与燕山期浅成-超浅成侵入作用有关的铁、铜、金、钼矿床成矿系列	金堆城
29		Mz_{2-3}-4	长江中下游与燕山期壳幔混源中酸性浅成侵入-喷发活动有关的铜、金、铁、锌矿床成矿系列	巷子口
30		Mz_{2-3}-5	滇川黔与燕山期海陆过渡相碳酸盐、碎屑岩建造有关的铅、锌、铁矿床成矿系列	方家屯
31		Mz_{2-3}-6	南岭与燕山期浅成-超浅成花岗闪长岩类有关的铅、锌、钨、钼、铜矿床成矿系列	大宝山、铜山岭
32		Mz_{2-3}-7	江南隆起东段与燕山期中酸性-酸性岩浆侵入-喷发活动有关的金、银、铅、锌、钨矿床成矿系列	七宝山、张十八
33		Mz_{2-3}-8	武功山、杭州湾与燕山期中浅成花岗岩类有关的铜、铅、锌、银、金、钨矿床成矿系列	康家湾
34		Mz_{2-3}-9	海拉尔燕山早期与中、酸性岩浆侵入有关的铁、铜、钨矿床成矿系列	乌鲁格吐山
35		Mz_3-1	滇黔桂北部与燕山晚期中浅成酸性花岗岩类的金、锑、汞、银、锰矿床成矿系列	益兰
36		Mz_3-2	玉山与燕山晚期陆相火山喷发-超浅成岩浆作用有关的铜、铅、锌、银、金、钨矿床成矿系列	银山
37	新生代	Q-1	昌宁、澜沧与新生代花岗岩浆侵入活动有关的铁、铜、铅、锌、银、锡矿床成矿系列	翁子
38		Q-2	盐源、丽江、金平与新生代构造-岩浆作用有关的金、铜、镍、铁矿床成矿系列	白马寨
39		Q-3	昌都喜马拉雅期与中-中酸性岩浆侵入作用有关的铜、钼、金矿床成矿系列	马拉松多

表 1-7 中国主要硒矿床与成矿时代（陈炳翰，2020）

序号	矿产地	矿种（除硒矿外其他共伴生矿种）	硒矿规模	硒矿类型	成矿时代
1	山西铜矿峪铜钼矿	铜、钼	大型	斑岩型	古元古代（2117±13）Ma
2	内蒙古阿巴嘎旗高尔旗、铅锌多金属矿	银、铅、锌	中型	热液型	早二叠世
3	内蒙古乌鲁格吐山铜钼矿	铜、钼	中型	斑岩型	燕山早期（177 Ma）
4	黑龙江多宝山铜钼矿	铜、钼	大型	斑岩型	奥陶纪（475.1±5.1）Ma
5	吉林赤柏松铜镍矿	铜、镍、碲	中型	岩浆型	晚古生代和早中生代
6	浙江平水铜矿	铜、锌、金、银	中型	岩浆型	青白口纪（主成矿期 899 Ma）/白垩纪（叠加期 93~120 Ma）
7	浙江银山银多金属矿	银、锌、铅、铜、铋、镉、钢	中型	热液型	燕山晚期
8	安徽狮子山铜矿	铜	中型	矽卡岩型	燕山期
9	江西张十八铅锌锑矿	银、铅、锌	大型	热液型	燕山期
10	江西城门山铜矿	铜、银	大型	矽卡岩型	燕山期
11	江西富家坞斑岩铜矿	铜	大型	斑岩型	中侏罗世（170 Ma）
12	江西金山金矿	金	中型	热液型	中侏罗世
13	青海德尔尼铜矿	铜、银	大型	热液型	石炭纪
14	青海铜峪沟铜矿	铜、铅、锌、锡、锗、镓、镉	大型	矽卡岩型	三叠纪
15	青海祁连山铜矿	铜、银	中型	热液型	奥陶纪
16	湖北鸡冠咀铜金矿	铜、金	中型	矽卡岩型	早白垩世
17	湖北徐家山锑矿	锑	中型	热液型	加里东期（323 Ma、348 Ma 和 402 Ma）
18	湖北白果园银钒矿区	银、钒	大型	沉积型	早震旦世
19	湖南天门山大坪-晓坪硒多金属矿	钼、钒、镍、磷、钇	大型	沉积型	寒武纪
20	湖南天门山杆子坪-汪家寨硒多金属矿	钼、钒、镍、磷	大型	沉积型	寒武纪
21	湖南铜山岭银多金属矿	银、镉、铅、锌、铜、铋、碲	中型	热液型	燕山中期
22	湖南七宝山铜多金属矿	铜、金、铁、硫	大型	矽卡岩型	燕山期
23	广东大宝山钨钼铜矿	钨、钼、铜	中型	矽卡岩型	燕山期
24	广西水岩坝烂头山钨矿	钨、砷、碲、铊	中型	矽卡岩型	燕山期（151 Ma）
25	四川里伍铜矿	铜、镉	大型	热液型	燕山期
26	贵州下高跃铅锌矿	铅、锌、银	中型	热液型	雪峰期（840 Ma）
27	西藏马拉松多斑岩型铜矿	铜	大型	斑岩型	喜马拉雅期（36.9 Ma）
28	陕西金堆城钼矿	钼	中型	热液型	燕山期
29	陕西大石沟钼矿	钼	中型	热液型	印支晚期
30	陕西华阳川铀多金属矿	锶、铅、铌、银、稀土	中型	岩浆型	印支期
31	甘肃金川铜镍矿	铜、镍、金、硫、锇、铱、钌、铑、铂、钯、碲	超大型	岩浆型	中-新元古代

续表

序号	矿产地	矿种（除硒矿外其他共伴生矿种）	硒矿规模	硒矿类型	成矿时代
32	甘肃碌曲拉尔玛金矿	汞、金	大型	热液型	燕山期（127 Ma）
33	新疆彩花沟钨铜硫铁矿	钨、铜、硫铁矿、银	中型	热液型	早泥盆世
34	新疆达坂城铜锌矿	铜、锌、银	中型	热液型	中二叠世

型，主要位于喀喇昆仑-三江造山带和扬子板块边缘。不同硒矿类型的成矿时代也有很大差异。岩浆型硒矿主要为铜镍硫化物型矿伴生的硒矿，成矿时代跨度较大，有中-新元古代、晚古生代、早中生代到新生代；矽卡岩型硒矿主要是矽卡岩型铜矿内伴生的硒矿，成矿时间集中在中生代；沉积型硒矿为独立形成的硒矿，主要赋存在新元古代、早古生代、晚古生代的层状、似层状、透镜状有机质硅质岩夹高碳质泥岩、页岩内；热液型硒矿主要为热液型铁矿、铜矿、铅锌矿的伴生硒矿，形成时代跨度大，从元古代、古生代、中生代到新生代；斑岩型硒矿形成时代从古元古代、早古生代、中生代到新生代（表1-5，表1-7）。

（二）硒的表生地球化学

硫与硒在内生作用中的地球化学行为非常相似且紧密共生，形成广泛的类质同象现象，但在外生（表生）作用中，硫与硒的地球化学行为则表现为明显的分离。最根本的原因是二者迁移能量的不同，硫的迁移能远大于硒（涂光炽等，2004）。

在硫化物矿垂直剖面上，自上而下依次分为氧化带、次生富集带、原生矿石带，其中氧化带（相当于渗透带）自上而下又可分为完全氧化亚带（铁帽）、淋滤亚带、次生氧化物富集亚带。从表1-8中可以看出，在厌氧的内生环境中，硒与硫共生；而在好氧的表生环境中，硫比硒更容易氧化而发生分离。在地表氧化介质环境中，硫化物中的硫被氧化为S^{4+}，以亚硫酸盐或硫酸盐形式存在，被地表水溶解进行远距离迁移。而地表硒化物中的硒仅氧化为零价硒，无法被氧化为亚硒酸盐或硒酸盐，即使个别地区如沙漠和半沙漠条件下，硒能够氧化为硒酸盐，也处于不稳定状态，易被重新还原为亚硒酸盐或零价硒。因此，硒酸盐矿物在自然界中无法存在（涂光炽等，2004）。零价硒不溶于水，不易于远距离传输。在次生风化作用中，硒极易转变为四价氧化物的黄硒矿，以及含水或不含水的硒盐矿物。四价硒易被铁氧化物和有机质吸附，特别是在低pH的酸性环境中，存在于铁帽中或以机械夹带存在于胶结带中，既不会被搬迁，也不会发生富集。

表1-8 硫与硒的氧化还原电位

硫反应	电位	硒反应	电位
$S \longrightarrow H_2SO_3$	−0.45 V	$Se \longrightarrow H_2SeO_3$	−0.74 V
$H_2SO_3 \longrightarrow H_2SO_4$	−0.17 V	$H_2SeO_3 \longrightarrow H_2SeO_4$	−1.15 V
$S^{2-} \longrightarrow S^0$	−0.14 V	$Se^{2-} \longrightarrow Se^0$	+0.4 V

氧化还原环境及 pH 影响硒的形态变化和吸附-解吸。地球上的环境变化影响硒的形态变化，特别是海洋中硒的形态变化，而硒形态变化影响硒在海水中的溶解度，导致海水中硒含量升高或降低。因为硒是生命必需元素，海洋中硒含量的变化可影响生命的形式，甚至与生物暴发或灭绝有关（Large et al.，2015）。最近研究表明，地球历史上 5 次物种大灭绝事件中有 3 次与海洋重要微量元素指数下降有关（Long et al.，2016），奥陶纪、泥盆纪和三叠纪时期（分别为约 4.43 亿、3.71 亿和 2.01 亿年前）的 3 次物种大灭绝事件到来之前的那些年里，均出现了海洋硒浓度陡然下降的情况，且只有硒元素显著降低，浓度下降至最低点（1～2 ng/L），不到现代海洋中硒浓度的 1%（155 ng/L）。例如，寒武纪末，海洋中的缺氧环境可能是造成全球三叶虫灭绝的重要因素：在氧气不足的条件下，海水中硒以+4 价或-2 价形式存在，这些低价硒在水中溶解度很差，生物有效性低，由于海洋中的生物生长需摄入硒却不可得，成为导致包括三叶虫在内的海洋生物灭绝的原因之一（Li et al.，2018）。

（三）岩石中的硒

岩石是环境中硒的主要来源，岩石中硒的形态和释放行为控制着土壤中硒的生物有效性。地壳中硒的最初来源与二叠纪的磷酸盐建造、成煤作用、火山喷发、硫化物热解流体的作用、岩浆分异作用及变质作用、地下水水岩作用等密切相关，其中火山喷发被认为是硒元素最重要的天然来源之一。Se^0 和 H_2Se 的沸点分别为 688℃和-42℃，SeO_2 在 300℃以上可以升华，因而硒在火山活跃期以挥发性高温气体放出，使硒在环境中的分布较为复杂、多样（程水源，2019）。岩石中的硒约占地壳中硒总量的 40%，主要分布在砂岩、硅岩和石灰岩中。几乎所有岩石中均存在硒，但一般含量都很低，并与岩石性质密切相关。变质岩、沉积岩和岩浆岩的硒含量分别为 0.031～0.131 mg/kg、0.028～0.118 mg/kg 和 0.059～0.108 mg/kg。岩浆岩中的硒主要为无机硒，其中基性岩中硒含量最高，其次是碱性岩和酸性岩。沉积岩中，页岩、深海碳酸盐岩硒含量高于砂岩和黄土（夏卫平和谭见安，1990）。沉积岩是构成地壳表层的主要岩石，覆盖地球表面约 75% 的面积，而且是农业土壤的主要母质，但由于沉积岩中较难赋存硒（平均硒含量 0.088 mg/kg），因此低硒环境的分布相对于高硒环境更为常见。

研究表明，土壤硒含量的高低与成土母质的岩石类型有很大相关性。王美珠和章明奎（1996）认为如果成土母质是二叠纪和寒武-奥陶纪的硅质页岩以及含碳的硅质页岩，则土壤硒含量高，其次为二叠纪的长兴组灰岩和玄武质火山灰岩等；如果第四纪沉积物如湖沼相沉积物、长江冲积物、黄淮海冲积物等为成土母质，则土壤硒含量低。蔡大为等（2020）的研究表明，贵州省不同成土母岩耕地表层土壤硒元素背景值为黑色页岩（0.61 mg/kg）＞玄武岩及辉绿岩（0.58 mg/kg）＝千枚岩（0.58 mg/kg）＞灰岩（0.54 mg/kg）＞泥、砂、砾（0.53 mg/kg）＞白云岩（0.48 mg/kg）＝泥（页）岩（0.48 mg/kg）＞花岗岩（0.42 mg/kg）＞变余砂（砾）岩（0.40 mg/kg）＞板岩（0.39 mg/kg）＞变余凝灰岩（0.38 mg/kg）＞砂岩（0.37 mg/kg）＞紫红色砂页岩（0.32 mg/kg）。火山岩和沉积岩中的硒含量分别可以高达 120 mg/kg 和 1 mg/kg，均高于地壳丰度（Fishbein，1983）。一般而言，黑色页岩、碳质岩、含碳质硅质岩和煤层等黑色岩系比较容易富集硒。研究显示，硒主要富集在黑色岩系中，并可以形成独立硒矿床，如渔塘坝硒矿床、拉尔玛 Se-Au 矿床和遵义 Ni-Mo-Se 矿床（温汉捷等，2019）；Se 也可以共（伴）生在铅锌矿床和砂岩型铀矿床中，其中铅锌矿床以浅成低温热液型和矽卡岩型中硒含量最高（Cook et al.，2009；Ye et al.，2011）。除岩石类型外，岩石沉积层位也是影响硒富集的关键因素。例如，美国白垩纪页岩平均含硒量为 2.0 mg/kg，日本古生代页岩含硒 0.24 mg/kg，苏联志留纪页岩含硒 3.3 mg/kg。美国农业部在 1935~1941 年对美国西部 10 个州和东海岸进行了海相沉积岩和土壤硒的调查工作，发现白垩纪皮埃尔（Pierre）页岩中硒含量最高达到 10 000 mg/kg，正是这套页岩发育的富硒土壤，最终导致南达科他州、内布拉斯加州和怀俄明州的牲畜在 1933 年发生硒中毒症（Presser and Ohlendorf，1987）。不同国家富硒土壤的形成机制各不相同，其中，美国富硒土壤主要由白垩纪页岩、凝灰岩、侏罗纪页岩、砂岩及三叠纪砂岩风化而成，加拿大富硒土壤主要由白垩纪页岩发育而来，哥伦比亚土壤中硒主要来源于黑色板岩，波多黎各富硒土壤的形成则与火山有关，英国和爱尔兰富硒土壤主要与富含碳质的页岩、板岩和火山岩有关，南非和澳大利亚的富硒土壤主要由白垩纪页岩、砂岩或灰岩风化而成，俄罗斯的侏罗纪砂岩则被认为是富硒土壤的成土母质（Fleming，1980）。

不同类型的岩石中，硒的结合形态也存在较大区别。对于富硒沉积岩层（煤和黑色页岩），煤中硒的结合态包括可交换态、硫化物结合态、有机物结合态、残渣态（Palmer and Lyons，1996；张军营等，1999）。页岩中硒的主要赋存相态为有机质和黄铁矿，这两者中所含的硒占页岩总硒的比例高达 90%，而且硒更倾向于进入黄铁矿内（Matamoros-Veloza et al.，2011）。在湖北恩施渔塘坝碳质页岩和碳质硅质岩中，主要以有机物结合态和硫化物/硒化物硒为主，其次为可交换态、水溶态和元素态硒，而残渣态硒含量较低，碳酸盐结合态硒可忽略不计，其中碳质页岩和碳质硅质岩中二者水溶态硒的平均含量仅 5.5% 和 5.7%；而恩施富硒碳质泥岩中硒主要以有机结合态、元素态和硫/硒化物态的形式存在。富硒碳质页岩和硅质岩中的水溶态硒显著高于碳质泥岩，元素态硒的比例较高（韩文亮等，2007）。与恩施二叠纪富硒炭碳质岩相比，贵州寒武纪牛蹄塘组富硒碳质硅质岩中硒以硫化物/硒化物态和有机物结合态为主，碳质页岩与镍钼矿层中硒以有机物结合态、残渣态和硫化物/硒化物态为主，斑脱岩中硒主要以有机物结合态、元素态和可交换态为主（朱建明等，2007）。紫阳富硒区富硒岩石中硒主要以硫化物/硒化物态、有机物结合态和残渣态为主，水溶态硒含量极低，一般少于 3%，而其水溶态和可交换态硒之和仅为 5%~8%（Tian et al.，2016）。

由于硒是一种典型的生物制约元素，生物质产生的煤中硒含量较高，与煤内源的有机或无机物质有很强的亲和力，与锗和硫的亲煤性相似，所以关于煤中硒的研究较多。硬煤和褐煤中硒的平均含量分别为（1.6±0.1）mg/kg 和（1.0±0.15）mg/kg，灰分中硒含量分别可达（9.9±0.7）mg/kg 和（7.6±0.6）mg/kg（Yudovich and Ketris，2006）。1896 年，硒首次在比利时的煤中被发现（Yudovich and Ketris，2006）。Goldschmidt 和 Hefter（1933）在约克郡无烟煤中也发现了硒。煤中的硒主要以硫化硒和有机硒形式存在，另外还有少量其他形态的硒。Goldschmidt 和 Strock（1935）研究表明富硒煤中含有丰富的黄铁矿。在乌兹别克斯坦侏罗纪褐煤中发现了天然硒（Se^0）、白硒铁矿（$FeSe_2$）、有机硒和含硒黄铁矿（Savel'ev and Timofeev，1977）。杨光圻等曾于湖北恩施渔塘坝富硒地层为二叠系茅口组顶部的碳质硅质岩中采集到硒含量高达 84.1 mg/g 的石煤（Yang et al.，1983）。在低硫煤中，硫化硒占总硒的 35%；而在高硫煤中，硫化硒形态的硒比重达到总硒含量的85%~89%（Kizilstein and Shokhina，2001）。中国煤中硒含量均值为3.91 mg/kg，与美国煤中硒的平均含量（4.1 mg/kg）相近，远远高于世界煤中硒的平均水平（1.51 mg/kg，n=19 154）（Yudovich and Ketris，2006；Wang et al.，2009）。此外，澳大利亚煤中硒含量为0.21~2.5 mg/kg，低于中国和美国煤中硒的含量（Plant et al.，2003）。

三、土壤中硒的来源及分布

硒在自然界中的分布非常广泛，地球上所有自然物质（包括空气、土壤、岩石、水体、动植物体）中均存在硒元素，但是含量很低（樊海峰等，2006）。地球主要外圈层中，岩石圈、大气圈（气态和微粒态）、土壤、海洋（溶液的）、陆相生物群、河流（溶解的含颗粒状）、浅层地下水、化石燃料矿床、极冰中的硒含量分别为3×10^{12} t、2000 t、1×10^8 t、2×10^8 t、7×10^4 t、1.4×10^4 t、800 t、1.4×10^8 t、4×10^5 t（姚林波等，1999）。

土壤圈中，世界土壤的硒含量范围为0.1~2.0 mg/kg，平均值为0.4 mg/kg（Fordyce，2005）。根据美国地质调查局的资料，美国土壤硒含量范围在0.1~5.32 mg/kg，最富硒地区分布在美洲大平原，最缺硒地区主要分布在美国东北部。英国土壤硒含量范围为0.1~4.0 mg/kg，其中硒含量小于1.0 mg/kg的土壤占95%（White and Broadley，2005）。同一国家不同区域硒含量差异较大，印度西北部地区表层土壤中的硒含量就存在着明显的地域差异，例如，斋普尔（Jainpur）地区土壤硒含量范围为2.3~11.6 mg/kg，而巴瓦（Barwa）地区土壤硒含量平均值为3.1 mg/kg（Bajaj et al.，2011）。在美国、加拿大和中国的一些干旱、半干旱地区，土壤硒含量可高达20~40 mg/kg。同时，中国、北美、新西兰、澳大利亚、瑞典、芬兰等国家和地区均报道有缺硒现象（Ylaranta，1983；Gissel-Nielsen et al.，1984；Gupta and Gupta，2000；Blazina et al.，2014）。山地国家如芬兰、瑞典和英国的苏格兰的土壤硒含量普遍不足，而世界上的页岩土和干燥地区则是富硒地区。英国、法国、印度、比利时、巴西、塞尔维亚、斯洛文尼亚、西班牙、葡萄牙、土耳其、波兰、德国、丹麦、斯洛伐克、奥地利、爱尔兰、希腊、荷兰、意大利、中国、尼泊尔、沙特阿拉伯、捷克共和国、克罗地亚、埃及、布隆迪和新几内亚等国均报道有硒缺乏地区（Zhu et al.，2009；Yin and Yuan，2012）；富硒地区有印度旁遮普（Bajaj et al.，2011）、中国湖北恩施地区和陕西紫阳地区（Feng et al.，2009；Qin et al.，2013；Cui et al.，2017）、巴西亚马孙的帕拉（Lemire et al.，2009）、日本、丹麦的格陵兰岛（Fordyce et al.，2005）、美国、委内瑞拉和加拿大（Yin and Yuan，2012；表1-9）。全球约80%的硒储量分布在秘鲁、中国、智利、美国、加拿大、赞比亚、菲律宾、扎伊尔、澳大利亚和新几内亚（Liu et al.，2011）。世界上将近40个国家的自然硒资源储量有限。世界卫生组织的数据显示，除了个别富硒地区的人存在摄入硒过量的问题外，全球大约有10亿人口面临硒营养不良问题（Wu et al.，2015）。世界卫生组织推荐成人硒摄入量为40~200 μg/d，而我国成人硒摄入量仅26.6 μg/d（Cao et al.，2001）。虽然中国是世界上硒储量排名第四的国家（仅次于加拿大、美国和比利时），但缺硒现象仍然发生在东北黑龙江省至西南云南省的地理低硒带，涉及22个省份（71.2%的土地）（中国环境监测总站，1990；Zhu et al.，2009；Sun et al.，2010）。

表1-9 世界土壤硒缺乏、低硒和富硒的地区分布（Gupta and Gupta，2017）

硒水平	居民硒摄入量/（μg/d）	地区分布
硒缺乏	<55	中国、尼泊尔、埃及、英国、法国、印度、芬兰
低硒	55～100	瑞士、新西兰、韩国、芬兰
富硒	100～200	美国、加拿大、委内瑞拉、日本、中国（湖北恩施、陕西紫阳）、比利时、印度（旁遮普）

《中华人民共和国地方病与环境图集》（1989）显示从东北三省至西南的云贵高原有一条低硒带，并且这条低硒带与克山病和大骨节病的分布高度相关。不过，王学求等（2021）的研究结果表明，在我国九大粮食主产区中，除河套平原土壤硒平均值（0.081 mg/kg）、中位数（0.053 mg/kg）、背景值（0.081 mg/kg）都低于WHO（1987）的最低临界值（0.1 mg/kg）和我国有关地方标准（DZ/T 0295—2016）的土壤缺硒临界值（0.125 mg/kg）以外，其他粮食主产区都高于这一临界值，而且，珠江三角洲平原、广西平原、成都平原、长江中下游平原是富硒区（>0.4 mg/kg）。我国富硒（>0.4 mg/kg）土地总面积达109万 km^2，占国土总面积约11.2%，贫硒土地总面积占国土面积30%左右。因此，王学求等（2021）认为我国粮食主产区总体上不缺硒。另外，我国表层土壤硒含量高于深层土壤，含量范围分别为0.010～16.40 mg/kg和0.082～10.74 mg/kg，中位数分别是0.171 mg/kg和0.128 mg/kg，算数平均值分别为0.235 mg/kg和0.183 mg/kg，背景值分别为0.184 mg/kg和0.136 mg/kg（王学求等，2021）。

有学者对我国硒的空间分布成因进行了分析报道。Blazina等（2014）认为中国东北三省至西南的云贵高原低硒带分布与季风气候有关。Sun等（2016）进一步提出表层土壤中硒的分布受大气沉降和挥发作用调控。瑞士Winkel教授团队根据土壤硒的大数据分析表明，土壤的母质、酸碱性、黏粒含量和有机物含量是决定硒水平最主要的几个参数，而干旱程度、植被的蒸腾作用和降水量是主要的气候和生物因素（Jones et al.，2017；孙国新等，2017）。王学求等（2021）通过覆盖全国的网格化土壤样品的采样与分析表明，我国低硒带呈不连续的片状，分布于内蒙古东部至青藏高原一带，与过去发现的"东北三省至西南云贵高原有一条低硒带"的结果存在差异，并认为硒的这一空间分布主要受地质背景、岩石类型、土壤类型和自然地理景观控制（Liu et al.，2021）。

（一）全球表层土壤硒分布及主要影响因素

全球范围内土壤中的硒含量目前无实测数据，Jones等（2017）采用机器学习等大数据分析方法，分析了全球范围内3.3万个土壤点位中的硒含量，预测了全球范围内表层土壤中的硒含量，同时分析了26个环境变量对土壤中硒含量的影响。表层土壤中的硒含量主要取决于两个主要的环境因素：一个是土壤自身的理化性质，另一个是气候和生物因素。总体趋势为热带、亚热带地区土壤中硒强烈富集，含量普遍偏高；温带地区土壤（山地褐土和暗棕壤）中硒含量较低，而干旱、半干旱地区的灰钙土和灰漠土中硒含量稍高。土壤的母质、酸碱性、黏粒含量和有机物含量这四个因素是决定表层土壤硒含量最关键的几个参数；而干旱程度、降水量和植被蒸腾作用是主要的气候和生物因素。利用这7个参数，Jones等（2017）预测了近期（1980～1999年）和未来（2080～2099年）全球表层土壤（0~30 cm）中的硒含量，这 7 个参数的重要性由高到低依次为：干旱程度、土壤酸碱性、降水量、植物蒸腾作用、黏粒含量、土壤母质和有机质含量。敏感性分析显示，在全球不同的区域，影响表层土壤硒含量的因素总体相同，即不论在什么环境条件下，表层土壤中硒含量主要受相同因素的影响。综合考虑气候变化因素的影响，与 20 世纪末相比，21 世纪末全球耕作土壤中的硒含量将下降约 5%。就中国而言，大部分的耕地会变得更加缺硒，仅在东北与朝鲜半岛交界带，以及四川盆地等少部分区域表层土壤硒含量呈上升趋势（孙国新等，2017）。另外，由于化石燃料的大量使用，大气中氧气含量逐年下降、二氧化碳含量逐年增加，造成温室效应，影响全球气候变化（Keeling and Manning，2014）。而全球气候变化必然会对降水分布及我国农业产生较大影响（Fischer and Knutti，2016；Piao et al.，2010），进一步影响粮食作物中的硒含量及人体硒摄入量。

（二）我国土壤硒分布概况

我国表层土壤中硒含量分布严重不均，从全国尺度看，具有很大的空间异质性，但总体上有规律可循，即西北和东南地区土壤中硒含量相对较高，从东北地区到西南地区表层土壤中硒的含量明显偏低，存在一条典型的低硒带，该低硒带主要位于年均降水量 400~800 mm 的气候范围内（Wang and Gao，2001）。我国是典型的大陆性季风气候，降水量主要受季风影响，降水量不同导致我国农业具有"东耕西牧，南稻北麦"的分布特点。400 mm 等降水量线是我国一条重要的地理分界线，是我国的半湿润和半干旱区的分界线，是森林植被与草原植被的分界线，是农耕文明与游牧文明的分界线，是种植业与畜牧业的分界线（即南部降水量大的区域属于种植业而北部降水量小的区域属于畜牧业）。土壤低硒带位于 400 mm 等降水量线南部，即位于农田区，这造成我国有 70% 以上耕地处于不同程度缺硒状态（孙国新等，2017），其中约 30% 耕地土壤严重缺硒。农田土壤中硒的缺乏导致作物中硒含量严重不足。

一段时间以来，我国土壤中硒含量研究主要集中在个别土壤硒含量较高区域（Yu et al.，2014），例如，湖北恩施和陕西紫阳是两个已知的富硒岩石出露区。近年来土壤调查发现了一些富硒土壤，主要分布在东南地区，黑龙江和青海也有少量富硒土壤。2021 年，中国地质学会公布了首批天然富硒土地，包括黑龙江、浙江、安徽、山东等地的 30 个地块。从全世界范围看，低硒或表现缺硒的土壤面积远大于高硒或硒毒土壤面积。低硒造成的人体疾病在南北半球各有一条分布带，基本在 30°以上的中高纬度地区（郑达贤等，1982）。这些地带在北半球分布更广，大部分地区年降水量在 400~1000 mm 范围内，以北温带湿润、半湿润气候和地中海型气候区为主。南半球主要分布在大陆南端的地中海气候区。成土母质对土壤的物理性状和化学组成有着重要的作用，同样是影响表层土壤硒含量的重要因素，世界范围内低硒区主要是灰化土、棕壤、暗棕壤、褐土、草甸黑土及与其性质相近的土壤（郑达贤等，1982）。结晶基岩和第四季沉积物是地球土壤的主要母质，其中沉积岩占 75%、变质岩和岩浆岩占 25%（Systra，2010）。我国土壤由 77.3% 的沉积岩风化而成。三大岩类硒含量也有显著不同，变质岩硒含量最高（0.031~0.131mg/kg），其次为沉积岩（0.028~0.118mg/kg），岩浆岩硒含量偏低（0.059~0.108mg/kg）。我国三大主要岩石中的硒含量分别为：变质岩 0.070 mg/kg；岩浆岩 0.067 mg/kg；沉积岩 0.047 mg/kg（夏卫平和谭见安，1990）。

全球范围内硒在地壳中的丰度为 0.05～0.09 mg/kg（Combs, 2001）。我国硒在地壳中的丰度为 0.058 mg/kg（夏卫平和谭见安, 1990）。显然，土壤母质中硒含量都很低，与硒在地壳中的丰度相当。而我国土壤中硒在不同地区含量为西北地区 0.19 mg/kg、中部低硒带 0.13 mg/kg、东南沿海 0.23 mg/kg（中华人民共和国地方病与环境图集编纂委员会, 1989）。显然，土壤中硒含量远高于土壤母岩硒含量，特别是西北和东南地区。土壤母质中硒含量比起土壤形成过程中其他因素引起的硒的变化量来说，贡献较小。将土壤硒含量分布图与土壤母质进行直观比较发现，土壤硒含量分布和母质分布完全不一致，说明土壤母质不是我国低硒带形成的主要成因（Sun et al., 2016）。仅仅依靠土壤母质无法解释表层土壤中硒含量的地带性分异（Blazina et al., 2014）。

（三）土壤中硒分布的气候成因

1. 土壤中硒的湿沉降输入

大尺度范围内，表层土壤中硒的另一个主要来源为大气沉降（Nriagu and Pacyna, 1988; Nriagu, 1989; Wen and Carignan, 2009）。大气中硒的主要来源是海洋中的硒挥发，占大气中总硒的 60%～80%（Blazina et al., 2017）。海水中硒被海洋微生物吸收，经微生物甲基化转化为气态硒化合物，大量的气态硒化合物被释放到大气中，以二甲基硒和二甲基二硒为主，这些挥发性硒易溶于水，经季风携带进入陆地系统中，通过湿沉降进入表层土壤（图 1-5），被认为是陆地硒的主要来源（Amouroux et al., 2001; Wen and Carignan, 2009）。据报道，欧洲沿海国家土壤中硒含量高于欧洲中部非沿海国家（Müller et al., 2012）。Weiss 等（1971）报道格陵兰冰层中平均硒含量为 0.019 μg/kg，以此估算降水量在 500～1000 mm 条件下，一万年间通过降水对土壤耕作层（20 cm）中硒的贡献为 0.2～0.8 mg/kg。我国东南地区属于典型的亚热带季风气候，土壤硒含量分布与降水量分布类似（Sun et al., 2016; 孙国新等, 2017）。Blazina 等（2014）研究我国黄土高原地区古气候显示，间冰期东亚季风引起的降水量与土壤中硒含量呈显著正相关，说明季风导致的降水对土壤中硒含量贡献较大。Sun 等（2016）的研究结果显示，我国雨水中硒浓度为 0.1～0.2 μg/L，降水量在 400～800 mm 条件下，硒年沉降为 40～160 μg/m²，

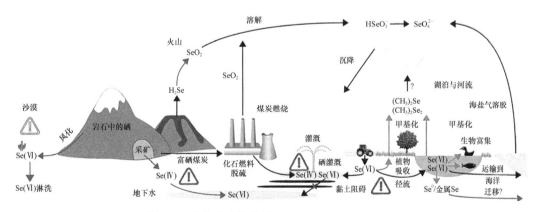

图 1-5　生物地球化学循环（Winkel et al., 2012）

蓝色箭头代表硒氧化过程，绿色箭头代表硒还原过程；空心叹号代表可能硒缺乏，实心叹号代表可能硒过量

当降水量达到 2000 mm 时，降水向土壤中硒的输入量达到 200~400 μg/m^2。东亚夏季季风造成的湿沉降，是我国东南地区土壤高硒分布的重要影响因素（Sun et al., 2016）。Feinberg 等（2020）估算，每年有 290~360 t 硒通过人为活动、火山喷发、海洋和陆地生物挥发进入大气中，绝大部分（96%）以气溶胶形式存在。而大气中硒的去除主要是湿沉降（81%），干沉降较少（19%）。据估计，每年全球硒湿沉降达 3500~1000 t（Ross, 1985）。

2. 土壤中硒的干沉降输入

我国西北干旱地区，由于受到山脉阻隔且远离海洋，不受东亚季风影响，造成该地区年降水量很小（<200 mm），一些戈壁或沙漠地区的年降水量甚至低于 100 mm（Qian and Lin, 2005）。而该地区土壤中硒含量却相对较高，显然，西北地区土壤中硒的输入与降水量关系不大。经对比显示，西北地区表层土壤硒分布与沙漠分布非常吻合。众所周知，东亚季风分为夏季季风和冬季季风，冬季季风由蒙古和西伯利亚与北太平洋大气压力差形成。亚洲冬季季风不仅是世界最强的冬季季风，而且是亚洲的主要尘源，占大气灰尘的 70%（Lawrence and Neff, 2009）。而气溶胶中硒含量远高于土壤母岩中硒含量（Ghauri et al., 2001）。中亚沙漠和戈壁上的矿物灰尘被冬季季风带入空中，进而主要通过干沉降重新回到土壤表面，其中有约 30% 沉降在沙漠周边地区。局地尺度灰尘的年沉降速率为 200 g/m（距尘源 0~10 km），区域尺度为 20 g/m（距尘源 10~1000 km），而全球尺度的沉降速率为 0.4 g/m（距尘源 >1000 km）。我国西北地区包含蒙古沙漠、塔克拉玛干沙漠等多个沙漠，西北高硒地区都紧邻沙漠尘源（Zhang et al., 1997）。所以靠近沙漠地区表层土壤中硒含量相对较高，而远离沙漠地区土壤硒含量较低。显然，干沉降提供了西北地区表层土壤中的硒，造成该地区表层土壤中硒含量相对较高。据报道，我国沙漠地区平均每年产生 8 亿 t 灰尘（5 亿~11 亿 t），占全球总灰尘量的一半（Zhang et al., 1997）。全球硒年干沉降量为 1700~2400 t，小于硒湿沉降量（3500~10 000 t）。

3. 土壤中硒的挥发损失

我国中部地区降水量相对较小（400~800 mm），由湿沉降进入土壤的硒较少；同时该区距离亚洲尘源相对较远，干沉降灰尘相对较少，对土壤硒的贡献也不大。即使这样，经多年积累，土壤中硒含量也应提高，为什么该地区硒含量显著低于西北和东南地区，形成典型的低硒带？一个不容忽视的原因是土壤硒挥发（图 1-6）。土壤硒挥发是硒生物地球化学循环的一个重要过程，同时也减少了土壤中硒的积累。生物甲基化形成的挥发性硒甲基化合物是该元素生物地球化学循环的主要驱动（图 1-6）（Winkel et al., 2015）。植物和微生物皆可甲基化硒，常见的挥发性硒化合物有二甲基硒（CH_3SeCH_3, DMeSe）、二甲基硫硒（CH_3SeSCH_3, DMeSeS）、二甲基联硒（$CH_3SeSeCH_3$, DMeDSe）（Meija et al., 2002）。在微生物体内已发现一种参与挥发硒生成的酶及相关基因，即巯基嘌呤甲基转移酶（Ranjard et al., 2002, 2003）。生物甲基化和挥发是土壤中硒减少的主要驱动因素，土壤微生物是硒挥发的主要驱动因子。

影响土壤硒挥发的主要因素包括温度（Chau et al., 1976; Zieve and Peterson, 1981; Frankenberger and Karlson, 1989）、土壤湿度（Abu-Erreish et al., 1968; Zieve and Pe-

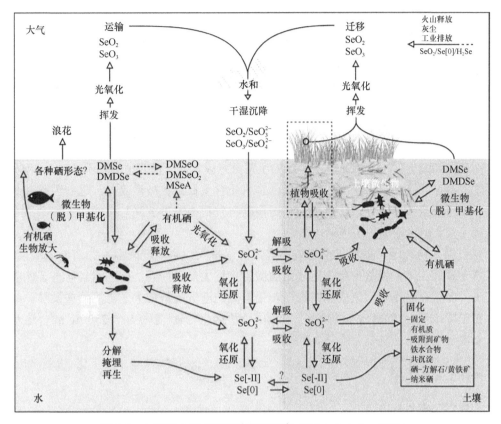

图 1-6　土壤和水体中硒的转化及挥发（Winkel et al.，2015）

terson，1981；Frankenberger and Karlson，1989）、有机碳含量（Abu-Erreish et al.，1968；Doran and Alexander，1977）、土壤硒含量和土壤微生物丰度（Winkel et al.，2015）等多种因素。目前关于硒挥发的研究主要集中在富硒地区土壤，个别研究表明富硒土壤中硒挥发通量达到 1300 μg/(m²·d)，富硒土壤中硒浓度可达到 1200 mg/kg，远高于世界平均土壤硒浓度（0.1～2 mg/kg）（Banuelos et al.，2005）。富硒土壤中挥发硒通量也会高于普通农田，不过富硒地区面积很小，无法代表广大的普通陆地低硒土壤的硒挥发通量。对普通土壤中硒的挥发研究甚少，数据严重缺乏（Haygarth et al.，1994；Blazina et al.，2014）。普通土壤中硒的挥发通量，以及植物和微生物对该通量的贡献目前仍未知，因此无法准确估算全球土壤硒挥发能力。据报道，湿地系统中硒挥发通量为 43.8 μg/(m²·a)（Banuelos and Lin，2009），而草地系统中硒挥发通量为 100～200 μg/(m²·a)（Barkes and Fleming，1974；Haygarth et al.，1994）。可能影响微生物生物量、生物多样性、丰度和活性的因素包括温度、降水量、土壤湿度、植被覆盖、有机质含量、土壤 pH 和硒生物有效性等，而这些因素在我国西北、中部和东南地区差异显著，造成土壤中微生物生物量、生物多样性和丰度不同，硒挥发损失差别巨大，对表层土壤中硒积累有重要影响。

4. 土壤中硒的净积累

土壤中硒的含量实际上是大气干湿沉降（输入量）与土壤硒挥发（输出量）的差值，即土壤净积累为硒干湿沉降减去硒挥发。我国中部的低硒带属于半湿润气候，该地区年平均气温相对较高，由于距离海洋和沙漠都比较远，湿沉降和干沉降提供的硒含量均相对较少，总体而言，硒沉降量较低。而该地区被森林覆盖，土壤有机质含量很高，土壤温度、湿度均适合微生物生长，因此该地区土壤微生物含量大、活性高，土壤中硒挥发损失相对较大。较大的土壤硒挥发损失与较小的土壤干湿沉降，造成土壤硒挥发与沉降相对平衡。经粗略估算，在我国年降水量400～800 mm的地区，土壤中硒挥发量与湿沉降造成的土壤硒输入量相当，没有或仅有少量硒在土壤中积累，使土壤硒含量接近土壤母质，保持较低水平，形成典型的低硒带。

在我国西北地区，干沉降是土壤硒的主要贡献因素，而西北地区属于沙漠、半沙漠，年平均气温低，降水量小，土壤湿度低，有机质含量低，植被覆盖差，这些因素限制了土壤中微生物量及其活性，造成该地区土壤硒挥发很低。干沉降与极低的土壤硒挥发能力造成了西北地区较高的土壤硒含量。

在我国东南地区，虽然年平均气温高，降水量大，硒湿沉降多，然而土壤湿度高、有机质含量高、植被覆盖密集，造成土壤微生物量大，活性高，这些因素促进了硒的挥发损失。但东亚夏季季风造成的强降雨，使得硒的湿沉降量远大于土壤硒挥发量。经估算，在年降水量为2000 mm的地区，土壤中硒的年净积累量为50～250 μg/m^2，这是该地区土壤硒积累而呈现高含量的主要原因。

四、硒的生物化学

1. 植物硒吸收与转运

目前还未发现植物体内特异性硒转运蛋白，但生理实验及生物信息学分析显示，特异性硒转运蛋白可能存在于硒超积累植物中（Schiavon et al.，2012；Harris et al.，2014）。硒酸盐主要通过硫酸盐转运途径进入植物体内。对于亚硒酸盐的植物吸收，其吸收通道依条件而定。当溶液中pH较低时（pH≤3.0），亚硒酸盐主要以中性分子H_2SeO_3的形式存在（图1-7），植物通过水通道吸收，参与亚硒酸盐吸收的载体为水通道家族的成员硅转运蛋白OsNIP2;1（Zhao et al.，2010）；在相对较高的pH环境中（pH 5.0），亚硒酸盐主要以$HSeO_3^-$的离子形式存在，此时亚硒酸盐的吸收主要通过磷酸盐转运蛋白进入植物体内（Zhang et al.，2014）；在pH 8.0的碱性条件下，SeO_3^{2-}是亚硒酸盐的主要存在形态（图1-7，图1-8），其可能通过阴离子通道进入植物体内，但阴离子通道抑制剂的添加并不能显著抑制亚硒酸盐的吸收，说明植物还存在其他未知的亚硒酸盐的吸收途径（Zhang et al.，2010）；环境中有机硒以硒代半胱氨酸（SeCys）和硒代甲硫氨酸（SeMet）为主，植物可通过氨基酸转运蛋白从外界环境中吸收有机态硒。植物吸收的无机硒通过木质部从植物根向上转运，转运过程与吸收的硒形态有关。植物极易吸收硒酸盐并将其转运到地上部；如果吸收亚硒酸盐，则大部分硒在植物根部首先同化，被转化为有机硒化合物再向上运输，木质部伤流液中只能检测到微量的亚硒酸盐（Li et al.，2008）。植物地上部的硒则主要通过韧皮部进行重分配。

2. 植物硒代谢途径

植物体内硒代谢主要包括硒酸盐的还原、硒代氨基酸的合成以及挥发性硒的形成

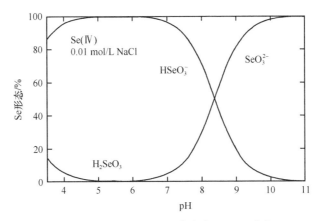

图 1-7 0.01 mol/L NaCl 溶液中 Se(Ⅳ) 形态

Se(Ⅳ) 形态未显示,因为一直以 SeO_4^{2-} 为主

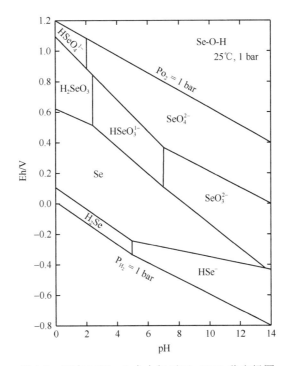

图 1-8 硒在 25℃、1 个大气压下 pH-Eh 稳定场图

(图 1-9)。硒酸盐被植物吸收之后运输到叶绿体内,并按硫同化途径进行转化。硒酸盐首先通过 ATP 硫酸化酶激酶(ATP sulfurylase,APS)激活形成 5-磷酸硒腺苷(adenosine phospho-selenate,APSe),之后通过 5-磷硫酸腺苷还原酶(adenosine 5-phosphosulfate reductase,APR)催化 APSe 还原产生亚硒酸盐(Sors et al.,2005)。形成的亚硒酸盐被进一步还原生成硒化物。参与该过程的酶目前仍不清楚,据推测可能是亚硫酸还原酶(sulfite reductase,SiR)。蛋白质中半胱氨酸(cysteine,Cys)残基能够稳定蛋白质与(类)金属辅基结合,Cys 残基被 SeCys 随机取代,硒被结合进入蛋白质。叶绿体内合成的部分 SeCys 能被转运到细胞质中并在硒代半胱氨酸甲基转移酶(SeCys methyltransferase,SMT)等的作用下形成挥发性二甲基二硒化物(dimethyldiselenide,DMeDSe),而 SeMet

则能在 S-腺苷-L-甲硫氨酸甲基转移酶（S-adenosyl-L-methionine：L-methionine S-methyl-transferase，MMT）等的催化下形成挥发性的二甲基硒化物（dimethylselenide，DMeSe）（Zhu et al.，2009；陈松灿等，2014）。

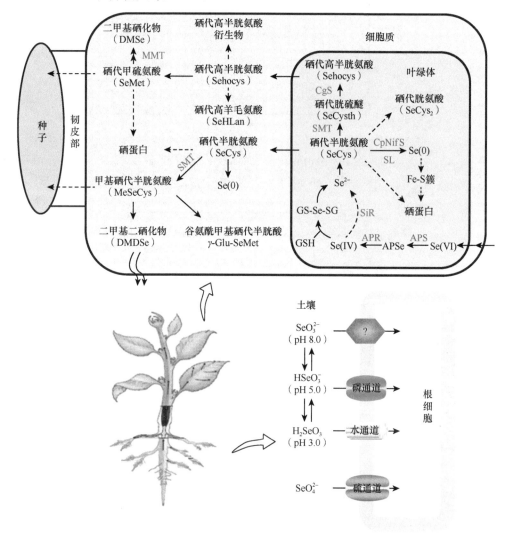

图 1-9　植物硒代谢（改自陈松灿等，2014）

除了上述代谢途径之外，Matich 等（2009）在转基因的烟草（*Nicotiana tabacum*）叶片中分离鉴定了 4 种新的含硒小分子，这意味着植物体内可能还存在其他硒代谢途径。对于植物硒代谢的探索不能仅局限于与硫代谢直接相关的途径，还可以在不同硒代谢产物形态的基础上，从其他代谢途径出发探究可能的植物硒同化机理（陈松灿等，2014）。

不同食物来源中硒含量差别较大（图 1-10）。粮食作物如水稻、小麦，因硒浓度相对较高、消费量较大（均远高于蔬菜水果），因此是人体中硒的重要来源之一。不同国家水稻和小麦中硒含量见图 1-11，从图中可见，美国、加拿大、澳大利亚生产的小麦中硒含量较高，而意大利、英国等欧洲国家小麦硒含量较低；美国、印度、泰国大米中硒含量较高，而中国、意大利、埃及等水稻硒含量偏低（Zhu et al.，2009；Williams et al.，2009）。

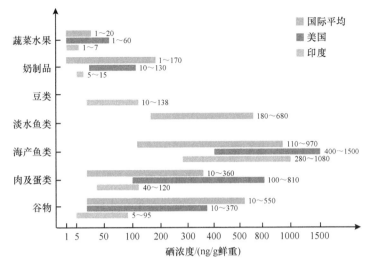

图 1-10　美国、印度和国际上不同来源食物中硒含量范围（Kumar and Priyadarsini，2014）

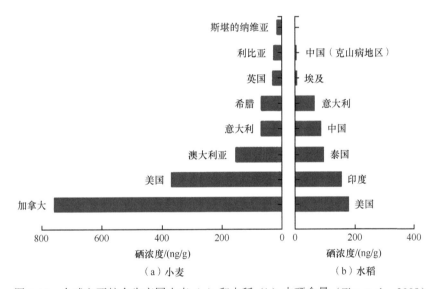

图 1-11　全球主要粮食生产国小麦（a）和水稻（b）中硒含量（Zhu et al.，2009）

水稻秸秆、稻壳、籽粒、米糠中硒浓度相互之间呈显著正相关，与土壤中硒含量也呈显著正相关，表明硒从根部吸收进入植物体内，从茎秆木质部向上运输进入籽粒中（Sun et al.，2010）。籽粒中硒含量高于秸秆和稻壳中，植物将更多的硒存储在籽实中。不同形态的硒在水稻叶片和籽粒之间通过韧皮部运输进行再分配，硒可能主要以有机硒的形式装载到大米籽粒中。有机硒通过韧皮部迅速转运并存储在水稻籽粒的胚及胚乳中；而无机硒转运过程相对较慢，而且重分配的硒滞留在维管束末端很难进入籽粒中（Carey et al.，2012）。

硒在植物根、茎、叶、种子等不同部位表现出特定的空间分布特征，且与硫分布特征相似（Pickering et al.，2000）。水稻中硒主要以有机硒的形式分布于糊粉层，也有少量硒分布在胚乳中，硒浓度由外部向中心逐渐降低，该分布特征与硫相似（Williams et al.，2009）。米糠中硒含量略高于脱糠后的精米（Sun et al.，2010），脱糠会造成水稻中硒含量损失，水稻籽粒中近 1/3 的硒（29.1%）经脱糠去除，造成营养流失（Yao et al., 2020）。小麦籽粒中硒在糊粉粒细胞内以及淀粉粒细胞周围的蛋白质基质中的浓度较高、分布比较均匀，这与硫的空间分布相同（Moore et al., 2010）。

第三节 土壤-植物系统中硒的研究现状与问题

一、研究现状

1. 土壤硒生物有效性

多种环境变量影响土壤中硒的有效性，这些变量在不同程度上影响着土壤中硒的分布与含量。例如，在纳米尺度上，不同黏土颗粒的矿物结构影响硒的分配；在大尺度上（100 km），气候条件的作用更加明显。不同的环境尺度下，这些环境变量的异质性对硒有效性的影响显著不同（图1-12）。在微米尺度上，含铁和铝氧化物的黏土矿物对硒有较强的吸附能力，黏土矿物是硅酸盐岩石长期风化形成，不同土壤中含量不同，吸附固定能力也有差异，土壤 pH 和 Eh 显著影响矿物对硒的吸附。矿物表面存在 pH 和 Eh 梯度，这影响黏土矿物对硒的吸附能力及硒的生物有效性。局地尺度上，生物（植物分布）和非生物（地形、土地利用方式）等因素显著影响土壤中的硒。藻类及微生物可挥发土壤中的硒、减少土壤硒含量，植物根际分泌物可调控土壤 pH，增加硒的生物有效性。另外，植物可改变土壤微环境，改善营养元素的有效性。例如，沙漠中植物可增加土壤湿度和有机质含量，这可以提高土壤硒的生物有效性；植物还可降低土壤风蚀和水蚀，减少土壤硒的流失。根据土壤上生长植物的不同，可在局地尺度或景观尺度影响土壤中硒的含量。土地利用方式的差异影响土壤理化性质，例如，退耕还林还草增加土壤有机质含量，从而增加土壤硒含量；土壤水分含量变化，影响土壤氧化还原条件，从而影响土壤中硒的形态转变，硒形态的变化又会影响硒的移动性和生物有效性。田间尺度上，地形对土壤物理、化学及生物过程有显著影响，田间尺度的微气候受坡度、朝向（阳坡或阴坡）、太阳辐照等因素影响，这些因素影响植物组成、土壤含水率、氧化还原、有机质含量等，进而影响土壤中硒的分布。区域尺度上，机器学习分析显示，气候变量（包括干旱、降水量）显著影响土壤硒含量，而土壤理化性质及地质成因对土壤中硒的影响较小（Jones et al., 2017）。

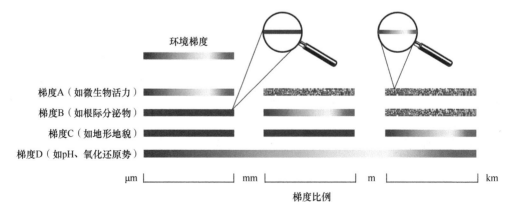

图 1-12 不同尺度下调控土壤硒含量的环境因素（Elizabeth et al., 2017）

2. 硒与植物微生物

豆科植物中有 25 种硒超积累植物。研究显示，根瘤增强植物氮的获取能力，不仅有利于植物生长，而且增加了硒的积累（Alford et al., 2012, 2014）。硒超积累豆科植物中，硒含量增加与根瘤的形成有关。硒超积累植物将硒转变为含硒氨基酸，根瘤中的微生物提供大量氨基酸合成所需的氮，造成含硒氨基酸的积累。很多根际细菌和真菌可以将亚硒酸盐还原为元素硒，除此之外，还可以将硒转变为挥发性硒化物，这些微生物调节根际土壤硒形态，对植物中硒的吸收和积累产生影响。接种根际微生物可增强植物对硒的积累（El Mehdawi et al., 2015；Yasin et al., 2015；Lindblom et al., 2013）。接种一些菌根真菌同样可以增加植物中硒的含量。硒超积累植物内生菌具有很强的硒耐受能力，可还原亚硒酸盐为元素硒，接种该内生菌促进了植物生长（Durán et al., 2014；Staicu et al., 2015）。植物内生细菌和真菌可促进植物生长，改变硒形态，影响宿主植物对硒的积累。

3. 植物积累硒的生态意义

植物中硒的积累，特别是硒超积累植物对硒的积累，可影响生态系统与周边环境的相互关系（图 1-13）。因为高浓度的硒对动物具有毒性，植物体内积累高浓度的硒可保护植物免受草食动物及病菌的侵犯（Freeman et al., 2006；Prins et al., 2011；Quinn et al., 2011），包括夜蛾、蝴蝶、蚜虫、螨虫、草原土拨鼠。例如，硒超积累植物沙漠圆锥花和二沟黄芪均将硒储存在嫩叶和生殖器官，包括叶片边缘、花粉和胚珠中，从而保护植物有价值的组织或食草动物可能的进食部位。有机硒和无机硒均能够提供保护作用，因为六价硒和硒代半胱氨酸对动物均有毒，六价硒产生自由基，硒代半胱氨酸影响蛋白质功能。除叶片外，硒超积累植物花粉中硒含量同样很高，造成花蜜中硒含量也很高，达到昆虫硒中毒浓度的 10 倍以上。

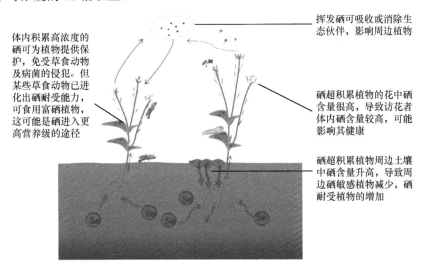

图 1-13　硒超富集植物的生态交互作用（Elizabeth et al., 2017）

因为硒超积累植物中硒浓度是其他植物中的 100 倍以上，可能影响周边植物的生态。硒超积累植物中硒浓度是土壤中的 1000 倍以上，即植物对硒进行了富集。硒超积累植物的残体进入土壤造成周边土壤中硒富集，同时改变了土壤中硒的形态。这些高浓度的硒可能对周边植物,特别是硒敏感植物的发芽和生长产生抑制作用(El Mehdawi et al., 2011),这是植物化感作用中元素相克的表现。硒超积累植物通过积累硒抑制周边硒敏感植物生长,增加硒耐受植物的增长,总体上影响植物的群落结构。

二、研究中存在的问题

土壤-植物系统中硒的生物地球化学转化、硒的生物有效性及测试方法需进一步完善。植物系统中通过氮、磷、硫及富硒有机肥或农业废弃物的施用,可强化植物中硒的积累(Dinh et al., 2019)。生物有效性的分析主要采用化学浸提法和薄膜梯度扩散技术(DGT)。梯度扩散薄膜技术既考虑了土壤固-液之间的动态平衡,同时削弱或消除了土壤有机质含量等理化性质对测定结果的影响,可有效地测量硒生物有效性;但 DGT 在实际应用中对技术和设备要求较高,目前应用还不十分普遍。

植物对硒的吸收、转运和代谢已经有大量的研究和报道。植物体内产生大量的含硒代谢物,特别是含硒蛋白等。这些含硒代谢物种类繁多,目前已确认有约 25 种含硒蛋白,仍有大量含硒物质未被确认。另外,这些含硒代谢物的功能仍未知。硒是植物特别是维管植物的有益元素,硒蛋白在植物体内的功能及作用有待于深入研究。

土壤有机质和黏土矿物是影响土壤中硒吸附-解吸、溶解-固定及生物有效性的关键因素。硒与黏土矿物有较高的亲和性,所以黏土含量高的土壤中,硒的生物有效性相对较低。pH 影响硒在黏土矿物上的吸附-解吸,因为 pH 影响黏土矿物表面的电荷。土壤有机质含有大量功能基团,对硒同样具有很强的吸附能力。据报道,土壤中硒约 20% 被矿物吸附,而 40%~50% 被土壤有机质吸附(Qin et al., 2013)。因此,土壤有机质含量显著影响土壤中硒的迁移和生物有效性。土壤有机质与硒的相互作用主要包括三个方面:①有机质吸附位点与硒形成复杂化合物;②有机质与其他金属化合物形成复杂化合物,之后又吸附硒,形成间接复合物;③微生物将硒结合进入蛋白质中。溶解性有机质由不同分子质量的有机物组成,对硒的迁移和生物有效性具有双重作用。一方面,大分子溶解性有机质与硒结合可降低硒的生物有效性;另一方面,小分子溶解性有机质与硒结合可增加硒的溶解,提高硒的生物有效性。超过 70% 的植物根际分泌物为亲水或疏水的小分子物质,这显著影响硒的溶解和生物有效性。农田中植物残体腐解主要产生小分子的溶解性有机质,这同样能够提高土壤中硒的生物有效性。

土壤中有机质同样影响土壤微生物群落结构和微生物多样性。而微生物是硒生物地球化学循环的主要驱动,多种微生物通过硒获得能量,维持自身生长,包括硒酸盐呼吸菌、亚硒酸盐呼吸菌、元素硒呼吸菌、纳米硒合成菌等,这些微生物广泛存在于土壤环境中。这些微生物代谢土壤中的硒,转变硒的形态,影响硒的吸附-解吸、迁移转化。植物根际存在大量的微生物,这些微生物与植物共生,通过活化土壤中元素,增强硒生物有效性,帮助植物获得营养物质。这些矿物-微生物-植物之间的复杂电子传递通路、微生物转化硒形态的方式,以及哪些微生物可提高硒生物有效性等仍缺乏明确的信息,亟须深入探索。

目前,我国土壤退化严重,特别是土壤酸化、土壤有机质缺乏和土壤微生物失衡尤为突出。我国正采取措施改善土壤质量、治理土壤酸化、提高土壤有机质含量和调节土壤微生物，而土壤 pH、有机质含量及微生物群落结构显著影响土壤中硒的含量及活性。通过治理退化土壤，可提高土壤中硒含量及硒生物有效性。

硒与植物及土壤微生物仍有很多方面需要深入探索，例如，硒超积累植物中是否具有特有的核心微生物菌群，影响植物硒的超积累？如果该菌群存在，是否可以用于硒污染土壤的修复或硒的生物强化？植物尤其是硒超积累植物的叶际微生物与硒的关系几乎没有研究，植物内生菌与硒的关系仍需深入探索。

植物特别是硒超积累植物中硒的生态效应方面的研究还比较初步，富硒植物或硒超积累植物如何影响植物营养及植物群落结构的研究目前取得了一些成果，但仍需进一步探索。另外，硒超积累植物改变土壤中硒的分配，使土壤硒在根际富集，其生态学作用目前还不明确。通过多学科交叉探索植物元素生态学，可拓展我们对生态的理解及植物元素的生态学意义。

地球上不同地区存在多个季风区，包括南亚季风、东亚季风、北澳季风、西北太平洋季风、南非季风、北非陆地季风、南美季风、北美季风，这些季风显著影响当地的降水量和硒的干湿沉降，势必影响土壤中硒的含量。未来将继续探讨这些季风对当地土壤中硒分布的影响，从而探索气候因素对土壤中硒分布的影响。另外，土壤中硒的挥发，尤其是常规农田土壤中微生物驱动的硒挥发仍缺乏系统研究，这对土壤中硒的积累、硒生物地球化学循环以及不同介质中硒通量的准确计算，都有重要的意义和价值。

参 考 文 献

蔡大为, 李龙波, 蒋国才, 等. 2020. 贵州耕地主要元素地球化学背景值统计与分析. 贵州地质, 37(3): 233-239.
陈炳翰, 丁建华, 叶会寿, 等. 2020. 中国硒矿成矿规律概要. 矿床地质, 39(06): 1063-1077.
陈松灿, 孙国新, 陈正, 等. 2014. 植物硒生理及与重金属交互的研究进展. 植物生理学报, 50(5): 612-624.
程水源. 2019. 硒学导论. 北京: 中国农业出版社.
翟秀静, 周亚光. 2009. 稀散金属. 合肥: 中国科学技术大学出版社.
樊海峰, 温汉捷, 凌宏文, 等. 2006. 表生环境中硒形态研究现状. 地球与环境, 34(2): 19-26.
冯彩霞, 刘家军, 刘燊, 等. 2002. 硒资源及其开发利用概况. 地质与资源, 11(3): 152-156.
韩文亮, 朱建明, 秦海波, 等. 2007. 恩施渔塘坝富硒碳质岩石中硒的形态分析. 矿物学报, 27(1): 89-95.
李家熙, 张光弟, 葛晓立, 等. 2000. 人体硒缺乏与过剩的地球化学环境特征及其预测. 北京: 地质出版社: 2-10.
刘家军, 翟德高, 王大钊, 等. 2020. Au-(Ag)-Te-Se 成矿系统与成矿作用. 地学前缘, 27(2): 79-98.
刘家军, 冯彩霞, 郑明华. 2001. 硒矿资源研究现状. 世界科技研究与发展, 23(5): 16-21.
刘英俊, 曹励明. 1987. 元素地球化学导论. 北京: 地质出版社: 1-281.
彭大明. 1996. 罕见硒矿中国蕴藏. 建材地质, 6: 9-14.
彭大明. 1997. 中国硒矿资源概述. 化工矿产地质, 19(1): 36-42.
宋成祖. 1989. 鄂西南渔塘坝沉积型硒矿化区概况. 矿床地质, 8(3): 83-89.
孙国新, 李媛, 李刚, 等. 2017. 我国土壤低硒带的气候成因研究. 生物技术进展, 7(5): 387-394.
涂光炽, 高振敏, 胡瑞忠, 等. 2004. 分散元素地球化学及成矿机制. 北京: 地质出版社.
王美珠, 章明奎. 1996. 我国部分高硒低硒土壤的成因初探. 浙江农业大学学报, (01): 89-93.
王学求, 柳青青, 刘汉粮, 等. 2021. 关键元素与生命健康: 中国耕地缺硒吗? 地学前缘, 28(3): 412-423.
王越. 2018. 生物源纳米硒对阿霉素抑制肝癌 HepG2 细胞移植瘤生长的增效与解毒作用研究. 太原: 山

西大学硕士学位论文.

温汉捷. 1999. 分散元素硒的有机成矿作用初析. 地球学报, 20(2): 190-194.

温汉捷, 肖化云. 1998. 硒矿物综述. 岩石矿物学杂志, 17(3): 260-265.

温汉捷, 周正兵, 朱传威, 等. 2019. 稀散金属超常富集的主要科学问题. 岩石学报, 35(11): 3271-3291.

夏卫平, 谭建安. 1990. 中国一些岩类中硒的比较研究. 环境科学学报, 10(2): 125-131.

肖鹏, 王红军, 叶逢春, 等. 2020. 稀散金属硒、碲回收工艺现状与展望. 金属矿山, 4: 52-60.

姚林波, 高振敏, 龙洪波. 1999. 分散元素硒的地球化学循环及其富集作用. 地质地球化学, 27(3): 62-67.

张军营, 任德贻, 许德伟, 等. 1999. 煤中硒的研究现状. 煤田地质与勘探, 27(2): 17-19.

郑宝山, 严良荣, 毛大钧, 等. 1993. 鄂西的硒资源及其开发战略研究. 自然资源学报, 8(3): 204-212.

郑达贤, 李日邦, 王五一. 1982. 初论世界低硒带. 环境科学学报, 2(3): 241-250.

中国环境监测总站. 1990. 中国土壤元素背景值. 北京: 中国环境科学出版社.

中华人民共和国国土资源部. 2016. 中华人民共和国地质矿产行业标准-土地质量地球化学评价规范(S) DZ/T 0295-2016. 北京: 地质出版社.

中华人民共和国地方病与环境图集编纂委员会. 1989. 中华人民共和国地方病与环境图集. 北京: 科学出版社: 40-117.

周令治, 陈少纯. 2008. 稀散金属提取冶金. 北京: 冶金工业出版社.

朱建明, 秦海波, 罗泰义, 等. 2007. 西南寒武、二叠富硒碳质岩中硒结合态的比较研究. 高校地质学报, 13(1): 69-74.

Abu-Erreish G M, Whitehead E I, Olson, O E. 1968. Evolution of volatile selenium from soils. Soil Sci, 106: 415-420.

Alford E R, Lindblom S D, Pittarello M, et al. 2014. Roles of rhizobial symbionts in Astragalus selenium hyperaccumulation. Am J Bot, 101: 1895-1905.

Alford E R, Pilon-Smits E A H, Marcus M A, et al. 2012. No evidence for a costof tolerance: selenium hyper-accumulation by Astragalus does not inhibit root nodule symbiosis. Am J Bot, 99: 1930-1941.

Amouroux D, Liss P S, Tessier E, et al. 2001. Role of oceans as biogenic sources of selenium. Earth Planet Sci Lett, 189: 277-283.

Bajaj M, Eiche E, Neumann T, et al. 2011. Hazardous concentrations of selenium in soil and groundwater in North-West India. J Hazard Mater, 189: 640-646.

Banuelos G S, Lin Z Q. 2009. Development and Use of Biofortified Agricultural Products. USA, Boca Raton: CRC Press: 297.

Banuelos G S, Terry N, Leduc D L, et al. 2005. Field trial of transgenic Indian mustard plants shows enhanced phytoremediation of selenium-contaminated sediment. Environ Sci Technol, 39: 1771-1777.

Barkes L, Fleming R W. 1974. Production of dimethylselenide gas from inorganic selenium by eleven soil fungi. Bull Environ Contam Toxicol, 12(3): 308-311.

Benovic J L, Shorr R G, Caron M G, et al. 1984. The mammalian beta 2-adrenergic receptor: purification and characterization. Biochemistry, 23(20): 4510-4518.

Biesalski H K. 2013. Hidden Hunger. Berlin Heidelberg: Springer-Verlag.

Blazina T, Läderach A, Jones G D, et al. 2017. Marine primary productivity as a potential indirect source of selenium and other trace elements in atmospheric deposition. Environ Sci Technol, 51(1): 108-118.

Blazina T, Sun Y, Voegelin A, et al. 2014. Terrestrial selenium distribution in China is potentially linked to monsoonal climate. Nature Communications, 5: 4717.

Bodnar M, Konieczka P, Namiesnik J. 2012. The properties, functions, and use of selenium compounds in living organisms. J Environ Sci Health Part C, 30: 225-252.

Cao Z H, Wang X C, Yao D H, et al. 2001. Selenium geochemistry of paddy soils in Yangtze River Delta. Environ Int, 26(5): 335-339.

Carey A M, Scheckel K G, Lombi E, et al. 2012. Grain accumulation of selenium species in rice (*Oryza sativa* L.). Environ Sci Technol, 46: 5557-5564.

Chasteen T G, Bentley R. 2003. Biomethylation of selenium and tellurium: Microorganisms and plants. Chem Rev, 103: 1-25.

Chau Y K, Wong P T S, Silverberg B A, et al. 1976. Methylation of selenium in the aquatic environment. Science, 192: 1130-1131.

Chauhan R, Awasthi S, Srivastava S, et al. 2019. Understanding selenium metabolism in plants and its role as a beneficial element. Critical Reviews in Environmental Science and Technology, 49(21): 1937-1958.

Chizhikov D M, Shchastlivyi V P. 1968. Selenium and Selenides. London: Collets Ltd.

Clark L C, Combs G F Jr, Turnbull B W, et al. 1996. Effects of selenium supplementation for cancer prevention in patients with carcinoma of the skin. A randomized controlled trial. Nutritional Prevention of Cancer Study Group. JAMA, 276(24): 1957-1963.

Combs G F. 2001. Selenium in global food systems. Br J Nutr, 85: 517-547.

Cook N J, Ciobanu C L, Pring A, et al. 2009. Trace and minor elements in sphalerite: A LA-ICPMS study. Geochimica et Cosmochimica Acta, 73(16): 4761-4791.

Cui Z, Huang J, Peng Q, et al. 2017. Risk assessment for human health in a seleniferous area, Shuang'an, China. Environmental Science and Pollution Research, 24(21): 17701-17710.

Dinh Q T, Wang M, Tran T A T, et al. 2019. Bioavailability of selenium in soil-plant system and a regulatory approach. Critical Reviews in Environmental Science and Technology, 49(6): 443-517.

Doran J W, Alexander M. 1977. Microbial transformations of selenium. Appl Environ Microbiol, 33: 31-37.

Dumont E, Vanhaecke F, Cornelis R. 2006. Selenium speciation from food source to metabolites: a critical review. Anal Bioanal Chem, 385, 1304-1323.

Durán P, Acuna J J, Jorquera M A, et al. 2014. Endophytic bacteria from selenium-supplemented wheat plants could be useful for plant-growth promotion, biofortification and *Gaeumannomyces graminis* biocontrol in wheat production. Biol Fertil Soils, 50(6): 983-990.

El Mehdawi A F, Cappa J J, Fakra S C. 2012. Interactions of selenium and non-accumulators during co-cultivation on seleniferous or non-seleniferous soil—the importance of having good neighbors. New Phytol, 194: 264-277.

El Mehdawi A F, Paschke M, Pilon-Smits E A H. 2015. Symphyotrichum ericoides populations from seleniferous and non- seleniferous soil display striking variation in selenium accumulation. New Phytol, 206: 231-242.

El Mehdawi A F, Quinn C F, Pilon-Smits E A H. 2011. Effects of selenium hyperaccumulation on plant-plant interactions: evidence for elemental allelopathy. New Phytol, 191: 120-131.

Elizabeth A H, Pilon-Smits, Lenny H E. 2017. Selenium in Plants: Molecular, Physiological, Ecological and Evolutionary Aspects. Switzerland: Springer International Publishing AG.

Fadlelmawla A A, Reddy K J, Vance G F. 1994. Geochemical processes affecting the mobility of dissolved selenium in surface coal mine backfill environment. Amexucan Wacser Resovrces Ssociation, (6): 1003-1009.

Fairweather-Tait S J, Bao Y, Broadley M R, et al. 2011. Selenium in human health and disease. Antioxid. Redox Signal, 14: 1337-1383.

FAO, WHO. 2001. Human Vitamin and Mineral Requirements. Rome: Food and Nutrition Division FAO.

Feinberg A, Stenke A, Peter T, et al. 2020. Constraining atmospheric selenium emissions using observations, global modeling, and bayesian inference. Environmental Science & Technology, 54(12): 7146-7155.

Feng R W, Wei C Y, Tu S X, et al. 2009. Effects of Se on the essential elements uptake in *Pterisvittata* L. Plant Soil, 325, 123-132.

Fernandez-Martinez A, Charlet L. 2009. Selenium environmental cycling and bioavailability: A structural chemist point of view. Rev Environ Sci Biotechnol, 8: 81-110.

Fischer E, Knutti R. 2016. Observed heavy precipitation increase confirms theory and early models. Nature Climate Change, 6: 986-991.

Fishbein L. 1983. Environmental selenium and its significance. Fundamental and Applied Toxicology, 3(5): 411-419.

Fleming G. 1980. Essential micronutrients II: iodine and selenium. In: Davis B E. Applied Soil Trace Elements. New York: Wiley: 199-234.

Fordyce F, Selinus O, Alloway B, et al. 2005. Essentials of medical geology. In: Fordyce F. Selenium Deficiency and Toxicity in the Environment. Edinburgh: Elsevier: 373-415.

Frankenberger Jr W T, Karlson U. 1989. Environmental factors affecting microbial production of dimethylselenide in a selenium-contaminated sediment. Soil Sci Soc Am J, 53: 1435-1442.

Freeman J L, Zhang L H, Marcus M A, et al. 2006. Spatial imaging, speciation and quantification of selenium in the hyperaccumulator plants *Astragalus bisulcatus* and *Stanleya pinnata*. Plant Physiol, 142: 124-134.

Gao J, Liu Y, Huang Y, et al. 2011. Daily selenium intake in a moderate selenium deficiency area of Suzhou, China. Food Chemistry, 126(3): 1088-1093.

George M W. 1998. Selenium and Tellurium. Virginia: U.S. Geological Survey: 6-13.

Ghauri B M, Mirza M I, Richter R, et al. 2001. Composition of aerosols and cloud water at a remote mountain site (2.8 kms) in Pakistan. Chemosphere: Global Science Change, 3(1): 51-63.

Gissel-Nielsen G, Gupta U C, Lamand M, et al. 1984. Selenium in soil and plants and its importance in livestock and human nutrition. Adv Agron, 37: 397-460.

Goldschmidt V M, Hefter O. 1933. Zur Geochemie des Selens. Nachr. Ges. Wiss. Göttingen, Math.-Phys. Kl., Fachgruppe IV: 245-252.

Goldschmidt V M, Strock L. 1935. Zur Geochemie des Selens: II. Nachr. Ges. Wiss. Göttingen, Math.-Phys. Kl., IV: 123-142.

Greenwood N N, Earnshaw A. 1997. Chemistry of the Elements. Oxford: Reed Educational and Professional Publishing Ltd.: 747-750.

Gupta M, Gupta S. 2017. An overview of selenium uptake, metabolism, and toxicity in plants. Frontiers in Plant Science, 7: 2074.

Gupta U C, Gupta S C. 2000. Selenium in soils and crops, its deficiencies in livestock and humans: Implications for management. Commun. Soil Sci Plant Anal, 31: 1791-1807.

Harris J, Schneberg K A, Pilon-Smits E A H. 2014. Sulfur-selenium-molybdenum interactions distinguish selenium hyperaccumulator Stanleya pinnata from non-hyperaccumulator *Brassica juncea* (Brassicaceae). Planta, 239: 479-491.

Haygarth P M, Fowler D, Sturup S, et al. 1994. Determination of gaseous and particulate selenium over a rural grassland in the U.K. Atmos. Environ, 28(22): 3655-3663.

Hoffman J E, King M G. 1997. Selenium and Selenium Compounds. New York: John Wiley & Sons Inc: 686-719.

John D A, Taylor R D. 2016. By-products of porphyry copper and molybdenum deposits. Econ Geol, 18: 137-164.

Johnson C C, Ge X, Green K A, et al. 2000. Selenium distribution in the local environment of selected villages of the Keshan Disease belt, Zhangjiakou District, Hebei Province, People's Republic of China. Appl Geochem, 15: 385-401.

Jones G D, Droz B, Greve P, et al. 2017. Selenium deficiency risk predicted to increase under future climate change. P Natl Acad Sci USA, 114(11): 2848-2853.

Keeling R F, Manning A C. 2014. Studies of recent changes in atmospheric O_2 content. In: Keeling R, Russell L. Treatise on Geochemistry, 5. Amsterdam: Elsevier: 385-404.

Kizilstein L Y, Shokhina O A. 2001. Geochemistry of selenium in coal: environmental aspect. Geokhimiya

[Geochemistry], 4: 434-440.

Large R R, Halpin J A, Lounejeva E, et al. 2015. Cycles of nutrient trace elements in the Phanerozoic ocean. Gondwana Research, 28(4): 1282-1293.

Lawrence C R, Neff J C. 2009. The contemporary physical and chemical flux of aeolian dust: A synthesis of direct measurements of dust deposition. Chem Geol, 267: 46-63.

Lemire M, Mergler D, Huel G, et al. 2009. Biomarkers of selenium status in the amazonian context: blood, urine and sequential hair segments. J Expo Sci Environ Epidemiol, 19: 213-222.

Li D, Zhang X, Hu D, et al. 2018. Evidence of a large $\delta^{13}C$carb and $\delta^{13}C$org depth gradient for deep-water anoxia during the late Cambrian SPICE event. Geology, 46(7): 631-634.

Li H F, McGrath S P, Zhao F J. 2008. Selenium uptake, translocation and speciation in wheat supplied with selenate or selenite. New Phytol, 178: 92-102.

Lindblom S D, Fakra S C, Landon J, et al. 2013. Inoculation of selenium hyperaccumulator Stanleya pinnata and related non-accumulator Stanleya elata with hyperaccumulator rhizosphere fungi—effects on Se accumulation and speciation. Physiol Plant, 150: 107-118.

Liu H L, Wang X Q, Zhang B M, et al. 2021. Concentration and distribution of selenium in soils of mainland China, and implications for human health. Journal of Geochemical Exploration, 220: 1-14.

Liu Y, Li F, Yin X B, et al. 2011. Plant-based biofortification: from phytoremediation to Se-enriched agriculture products. In: Sharma S K, Mudhoo A. Green Chemistry for Environmental Sustainability. BocaRaton: CRC Press: 341-356.

Long J A, Large R R, Lee M S Y, et al. 2016. Severe selenium depletion in the Phanerozoic oceans as a factor in three global mass extinction events. Gondwana Research, 36: 209-218.

Matamoros-Veloza A, Newton R J, Benning L G. 2011. What controls selenium release during shale weathering? Applied Geochemistry, 26(supp-S): S222-S226.

Matich A J, McKenzie M J, Brummell D A, et al. 2009. Organoselenides from *Nicotiana tabacum* genetically modified to accumulate selenium. Phytochemistry, 70: 1098-1106.

Mehdi Y, Hornick J L, Istasse L, et al. 2013. Selenium in the environment, metabolism and involvement in body functions. Molecules, 18: 3292-3311.

Meija J, Montes-Bayon M, Le Duc, et al. 2002. Simultaneous monitoring of volatile selenium and sulfur species from se accumulating plants (wild type and genetically modified) by gc/ms and GC/ICPMS using solid-phase microextraction for sample introduction. Anal Chem, 74: 5837-5844.

Michalke B. 2018. Molecular and Integrative Toxicology: Selenium. Switzerland: Springer International Publishing AG, part of Springer Nature.

Minaev V S, Timoshenkov S P, Kalugin V V. 2005. Structural and phase transformations in condensed selenium. Journal of Optoelectronics and Advanced Materials, 7(4): 1717-1741.

Moore K L, Schroder M, Lombi E, et al. 2010. NanoSIMS analysis of arsenic and selenium in cereal grain. New Phytologist, 185(2): 434-445.

Müller A, Bertram A, Moschos A. 2012. Seasonal and national differences in the selenium supply of horses across Europe. Tierarztl Prax Ausg G Grosstiere Nutztiere, 40(3): 157-166.

Nriagu J O, Pacyna J M. 1988. Quantitative assessment of worldwide contamination of air, water, and soils by trace metals. Nature, 333: 134-139.

Nriagu J O. 1989. A global assessment of natural sources of atmospheric trace metals. Nature, 338: 47-49.

Palmer C A, Lyons P C. 1996. Selected elements in major minerals from bituminous coal as determined by INAA: implications for removing environmentally sensitive elements from coal. International Journal of Coal Geology, 32(1-4): 151-166.

Patai S. 1987. The Chemistry of Organic Selenium and Tellurium Compounds. Great Britain: The Bath Press.

Piao S, Ciais P, Huang Y, et al. 2010. The impacts of climate change on water resources and agriculture in China. Nature, 467: 43-51.

Pilon-Smits E A H, LeDuc D L. 2009. Phytoremediation of selenium using transgenic plants. Curr Opin Biotechnol, 20: 207-212.

Plant J A, Kinniburgh D G, Smedley P L, et al. 2004. Arsenic and selenium. In: Holland H D, Turekian K K eds. Treatise on geochemistry 9, Environmental geochemistry. Amsterda: Elsevier: 17-66.

Preedy V R. 2015. Selenium Chemistry, Analysis, Function and Effects. Cambridge: RSC.

Presser T S, Ohlendorf H M. 1987. Biogeochemical cycling of selenium in the SanJoaquin Valley, California, USA. Environmental Management, 11(6): 805-821.

Prins C N, Hantzis L J, Quinn C F, et al. 2011. Effects of selenium accumulation on reproductive functions in *Brassica juncea* and *Stanleya pinnata*. J Exp Bot, 62: 5233-5239.

Qian W, Lin, X. 2005. Regional trends in recent precipitation indices in China. Meteorol Atmos Phys, 90: 193-207.

Qin H, Zhu J, Liang L, et al. 2013. The bioavailability of selenium and risk assessment for human selenium poisoning in high-Se areas, China. Environment International, 52: 66-74.

Quinn C F, Prins C N, Gross A M. 2011. Selenium accumulation in flowers and its effects on pollination. New Phytol, 192: 727-737.

Ranjard L, Nazaret S, Cournoyer B. 2003. Freshwater bacteria can methylate selenium through the thiopurine methyltransferase pathway. Appl Environ Microb, 69(7): 3784-3790.

Ranjard L, Prigent-Combaret C, Nazaret S, et al. 2002. Methylation of inorganic and organic selenium by the bacterial thiopurine methyltransferase. J Bacteriol, 184(11): 3146-3149.

Ren D, Zhao F, Wang Y, et al. 1999. Distributions of minor and trace elements in Chinese coals. International Journal of Coal Geology, 40(2-3): 109-118.

Ross H B. 1985. An atmospheric selenium budget for the region 30° N to 90° N. Tellus B: Chemical and Physical Meteorology, 37(2): 78-90.

Rotruck J T, Pope A L, Ganther H E, et al. 1973. Selenium: Biochemical role as a component of glutathione peroxidase. Science, 79(4073): 588-590.

Rudnick R L, Gao S. 2003. Composition of the Continental Crust. Treatise Geochemistry: Elsevier Ltd: 1-64.

Santhosh Kumar B, Priyadarsini K I. 2014. Selenium nutrition: How important is it? Biomedicine & Preventive Nutrition, 4: 333-341.

Saunders J A, Brueseke M E. 2012. Volatility of Se and Te during subduction-related distillation and the geochemistry of epithermal ores of the western United States. Economic Geology, 107(1): 165-172.

Savel'ev V F, Timofeev N I. 1977. On selenium in coal deposits. Sb. nauch. tr. Tashkentsk, 530: 101-106.

Schiavon M, Pittarello M, Pilon-Smits E A H. 2012. Selenate and molybdate alter sulfate transport and assimilation in *Brassica juncea* L. Czern: Implications for phytoremediation. Environ Exp Bot, 75: 41-51.

Schilling K. 2010. Stable Isotope Fractionation of Selenium by Biomethylation in Soil. German, Mainz Johannes Gutenberg-University Mainz.

Schwarz K, Foltz C M, 1957. Selenium as an integral part of factor-3 against dietary necrotic liver degeneration. Journal of the American Chemical Society, 79: 3292-3293.

Simon G, Essene E J. 1996. Phase relation among selenides, sulfides, tellurides and oxides: Ⅰ. Thermadynamic data and calculated equilibria. Econ Geol, 91: 1183-1208.

Simon G, Kesler S E, Essene E J. 1997. Phase relation among selenides, tellurides and oxides: Ⅱ. Application to selenide-bearing ore deposits. Econ Geol, 92: 468-484.

Sindeeva N D. 1964. Mineralogy and Types of Deposits of Selenium and Tellurium. New York: Interscience Publishers.

Sors T G, Ellis D R, Salt D E. 2005. Selenium uptake, translocation, assimilation and metabolic fate in plants. Photosynth Res, 86: 373-389.

Staicu L C, van Hullebusch E D, Lens P N L, et al. 2015. Electrocoagulation of colloidal biogenic selenium. Environ Sci Pollut Res, 22: 3127-3137.

Sun G X, Liu X, Williams P N, et al. 2010. Distribution and translocation of selenium from soil to grain and its speciation in paddy rice (*Oryza sativa* L.). Environ Sci Technol, 44(17): 6706-6711.

Sun G X, Meharg A A, Li G, et al. 2016. Distribution of soil selenium in China is potentially controlled by deposition and volatilization? Scientific Reports, 6: 20953.

Sun G X, Van de Wiele T, Alava P, et al. 2017. Bioaccessibility of selenium from cooked rice as determined in a simulator of the human intestinal tract (SHIME). Journal of the Science of Food and Agriculture, 97(11): 3540-3545.

Systra Y J. 2010. Bedrock and quaternary sediment geochemistry and biodiversity in Eastern Fennoscandia and Estonia. Forestry Studies Metsanduslikud Uurimused, 53: 35-52.

Tabelin C B, Igarashi T, Villacorte-Tabelin M, et al. 2018. Arsenic, selenium, boron, lead, cadmium, copper, and zinc in naturally contaminated rocks: A review of their sources, modes of enrichment, mechanisms of release, and mitigation strategies. Science of the Total Environment, 645: 1522-1553.

Tan J A, Zhu W Y, Wang W Y, et al. Selenium in soil and endemic diseases in China. Science of the Total Environment, 284: 227-235.

Taylor S R, Mclennan S M. 1995. The geochemical evolution of the continental crust. Reviews of Geophysics, 33(2): 293-301.

Tian H, Ma Z, Chen X, et al. 2016. Geochemical characteristics of selenium and its correlation to other elements and minerals in selenium-enriched rocks in Ziyang County, Shaanxi Province, China. Journal of Earth Science, 27(5): 763-776.

Wang L, Ju Y, Liu G, et al. 2009. Selenium in Chinese coals: distribution, occurrence, and health impact. Environmental Earth Sciences, 60(8): 1641-1651.

Wang Z, Gao Y. 2001. Biogeochemical cycling of selenium in Chinese environments. Applied Geochemistry, 16(11-12): 1345-1351.

Wedepohl K H. 1995. The composition of the continental crust. Geochimica Et Cosmochimica Acta, 59(7): 1217-1232.

Weiss H V, Koide M, Goldberg E D. 1971. Selenium and sulfur in a greenland ice sheet: relation to fossil fuel combustion. Science, 172(3980): 261-263.

Wen H, Carignan J. 2009. Ocean to continent transfer of atmospheric Se as revealed by epiphytic lichens. Environ Pollut, 157: 2790-2797.

White P J, Broadley M R. 2005. Biofortifyingcrops with essential mineral elements. Trends in Plant Science, 12(10): 586-593.

White P J, Brown P H. 2010. Plant nutrition for sustainable development and global health. Ann Bot, 105: 1073-1080.

WHO (World Health Organization). 1987. Environmental Health Criteria: Selenium Environmental Health Criteria: 1-110.

Williams P N, Lombi E, Sun G X, et al. 2009. Selenium characterization in the global rice supply chain. Environ Sci Technol, 43: 6024-6030.

Winkel L H E, Vriens B, Jones G D, et al. 2015. Selenium cycling across soil-plant-atmosphere interfaces: A critical review. Nutrients, 7: 4199-4239.

Winkel L H, Johnson C A, Lenz M, et al. 2012. Environmental selenium research: from microscopic processes to global understanding. Environ Sci Technol, 46(2): 571-579.

Wu Z L, Banuelos G S, Lin Z Q, et al. 2015. Biofortification and phytoremediation of selenium in China. Front Plant Sci, 6(1): 136.

Yang G Q, Wang S Z, Zhou R H, et al. 1983. Endemic selenium intoxication of humans in China. American Journal of Clinical Nutrition, 37(5): 872-881.

Yao B M, Chen P, Sun G X. 2020. Distribution of elements and their correlation in bran, polished rice, and whole grain. Food Sci Nutr, 8(2): 982-992.

Yasin M, El-Mehdawi A F, Anwar A, et al. 2015. Microbial-enhanced selenium and iron biofortification of wheat (*Triticum aestivum* L.)—applications in phytoremediation and biofortification. Int J Phytoremediation, 17: 341-347.

Ye L, Cook N J, Ciobanu C L, et al. 2011. Trace and minor elementsin sphalerite from base metal deposits in South China: A LA-ICPMS study. Ore Geology Reviews, 39(4): 188-217.

Yin X B, Yuan L X. 2012. Phytoremediation and Biofortification: Two Sides of One Coin. Berlin: Springe: 1-31.

Ylaranta T. 1983. Sorption of selenite and selenate added in the soil. Ann Agr Fenn, 22: 9-39.

Yokota A, Shigeoka S, Onishi T, et al. 1988. Selenium as inducer of glutathione peroxidase in low-CO_2-grown *Chlamydomonas reinhardtii*. Plant Physiol, 86: 649-651.

Yu T, Yang Z, Lv Y, et al. 2014. The origin and geochemical cycle of soil selenium in a Se-rich area of China. Journal of Geochemical Exploration, 139: 97-108.

Yudovich Y E, Ketris M P. 2006. Selenium in coal: A review. International Journal of Coal Geology, 67(1-2): 112-126.

Zhang L H, Hu B, Li W. 2014. OsPT2, a phosphate transporter, is involved in the active uptake of selenite in rice. New Phytol, 201: 1183-1191.

Zhang L H, Yu F Y, Shi W M, et al. 2010. Physiological characteristics of selenite uptake by maize roots in response to different pH levels. J Plant Nutr Soil Sci, 173: 412-422.

Zhang X Y, Arimoto R, An Z S. 1997. Dust emission from Chinese desert sources linked to variations in atmospheric circulation. Atmospheres J Geophys Res, 102, D23: 28041-28047.

Zhao X Q, Mitani N, Yamaji N, et al. 2010. Involvement of silicon influx transporter OsNIP2;1 in selenite uptake in rice. Plant Physiol, 153: 1871-1877.

Zhu Y G, Pilon-Smits E A H, Zhao F J, et al. 2009. Selenium in higher plants: understanding mechanisms for biofortification and phytoremediation. Trends in Plant Science, 19: 436-442.

Zieve R, Peterson P J. 1981. Factors influencing the volatilization of selenium from soil. Sci Total Environ, 19: 277-284.

第二章 土壤中硒的含量、形态及其有效性

硒（Se）是哺乳动物及部分藻类必需的微量营养元素，其丰缺与人体健康密切相关。硒大量存在于地球表面岩石中，且依据在环境中浓度的不同而具有双重生物学效应（有益/毒害）。人体摄入的硒主要源于日常饮食，而膳食硒主要通过食物链源于土壤（Emmanuelle et al.，2012）。因此，土壤硒是人和动物体内硒的基本来源，而植物是土壤硒的重要吸收和转移者（Broadley et al.，2006）。大量研究表明，硒的缺乏或者毒害作用具有明显的地带性，且与当地土壤中硒的含量及其有效性显著相关（Supriatin et al.，2015）。

土壤有效硒是指土壤中能够被植物直接吸收利用的硒，是影响植物硒含量的主要决定因子（Fordyce，2013），可通过土壤中硒的化学形态，或者通过可提取硒（操作性定义）与植物硒吸收的关系来确定（朱永官等，2018；周菲等，2022）。许多研究均已证实土壤硒的总量无法很好地表征和评价硒的生物有效性，而土壤中硒的赋存形态及其价态组成才是决定其溶解性、迁移性、有效性和毒性的关键（Bawrylak-Nowak，2013；Tolu et al.，2011）。

土壤中的硒常以不同的化学形态及价态存在，且受土壤理化性质及环境条件等影响，不断地与有机质、黏土矿物、铁锰氧化物等土壤组分发生着吸附-解吸、沉淀-溶解、氧化-还原、配位作用等一系列环境化学过程，使其形态和价态常常处于动态变化之中，进而影响其有效性（Chen et al.，2010）。土壤不仅是生物圈之间的联系枢纽，亦是人类赖以生存的基础，而其中硒的含量及形态/价态组成（即有效性）是决定植物硒吸收的关键，且通过食物链进一步影响人体健康。因此，研究土壤中硒的含量、形态/价态的分布特征，充分了解土壤中硒的有效性及其影响因子，并基于此提出土壤硒有效性的调控措施，避免硒供应不足或者过量引发疾病，对硒的生物强化和硒污染地区的植物修复均具有重要意义。

第一节 土壤中硒的来源及其含量

一、土壤中硒的来源

硒为稀有分散元素，介于氧族元素硫和金属元素碲之间（原子序数 34）。地壳中硒的含量很低，且受到成土母质和地球化学环境因素的影响，其在全球土壤环境中的分布极不均匀，甚至呈点状分布。

土壤中硒的来源分为天然源及人为源。成土母质/母岩是土壤中硒的主要天然来源，但不同矿石硒含量差异较大，如沉积岩形成的土壤中硒含量高于火成岩和变质岩等（Long and Luo，2017），高硒含量岩石如碳质板岩 [13～42 mg/kg（Luo et al.，2004）]、黑色页岩 [0.35～25.08 mg/kg（Tian and Luo，2017）] 和碳质硅质岩 [12.9～50.9 mg/kg（Feng et al.，2012）] 的自然风化过程是硒进入土壤的重要途径（Qin et al.，2012）。同时，

煤炭开采、火力发电、矿物肥料/富硒有机物料的使用等同样是导致岩石矿物或含硒农业废弃物中的硒进入土壤的关键（Huang et al.，2009），也是土壤中硒的重要人为来源。

此外，海洋中硒可经过转化以二甲基硒和其他挥发性硒化合物的形式进入大气，并经过沉降作用进入土壤，这也是土壤中硒的一个重要来源（郭莉等，2012）。大气中硒的滞留时间大约是5d，大气中颗粒态硒和水溶含氧阴离子硒主要通过湿沉降作用（约占81%，干沉降约占19%）进入土壤（Feinberg et al.，2020），据统计，每年约有6600 t的硒通过大气沉降进入土壤（杨兰芳，2000）。

二、世界土壤硒含量

土壤中硒的含量取决于土壤母质的组成，岩石的风化是土壤中硒的主要来源。世界范围内大部分土壤硒含量为0.1~2.00 mg/kg，平均约为0.40 mg/kg；缺硒地区土壤硒含量可以低至0.01 mg/kg（Fordyce，2007），但部分硒中毒地区土壤硒含量最高可达1200 mg/kg（Fordyce，2013；Jones and Winkel，2016）。此外，缺硒地区可以紧邻高硒含量地区（Fernández-Martínez and Charlet，2009）。

不同地区或国家由于成土母质、气候（温度和降水）和地形条件（坡度和海拔）等的不同，土壤硒含量差异较大（表2-1）。世界范围内土壤硒含量高的地区主要包括：美国的怀俄明州、南达科他州（20~40 mg/kg；Gerla et al.，2011）和加利福尼亚州（40~70 mg/kg；Sharma et al.，2009），中国的恩施和紫阳，爱尔兰（Séby et al.，1998），哥伦比亚（Minich，2022），印度（Dhillon and Dhillon，2003），委内瑞拉（Combs，2001）等。高硒含量的母岩和大气中硒的沉降是这些地区土壤硒含量偏高甚或达到中毒水平的主要原因（Oldfield，2002）。而欧洲许多地区土壤硒含量相对较高则归因于海洋的自然沉降（如爱尔兰、英国、荷兰）或酸沉降（如德国、斯洛伐克、波兰和捷克共和国）（Haug et al.，2007）。

表2-1 世界不同国家或地区土壤硒含量的分布

国家/地区	样本数	硒含量/（mg/kg）	均值/（mg/kg）	参考文献
日本	180	0.05~2.80	0.51	Yamada et al.，2009
瑞典	5170	0.05~13.3	0.30	Shand et al.，2012
英国	—	0.10~4.00	—	Broadley et al.，2006
芬兰	—	—	0.21	Eurola et al.，2003
比利时	539	0.14~0.70	0.33	De Temmerman et al.，2014
伊朗	17	0.04~0.45	0.23	Nazemi et al.，2012
英国（苏格兰）	661	0.06~19.2	1.04	Shand et al.，2012
挪威西部	485	0.06~2.73	0.30	Wu and Lag，1988
荷兰	83	0.12~1.97	0.58	Supriatin et al.，2015
印度北部	15	0.21~0.55	0.37	Yadav et al.，2005
孟加拉国	24	—	0.044	Spallholz et al.，2008
加拿大	—	0.1~6.1	—	Rosenfeld and Beath，1964
美国加州	—	42.76~78.09	—	Wu et al.，1993

续表

国家/地区	样本数	硒含量/（mg/kg）	均值/（mg/kg）	参考文献
美国南达科他州	24	20～40	—	Gerla et al.，2011
美国密西西比	8	—	18.94	Shaheen et al.，2017
巴西圣保罗	58	0.09～1.61	0.19	Gabos et al.，2014
埃及尼罗河	8	—	38.7	Shaheen et al.，2017
西班牙东南部	490	0.01～2.78	0.40	Perez-Sirvent et al.，2010
俄罗斯奥伦堡州	525	0.12～0.37	0.24	Skalny et al.，2019
均值	—	0.1～2.00	0.40	Jones and Winkel，2016

谭建安（1989）从硒的营养角度出发定义了土壤中总硒含量的界限值：＜0.125 mg/kg 为缺硒地区，0.125～0.175 mg/kg 为少硒地区，0.175～0.45 mg/kg 为足硒地区，0.45～2.0 mg/kg 为富硒地区，2.0～3.0 mg/kg 为高硒地区，3.0 mg/kg 以上为硒中毒地区。世界范围内约有 40 个国家和地区缺硒，其主要分布在南、北半球一条大致纬向性的带上，基本上位于 30°以上的中高纬度地区，如芬兰、塞尔维亚中部、刚果部分地区（Reilly，1996），而中国从东北到西南的整个条带都是极度缺硒的地区（Sun et al.，2016）。整体来看，世界范围内缺硒土壤的面积远大于富硒土壤（Dhillon and Dhillon，2003）。

世界范围内土壤硒中毒与缺乏的地区有可能紧邻，如英国土壤硒含量从 0.1 mg/kg 到 4.0 mg/kg 不等（Broadley et al.，2006）。目前，对于世界范围内土壤硒的分布情况了解仍较为有限，且由于硒在土壤中变异性较大，因而对很多地区土壤硒资源的认知仍然是空白，但可以肯定的是，世界多数地区食物中的硒含量并未达到理想水平（Dinh et al.，2018）。作物缺硒可大致归因于三种情况：一是土壤中固有的硒含量低；二是土壤中硒的有效性低或难以被植物吸收利用；三是一些地区（如新西兰）母岩中硒含量低和植物对硒的吸收能力弱的叠加作用（Haug et al.，2007）。

三、中国土壤硒含量

我国地域幅员辽阔，地理环境十分复杂。受此影响，不同地区表层土壤硒的含量分布同样存在显著差异（Yu et al.，2016），既有黑龙江（0.147 mg/kg；徐强，2016）、西藏（0.15 mg/kg；Zhang et al.，2002）和河北张家口（0.21 mg/kg；李振宁，2010）等缺硒地区，也有湖北恩施（36.69 mg/kg；Yuan et al.，2012）和陕西紫阳（79.08 mg/kg；Tian et al.，2016）等硒中毒地区。Dinh 等（2018）和 Tan 等（2002）的研究均表明，我国表层土壤硒含量范围为 0.005～79.08 mg/kg，平均为 0.29 mg/kg。一般农田土壤中硒含量变幅较大，但大部分在 0.2～0.3 mg/kg 范围内，平均约 0.25 mg/kg（廖自基，1992）。

我国不同地区表层土壤硒含量依次呈现出西北＞南方＞中部＞东部＞西南＞东北＞华北的规律（Dinh et al.，2018），存在一个从东北到西南的土壤硒缺乏和临界缺乏的地带（Tan，1989）；但最新的研究发现，即使这些典型的缺硒带上的一些有机质含量比较高、淋溶能力比较强的碳酸盐母质，也存在一些呈带状或点状分布的富硒地区（Liu et al.，2021a），但仍有约 51% 的土地面积存在不同程度的缺硒（Dinh et al.，2018）。

富硒地区主要分布在新疆焉耆、贵州开阳、湖南桃源、江西龙南、重庆江津、青海平安等地区（Dinh et al.，2018）。不同富硒地区，由于母质和成土因子的不同，土壤硒含量差异很大。我国一些典型富硒地区土壤中硒分布情况见表 2-2。

表 2-2 我国一些典型富硒地区土壤硒含量的分布

地区	样本数	硒含量/（mg/kg）	均值/（mg/kg）	参考文献
湖南桃源	111	0.18~7.05	0.76	Ni et al.，2016
贵州开阳	—	0.46~2.31	1.42	仝双梅等，2013
浙江永嘉	51	0.157~0.633	0.382	Xu et al.，2018
江西丰城	699	0.13~0.69	0.33	朱青等，2020
江苏如皋	203	0.07~0.17	0.13	Sun et al.，2009
陕西紫阳	37	0.50~16.96	4.78	Du et al.，2018
湖北鱼塘坝	161	0.41~42.3	4.75	Zhu et al.，2008
湖北恩施	19 734	0.03~86.59	1.12	Li et al.，2020
重庆江津	156	0.039~1.110	0.315	Liu et al.，2021a
重庆黔江	8 789	0.02~14.94	0.43	王锐等，2020
青海平安	573	0.076~1.88	0.369	喻大松，2015
广西三江	2 751	0.10~14.41	0.63	罗海怡等，2021
广西永福	104	0.27~1.40	0.80	Shao et al.，2018
广西马山	492	0.20~3.54	0.76	张春来等，2021
广东惠来县	321	0.02~1.41	0.50	王涵植等，2020
宁夏石嘴山	8 832	0.02~1.8	0.26	王志强等，2022
河南新密	386	0.06~3.99	0.44	毛香菊等，2021

第二节 土壤中硒的形态

土壤硒是否可被植物吸收利用取决于其存在的化学形态和价态，且很多研究均已证实对于土壤硒形态和价态的研究比总量更有意义（Peng et al.，2017）。土壤中的硒在自然条件下常以不同的化学形态和价态存在，且不同硒的形态或价态随环境条件改变而发生相互转化（Nakamaru and Altansuvd，2014；Wang et al.，2013）。

一、土壤中硒的价态

硒属于变价元素，以无机或者有机形式存在于土壤中。无机硒包括+6 价硒［硒酸盐 Se(Ⅵ)］、+4 价硒［亚硒酸盐 Se(Ⅳ)］、0 价［元素硒 Se(0)］及-2 价［硒化物 Se(-Ⅱ)］（Tamás et al.，2010）。在通气良好的碱性土壤中，硒主要以 Se(Ⅵ) 的形式存在于土壤溶液中，或通过静电吸附作用结合在土壤颗粒表面（郭璐等，2013；Wang et al.，2012）；而在酸性和还原性土壤中，硒酸盐常被还原为亚硒酸盐，因此主要以 Se(Ⅳ) 的形式存在于土壤溶液中、吸附于土壤颗粒表面或与土壤固相组分（如铁锰氧化物、水合氧化物及有机质等）结合（瞿建国等，1998；Li et al.，2017）。Se(0) 常被有机质包裹或以残渣态形式存在于土壤中（Kausch and Pallud，2013）。无机结合态的 Se(-Ⅱ) 多以沉淀形式存在；而有机结合态的 Se(-Ⅱ) 则存在于土壤溶液中，以挥发态（二甲基硒）脱离土壤或与土壤固相结合存在于有机质中（王丹，2019；Zhang and Moore，1996）。此外，硒化氢（H_2Se）

是自然界中微生物作用的产物（Thiry et al.，2012）。

二、土壤中硒的形态分级

有关土壤中硒的形态分级曾有简要的归纳和总结：一是按价态（并考虑配位官能团或物理状态）分级，包括元素态硒（Se^0）、硒化物（Se^{2-}）、硒酸盐（SeO_4^{2-}）、亚硒酸盐（SeO_3^{2-}）、有机态硒化物和挥发态硒；二是按操作定义，即按与土壤组分的结合方式来划分，包括吸附型、铝型、铁型和钙型等（朱永官等，2018）。在操作定义中，也有根据硒与土壤组分结合程度的强弱，使用不同的浸提剂和分级程序将土壤中硒通过连续提取划分为不同的形态。目前普遍采用的土壤硒形态分级方法为五步连续浸提法（Wang et al.，2012），即根据浸提的难易程度将土壤硒依次划分为水溶态（SOL-Se）、可交换态及碳酸盐结合态（EXC-Se）、铁锰氧化物结合态（FMO-Se）、有机结合态（OM-Se）和残渣态（RES-Se）。又由于有机硒在硒的地球生物化学循环中起着重要的源和库的作用（Qin et al.，2012），Abrams 和 Burau（1989）按照酸溶性将有机结合态硒进一步分为腐殖酸硒和富啡酸硒，富啡酸硒再细分为亲水性硒和憎水性硒。

由于不同类型土壤组成和性质的差异，不同土壤中各形态硒含量的占比各异。

（一）天然富硒土壤中硒的形态分布

富硒土壤中水溶态和可交换态硒含量均较低，且随土壤类型变异较大，较为稳定的有机结合态及残渣态硒是富硒土壤硒的主要存在形态。陕西紫阳硒中毒土壤中的硒主要以相对稳定的残渣态、铁锰氧化物结合态及有机结合态存在，三者之和约占土壤总硒的90%（Wang et al.，2012；Zhou et al.，2021），而水溶态和可交换态硒占总硒的百分比不足1%（Wang et al.，2012）；青海平安富硒地区土壤中有效硒含量占总硒的比例不足3%（喻大松，2015），青海东碱性土壤中水溶硒和可交换态硒占总硒的比例为6.38%（张亚峰等，2019）；湖北恩施高硒地区土壤碳酸盐及铁锰氧化物结合态、有机物结合态和残渣态硒分别占总硒的17%、26%和42%，稳定态硒的总量占了85%（吴少尉等，2004），最近 Li 等（2020）在恩施的研究表明，土壤水溶态和可交换态硒的总含量占总硒的比例仅为2.54%；浙江省金华市富硒地区土壤中硒主要为有机结合态（占总硒的64%），其次为残渣态硒（占总硒的16%），水溶态和可交换态硒在总硒中的占比均不足3%（黄春雷等，2013）；天津市蓟州富硒区土壤中有机结合态硒占比（44%）最高，其次为残渣态（约占35%），而水溶态、可交换态及碳酸盐结合态硒所占比例不足5%（谢薇等，2019）。Liu 等（2021b）研究发现，有机结合态及残渣态硒是陕西墡土及广西红壤中最主要的硒形态，两者之和在总硒中的占比分别为84.4%和76.9%。与此一致，Séby 等（1998）发现爱尔兰富硒土壤水溶态硒仅占总硒的2%，而有机结合态和残渣态硒分别占50%和30%。此外，矿区受硒污染土壤中60%以上的硒以残渣态形式存在，而有效态硒含量不足总硒的5%（Bujdos et al.，2005）。

（二）非富硒土壤中硒的形态分布

非富硒地区土壤硒形态同样主要以残渣态和有机结合态形式存在。Zhang 和 Moore（1996）报道湿地土壤中79%的硒以元素态和有机结合态存在，而水溶态硒和可交换态硒含量在总硒中的占比低于20%；上海不同地区土壤中有机结合态和残渣态硒含量占土壤总硒的80%以上，其余形态仅占12%~21%（瞿建国等，1998）；江苏几种低硒土壤中有机物-硫化物结合态及元素态硒约占土壤总硒含量的35%~56%，其次是残渣态（21%~40%），而水溶态硒仅占4%~9%（张艳玲等，2002）；浙江瑞安农田耕层土壤中有机结合态硒含量占总硒比例最高（约25%），硒多以残渣态形式存在，水溶态及可交换态硒仅占土壤总硒的1.1%左右（潘金德等，2007）；新疆库尔勒市绿地土壤中硒以有机结合态为主（占总硒的40%以上），其次为铁锰氧化物结合态（约20%），残渣态占15%，而水溶态硒占比不到总硒的5%（马玉锦，2009）。齐艳萍等（2012）报道黑龙江黑土中有机结合态硒的含量最高（58%~65%），残渣态硒次之（30%），而水溶态和可交换态硒含量在总硒中的占比不足15%；郎春燕等（2013）同样发现，四川水稻土中残渣态及有机结合态硒在土壤总硒中的占比超过80%。

综上所述，天然富硒和非富硒土壤中硒形态均以稳定态（五步连续浸提法中的残渣态、铁锰氧化物结合态及有机结合态）为主（占总硒80%以上）；富硒土壤中水溶态和可交换态硒（有效硒）占总硒的比例多小于5%，多数地区在2%~3%；非富硒地区土壤有效硒占总硒比例为1.1%~20%，且随土壤和环境条件变异较大。

第三节 土壤中硒的化学转化过程

土壤溶液中的硒是植物吸收利用硒的直接来源，但植物不断从土壤溶液中摄取硒的过程会影响土壤原有的不同形态硒之间的平衡。然而，土壤中的硒随环境条件不断发生溶解-沉淀、氧化-还原、配位作用、吸附-解吸、甲基化作用等转化过程，即当土壤溶液中的硒被植物根系吸收后，由于浓度梯度的形成，导致与土壤固相结合的硒会被重新释放出来补充到土壤溶液中，以维持硒在土壤固相、土壤溶液及植物根系间的动态平衡（图2-1）。因此，硒在土壤中的化学转化对土壤硒的供给或生物有效性的评价十分重要。

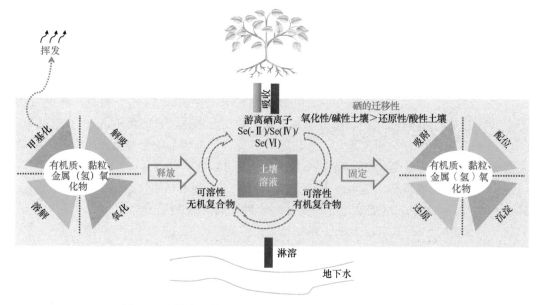

图2-1 土壤中硒的化学转化过程（改自Dinh et al.，2019）

一、溶解-沉淀

溶解-沉淀反应是直接影响土壤溶液中硒形态组成，进而影响其有效性的重要过程（Keskinen et al.，2011）。沉淀反应主要通过土壤中无机阳离子对硒的固定作用将土壤硒转化为植物难以吸收利用的形式（Séby et al.，1998）。各种理化性质中，土壤 pH 在硒的沉淀过程中起重要作用，在低 pH 条件下，Se(Ⅳ) 在硫化物溶液中被还原为 Se(0) 或 Se-S 沉淀物，且 pH 越小，沉淀作用越显著（Jung et al.，2016；Geoffroy and Demopoulos，2011）。此外，在还原条件下，沉淀作用形成的金属硒化物（如 FeSe 和 $FeSe_2$）[公式（2-1）～公式（2-2）]（何振立等，1998）或元素硒[公式（2-3）～公式（2-8）]（Kang et al.，2013；Ma et al.，2014）也限制了土壤中硒的溶解度（Hibbs et al.，2017；Séby et al.，1998），且其与土壤中铁硫化物的含量密切相关。此外，一些细菌和古细菌可以利用硒酸盐或亚硒酸盐作为代谢的末端电子受体，同样可将硒还原成元素硒或金属硒化物（Hageman et al.，2013），进而降低土壤硒的溶解度及其有效性。

$$Se^{2-} + Fe^{2+} = FeSe \tag{2-1}$$

$$2Se^- + Fe^{2+} = FeSe_2 \tag{2-2}$$

$$FeSe + 2Fe^{3+} = Se(0) + 3Fe^{2+} \tag{2-3}$$

$$FeSe_2 + 2Fe^{3+} = 2Se(0) + 3Fe^{2+} \tag{2-4}$$

$$HSeO_3^- + 2FeSe + 5H^+ = 3Se(0) + 2Fe^{2+} + 3H_2O \tag{2-5}$$

$$SeO_4^{2-} + 3FeSe + 8H^+ = 4Se(0) + 3Fe^{2+} + 4H_2O \tag{2-6}$$

$$HSeO_3^- + 2FeSe_2 + 5H^+ = 5Se(0) + 2Fe^{2+} + 3H_2O \tag{2-7}$$

$$SeO_4^{2-} + 3FeSe_2 + 8H^+ = 7Se(0) + 3Fe^{2+} + 4H_2O \tag{2-8}$$

有机质的微生物降解（Eswayah et al.，2016）和土壤矿物质（如氢氧化铁）的还原溶解（Shaheen et al.，2017）均会导致与其结合的硒被重新释放到土壤溶液中，从而提高土壤硒的有效性。此外，有机物料还田后产生的溶解性有机碳（DOC）由小分子有机酸和大分子腐殖物质组成，其中小分子有机酸可以溶解被土壤固定的硒，防止其被铁或铝氧化物/氢氧化物吸附而产生共沉淀（Dinh et al.，2017）。由此可见，微生物和小分子有机酸在土壤硒的溶解过程中也起着重要作用（Fan et al.，2018）。

二、氧化-还原

自然条件下，硒倾向于还原作用多于氧化作用，而微生物活性、竞争性电子受体和环境条件共同决定着硒的还原速率（Eswayah et al.，2016）。厌氧和好氧环境条件下均会发生硒的微生物还原作用（Bolan et al.，2014）。在厌氧条件下，一些细菌[如砷硒芽孢杆菌（*Bacillus arseniciselenatis*）、巴氏硫螺菌（*Sulfurospirillum barnesii*）和硒酸索氏菌（*Thauera selenatis*）]能够将 Se(Ⅵ) 还原为 Se(Ⅳ)，并进一步还原为 Se(0)（Dinh et al.，2019；Park et al.，2011）。Se(Ⅳ) 和 Se(Ⅵ) 在硒的微生物同化过程中被运输到细胞中，并通过同化还原进一步被转化为 Se(-Ⅱ)（Eswayah et al.，2016）；而除同化还原外，硒的异化还原同样是一个非常重要的过程（Wilber，1980）。厌氧条件下，Se(Ⅳ) 可在异化反应中被化学还原剂（如硫化物或羟胺）或生化还原剂（如谷胱甘肽还原酶）还原为 Se(0)

（Kausch and Pallud，2013）。此外，施入硫酸盐（SO_4^{2-}）和有机物料或 pH 较高的铁氧化物共存时，有助于将氧化态的 Se(Ⅵ) 还原为 Se(0) 或 Se(-Ⅱ)（Olegario et al.，2010）；而硝酸盐（NO_3^-）的共存会显著降低 Se(Ⅵ) 的还原，并使其保持为氧化态（Bassil et al.，2018）。

与还原过程相比，土壤中硒的氧化过程较为少见，但 Akiho 等（2010，2012）的研究证实 Se(Ⅳ) 能够被氧化为 Se(Ⅵ)。硒的氧化过程主要是由微生物介导，但也可由非生物因素介导（Bassil et al.，2018）。Torma 和 Habashi（1972）、Sarathchandra 和 Watkinson（1981）的研究均表明，在有氧条件下土壤微生物能够氧化 Se(0) 和 Se(Ⅳ)，而 Akiho 等（2010，2012）在过二硫酸根离子介导和高温条件下也得到类似的结果；也有 Se(Ⅳ) 的非生物氧化过程发生在合成氧化锰和水钠锰矿的表面的报道（Scott and Morgan，1996）。尽管如此，土壤中硒的还原过程还是比氧化过程更常见，但这两个过程均主要由微生物介导。此外，Se(Ⅳ) 的生物氧化速率低于 Se(Ⅵ) 的生物还原速率（Bassil et al.，2018），这也解释了在有氧条件下 Se(Ⅳ) 能持续存在及其在自然环境中能与 Se(Ⅵ) 共存的原因。

三、配位作用

配位反应对土壤硒的固定起重要作用。硒可与土壤有机质中的官能团和铁/铝氧化物发生螯合/络合反应，且其对 Se(Ⅳ) 的吸附作用大于 Se(Ⅵ)，因而外源亚硒酸盐较硒酸盐更易被土壤固定（Dinh et al.，2017；王丹，2019）。Fujita 等（2005）的研究表明，水溶液中的亚硒酸盐在 1 个月内能够被沉积物完全固定，而硒酸盐仍有 80% 残留在溶液中。因此，亚硒酸盐在吸附能力强的土壤中有较高的固液分配系数（K_d）（Li et al.，2015；Premarathna et al.，2010）。此外，土壤微生物可以进一步促进被固定的 Se(Ⅳ) 的还原，并将其转化为 Se(0) 或 Se(-Ⅱ)（Darcheville et al.，2008），提高了土壤对 Se(Ⅳ) 的固定作用。

有机物的官能团如羟基、羧基和羰基等及硒的含氧阴离子均带负电荷。因此，硒和有机物可通过带正电的金属桥键形成三元络合物（Se-HM-OM）（Dinh et al.，2017；Martin et al.，2017）。Tam 等（1995）发现，在羟基氧化铁存在时，Se(Ⅳ) 会被胡敏酸（HA）固定；也有报道称，当富啡酸（FA）存在时，无定形铁氧化物对 Se(Ⅳ) 的吸附作用增大（Tao et al.，2000）。此外，许多关于 Se(Ⅳ) 与 HA 通过 Fe(Ⅲ) 桥键相互作用的研究也进一步证实了土壤有机质与含氧硒阴离子结合形成三元配合物（Catrouillet et al.，2016；Martin et al.，2017）。

四、吸附-解吸

在好氧环境中，硒主要以 SeO_3^{2-} 和 SeO_4^{2-} 两种水溶态形式（植物吸收利用的主要形态）存在（王丹，2019）。外源硒酸盐进入土壤后，主要通过静电作用和物理结合形成土壤外层复合体（图 2-2）（Sharma et al.，2015），而亚硒酸盐则主要通过微孔扩散进入土壤内层晶格，与土壤铁铝锰氧化物、有机质等以共价键或离子键结合形成较为紧密的内层配合物（图 2-3）（Bruggeman et al.，2007；Sharma et al.，2015）。内层吸附比外层吸附作用更强，导致亚硒酸盐在土壤中的迁移性、生物有效性显著下降（He et al.，2018）。

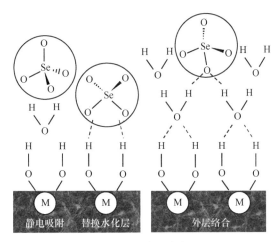

图 2-2　硒酸盐的表面吸附形式（改自 Jordan et al.，2013）

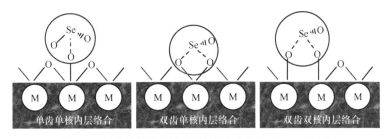

图 2-3　亚硒酸盐的表面吸附形式（改自 Phan，2017）

土壤黏土矿物对阴离子硒的吸附具有决定性作用。Dhillon 和 Dhillon（1999）的研究表明，土壤中黏土矿物的含量与亚硒酸盐的吸附量呈显著相关。无定形铁是土壤中活性最大的铁氧化物，也是土壤中少数带正电的胶体矿物，其对土壤中硒等阴离子有较强的吸附作用，可与硒形成较稳定的内层配合物（Tolu et al.，2014a；王松山等，2011）。而土壤有机质具有较大的比表面积和较强的螯合能力，且腐殖质可与土壤中的黏土矿物、氧化物等结合形成有机-无机复合胶体，从而提高土壤颗粒的比表面积和表面活性（胡宁静等，2007）。因此，土壤吸附硒酸盐的能力具有随着有机质含量的增加而增大的趋势（冯璞阳等，2016）。

此外，土壤硒［尤其是 Se(Ⅳ)］的吸附还受土壤溶液中共存竞争性阴离子（如 PO_4^{3-}、SO_4^{2-}、NO_3^-）的显著影响（Li et al.，2015），而这种影响的结果可以是促进、抑制或无影响（Lopez-Hernandez et al.，1986）。目前关于 SO_4^{2-} 和 PO_4^{3-} 等竞争阴离子对硒的吸附影响的研究较多，许多研究表明施用磷肥会降低土壤对亚硒酸盐的吸附（严佳，2014；Nakamaru and Sekine，2008），提高土壤有效硒含量。而有关 SO_4^{2-} 对亚硒酸盐的影响，不同研究者的结果不尽相同。例如，Dhillon（2000）发现，共存的 SO_4^{2-} 降低了土壤对硒的吸附，而 Lee 等（2011）和张华等（2009）均发现，SO_4^{2-} 对土壤硒的吸附基本无影响，这些相反的结果可能与供试的土壤性质不同（如土壤的矿物组成）有关。但总体而言，PO_4^{3-} 由于具有较高的亲和力，其对硒的吸附影响大于亲和力较低的 SO_4^{2-}（Goh and Lim，2004）。也有研究报道在 pH 范围为 6～8 的条件下，CO_3^{2-} 可以促进 SeO_4^{2-} 在氧化铝上的吸附（Wijnja and Schulthess，2000）。满楠（2013）的研究也表明，CO_3^{2-} 对土壤硒的吸附具有间接的正影响。

此外，土壤表面的吸附作用是控制土壤溶液中硒浓度的重要因素（Goldberg，2011）。解吸同样对土壤硒形态转化起至关重要的作用，且解吸率越低，越不利于植物对硒的吸收。Bar-Yosef 和 Meek（1987）发现，土壤 pH 从 5.6 提高到 8.7 会导致土壤液相中硒浓度的增加，而 pH 从 5.6 下降至 3.6 时，溶液中硒的浓度随之降低。Dhillon 和 Dhillon（1999）发现土壤中硒的解吸作用与土壤酸碱度有关，在酸性土壤中，KH_2PO_4 对硒的解吸作用大于 KCl，而碱性土壤则恰好相反。此外，Li 等（2015）的研究表明，土壤中硒的解吸率与金属（氢）氧化物含量呈负相关，而与土壤 pH（4.89～8.51）呈正相关。

由此可见，土壤中硒的吸附-解吸是一个复杂的过程，受到硒的种类、土壤组成（类型）、共存阴离子、pH 等许多因素的影响。

五、挥发作用

甲基化是土壤硒挥发和从土壤中去除硒的关键过程（Lin and Terry，2003）。土壤硒甲基化包括化学和生物两大机制，其中生物甲基化过程占主导，部分细菌、微藻、真菌和植物也可进行硒的生物甲基化（Terry et al.，2000；Neumann et al.，2003）。

Suzuki 等（2006）认为硒可能发生如下的甲基化反应[式（2-9）]，而 Ohta 和 Suzuki（2008）却提出二甲基硒化物才是土壤硒甲基化过程的最终产物。土壤硒的存在形态、微生物活性、有机物含量、温度、土壤含水量等条件的差异可能是导致甲基化过程最终产物不同的原因（Dungan and Frankenberger，1999）。Masscheleyn 等（1991）的研究表明，在氧化和适宜的还原条件（Eh：0～500 mV；pH 7～7.5）下，河流沉积物中可以检测到甲基硒化物。

$$H_2Se \rightarrow CH_3SeH \rightarrow (CH_3)_2Se \rightarrow (CH_3)_3Se^+ \qquad (2-9)$$

去甲基化过程与硒的甲基化过程相反。硒的去甲基化过程在好氧和厌氧条件下均可发生。Bolan 等（2014）发现，在厌氧沉积物中，二甲基硒化物中的甲基会被微生物（产甲烷菌和硫酸盐还原菌）快速去除，而 H_2Se 是这一过程的主要产物（Eswayah et al.，2016）。在好氧条件下，二甲基硒化物经去甲基化后生成 Se(Ⅵ)（Bolan et al.，2014）。

硒在土壤各组分间的分配是一个动态平衡的过程，外源施硒或发生硒污染后，进入土壤的硒经过多种途径的转化作用建立起新的平衡。然而，现有关于土壤中硒的行为过程的研究多集中于外源无机硒在土壤中的形态和价态转化，而有关自然条件下硒的行为过程的研究涉及很少，需进一步探究。

第四节　土壤硒的有效性及其影响因素

一、土壤中硒的有效性及其表征

（一）土壤硒有效性的概念

一般情况下，有效性的概念是指特定生物体在环境单元中可能利用的物质量，在不同学科之间有着广泛的应用。土壤养分有效性在土壤农业化学和植物营养学中有着悠久的历史和现实意义；而土壤环境研究中所涉及的污染和有益元素的有效性主要由土壤养分有效性的概念延伸而来，一般是用与植物吸收有良好相关性的土壤元素/化合物的可提取态来衡量。在水环境中，元素/化合物的有效性用于描述污染物在生物传输或生物反应中被利用的程度（Hamelink et al.，1994）。一些研究者随后将这个概念逐步引入到土壤和沉积物的重金属风险评价中（Lanno et al.，2004）。一般来说，金属按其在土壤中的活性可以分为：有效态、潜在有效态和无效态（Dinh et al.，2019）。

土壤有效硒是指土壤中可被植物直接吸收利用的硒的量，是影响植物硒含量的主要决定因子，而其在总硒中的占比（即硒的有效性）反映了土壤中硒的供给能力（Terry et al.，2000）。黄杰等（2018）和罗军强等（2016）的研究均表明，与土壤总硒相比，土壤有效硒含量与小麦籽粒及黄豆硒含量有更好的正相关关系。而且，大量研究表明，土壤硒的总量不能表征和评价硒的生物有效性，而硒的赋存形态，尤其是有效硒含量才是决定其迁移性和毒性的关键（Bawrylak-Nowak，2013；Tolu et al.，2011）。不同地区土壤有效硒含量受土壤类型、土壤性质、气候等因素影响，差别很大，相关方面需要进一步的研究（表 2-3）。

表 2-3 中国部分地区土壤水溶态硒含量特征（王志强等，2022）

地区	总硒/（mg/kg）	水溶态硒/（μg/kg）	水溶态占比/%
宁夏石嘴山	0.23	11.21	4.88
黑龙江海伦	0.29	8	4.68
江西鄱阳湖	0.308	11	3.60
天津蓟州	0.37	15	4.60
新疆焉耆盆地	0.39	7	2.05
青海东部	0.44	11.51	3.32
安徽庐江	0.45	5	1.40
湖北恩施沙地乡	1.88	13.3	0.71

（二）土壤有效硒的测定

由于土壤是一个多相非均质的复杂体系，不同类型土壤，其组成和性质差异很大，迄今国内外尚没有一个普遍适用的土壤有效硒测定方法。常用的土壤有效硒含量测定方法包括化学浸提法和薄膜梯度扩散技术（DGT 法）两类。

1. 化学浸提法

化学浸提法是根据不同形态硒的生物有效性差异，用一种或者几种组合的化学试剂一次或分步浸提将其分离测定的方法（Wang et al.，2012）。化学浸提法的核心是浸提剂的选择及提取条件的建立。不同浸提剂的提取机制不同，对应提取的化学形态也各异。

1）单一浸提法

单一浸提法常用的化学浸提剂包括水、盐类浸提剂（KH_2PO_4、NaH_2PO_4、KCl、$CaCl_2$、K_2SO_4、$NaHCO_3$ 等）、碱性浸提剂（$NaOH$、$NH_3 \cdot H_2O$ 等）及有机络合剂（EDTA、AB-DTPA 等）四种类型（周菲等，2022）。其中，用水作为浸提剂提取土壤有效硒在芬兰的硒生物强化实践中应用多年（Fordyce et al., 2000），而其他浸提剂也被广泛用于土壤有效硒的测定（Wang et al., 2012; Supriatin et al., 2015）。表 2-4 汇总了目前使用的单一化学浸提法的提取形态及其原理。

表 2-4 不同单一化学浸提法的浸提形态及其原理（周菲等，2022）

浸提剂	固液比	时间	浸提目标	浸提原理
水	1∶10 1∶30	30 min 24 h	水溶态	水的溶解作用
热水 （80℃去离子水）	1∶5	1 h	水溶态、有机结合态	弱结合态的亚硒酸盐和有机硒在高温下释放
0.0005 mol/L $CaCl_2$ 0.01 mol/L $CaCl_2$	1∶30 1∶10	24 h 2 h	水溶态	水的溶解及 Cl^- 的离子交换作用
0.1 mol/L KH_2PO_4-K_2HPO_4 （PBS）	1∶5	1 h	水溶态、可交换态、铁锰氧化物结合态	$H_2PO_4^-$ 与硒的含氧阴离子的配位交换作用
0.5 mol/L $NaHCO_3$（pH 8.5）	1∶10	1 h	水溶态、腐殖酸结合态	CO_3^{2-} 与铁、铝、钙形成沉淀，释放出吸附态的硒酸盐、亚硒酸盐
0.1 mol/L NaOH	1∶10	1 h	有机结合态	OH^- 置换作用释放硒酸盐和亚硒酸盐
0.05 mol/L EDTA 0.005 mol/L DTPA	1∶5 1∶2	12 h 15 min	铁锰氧化物结合态、有机结合态	与吸附亚硒酸盐的金属阳离子的配位作用
1 mol/L HNO_3 0.43 mol/L HNO_3	1∶30	24h	提取铁锰氧化物、硫化物和钙的化合物等	硒酸盐和亚硒酸盐离子的质子化作用

磷酸盐缓冲溶液不仅能提取出土壤中的水溶态、可交换态及部分有机结合态的硒（Dinh et al., 2018; Kulp and Pratt, 2004），且使用缓冲溶液能够避免低 pH（约 4.8）KH_2PO_4 造成的土壤中硒的再分配问题（Hagarová et al., 2005）。大量研究表明，磷酸盐缓冲溶液提取的有效硒含量与对应的作物硒含量间相关性最高（Wang et al., 2012; Wang et al., 2019a）。因此，推荐 0.1 mol/L KH_2PO_4-K_2HPO_4 作为土壤有效硒测定的浸提剂（周菲等，2022）。

2）连续浸提法

连续浸提法是根据提取能力的不同，使用多种选择性浸提剂将土壤中不同结合形态的元素依次提取出来的方法。与单一浸提法相比，连续浸提法能提供硒在土壤中的形态分布特征，因此可以更全面地评价土壤中硒的移动性、有效性和潜在毒性（Tolu et al., 2011），但连续浸提法测得的硒形态因浸提剂的选择及分级操作程序的不同而异（表 2-5）。

表 2-5 不同连续化学浸提法及硒形态分级（周菲等，2022）

浸提方法	硒形态分级
三步分级法	吸附型硒、铝型硒、铁型硒 水溶态、可交换态、有机结合态
四步分级法	吸附态、碳酸盐结合态、铁锰氧化物结合态、有机物结合态 弱酸提取态、可还原态、可氧化态、残渣态

续表

浸提方法	硒形态分级
五步分级法	水溶态、可交换态及碳酸盐结合态、有机结合态、铁锰氧化物结合态、残渣态
	有效硒、缓效硒、无定形氧化闭蓄态硒、游离氧化闭蓄态硒、残渣态
六步分级法	水溶态、可交换态及碳酸盐结合态、铁锰氧化物结合态、有机及硫化物结合态、元素态、残渣态
	水溶态、可交换态、酸溶态、有机及硫化物结合态、元素态、残渣态
	水溶态、可交换态、酸溶态、有机结合态、硫化物结合态、残渣态
七步分级法	水溶态、可交换态、有机结合态、元素态、酸溶态、有机/无机硒化物、残渣态
八步分级法	水溶态、可交换态、碳酸盐结合态、有机结合态、易还原氧化物结合态、无定型氧化物结合态、晶体氧化物结合态、无定型铝硅盐结合态

目前，普遍采用的土壤硒形态分级方法为五步连续浸提法（Wang et al., 2012），即根据浸提的难易程度将土壤硒划分为水溶态、可交换态及碳酸盐结合态、铁锰氧化物结合态、有机结合态和残渣态。

2. 薄膜梯度扩散技术

土壤中元素的生物有效性是一个不断变化的动态过程的结果（Richard, 2003），伴随作物根系对硒的吸收，根系附近的土壤溶液中硒含量下降，促使被土壤固持的部分硒释放出来补充到土壤溶液中（Wang et al., 2009），这个过程可以概括为两个截然不同的阶段，即以物理化学作用为基础的解吸过程和以植物生理学作用为驱动的吸收过程（Mccarty and Mackay, 1993）。以往采用的化学浸提法大都忽略了由土壤-植物体系引发的动态释放过程对土壤硒有效性的影响，而薄膜梯度扩散技术（diffusive gradients in thin-films, DGT）既考虑了土壤颗粒和土壤溶液之间的静态平衡过程，又包括了土壤固相吸附态重金属释放补给土壤溶液的动态过程（Davison and Zhang, 2012），能较好地反映自然条件下植物对重金属元素的复杂吸收过程。除此之外，此法还削弱甚至消除了土壤自身的性质及有机质含量对测定结果的影响（Williams et al., 2011）。

DGT装置核心由滤膜、扩散膜和吸附膜三部分构成，依次起着防止外界颗粒物进入、控制目标离子的扩散通量，以及快速、稳定地吸附固定目标离子的作用。随着土壤溶液中的金属离子被吸附膜固定，其在土壤溶液中的离子浓度下降，土壤颗粒表面吸附的非稳态金属会解吸并释放到溶液中（Harper et al., 2000）。可见，与传统的化学浸提法相比，DGT技术不仅操作简便，还能很好地模拟植物根系对土壤中金属元素的动态吸收过程（周菲等, 2022）。目前有关DGT方法对不同外源硒种类（Peng et al., 2019）、植物种类（彭琴等, 2017; Peng et al., 2017）、土壤类型（Peng et al., 2020）、有机肥施用（Dinh et al., 2021）中硒生物有效性的可靠性和准确性评价均做了一些工作，证明DGT是一种具有发展前景的测定土壤有效硒含量的技术。但由于DGT装置吸附膜的离子选择性因材质而异，且价格昂贵，因此在很大程度上限制着其在实际中的使用。

（三）土壤硒有效性的评价

虽然众多学者针对不同土壤及植物类型提出了对应的土壤有效硒测定方法，但目前对于土壤-植物体系中硒有效性的评价仍缺乏统一方法。现有的评价方法概括起来主要包括化学浸提法测定结果、参数评价法及植物指示法三种。

化学浸提法测定结果即直接根据测得的土壤硒含量或有效态硒的分布情况进行评价即可。

参数评价法以连续浸提法测定的土壤各形态硒含量为基础，通过数学运算得到的特征参数来评价土壤硒的有效性。常用的参数包括土壤结合强度（I_R 值）(Han et al., 2004) 和移动系数（MF 值）(Kabala and Singh, 2001)。I_R 值能够定量描述土壤重金属与土壤组分的相对结合强度；而 MF 值被定义为水溶态和可交换态硒含量占土壤总硒含量的百分比。一般来说，I_R 值越低，土壤硒有效性越高，硒多以水溶态、可交换态等不稳定态存在；MF 值越大，土壤硒有效性越高。I_R 值和 MF 值均可在一定程度上用来评估土壤硒的生物有效性（Terry et al., 2009; Davison and Zhang, 2012），但是要把硒的 5 个形态同时考虑进去，测定过程比较繁琐，需要进一步完善和验证；另外，酸性土壤中碳酸盐结合态划分的科学性和合理性需要进一步探讨。

测定土壤有效态硒含量最根本的目的是为了更好地反映植物吸收硒的能力。因此，土壤中硒的有效性也可以通过植物体硒含量来表征（Keskinen et al., 2010）。植物指示法，本质是一种植物的提取法，该法通过植物栽培试验，以植物受害症状、吸收元素的含量及根际微生物数量变化来指征土壤元素的生物有效性。Wang 等（2012）和 Peng 等（2017）的研究结果均表明，作物硒含量均可较好地表征土壤硒有效性。但植物指示法不仅生长周期长，而且还受供试植物种类（植物本身的生物特性）、品种、部位、生育期甚或环境条件的限制，使得不同的研究结果间往往难以比较，限制了其在实际中的应用（孙歆等，2006）。

综上所述，尽管土壤硒的生物有效性在硒的生物强化和生态风险评价中十分重要，研究者们对相关方面也做了大量工作，但是有关评价的方法还需要根据土壤、植物类型等进行进一步完善。

二、影响土壤硒有效性的因素

土壤有效态硒含量是决定作物硒吸收及积累的关键，土壤中硒的形态及价态分布是影响土壤硒有效性的最直接因素（Winkel et al., 2015），而土壤理化性质（如 pH、Eh、质地、有机质、共存离子等）又会通过改变土壤中硒的形态及价态分布间接影响土壤硒的有效性（Fan et al., 2018; Li et al., 2015; Liu et al., 2021b）。此外，农艺措施（Longchamp et al., 2013）及气候条件（Jones et al., 2017）等可通过改变土壤硒形态/价态、促进/抑制作物生长或改变土壤理化性质等途径影响土壤硒的生物有效性，因为土壤中硒有效性是一个多因子综合作用的结果（图 2-4），也随环境条件处于动态变化中。

（一）土壤中硒的形态和价态

大量研究表明，土壤总硒含量常常与作物硒吸收量间无显著相关性（姬华伟等，2021；Peng et al., 2017），这是因为岩石风化进入土壤的硒会以不同的化学形态和价态存在，且土壤中不同形态/价态硒的溶解性、移动性和生物有效性差异很大，因此植物对土壤硒的吸收受到土壤中不同硒形态及价态的影响（刘娜娜，2021；Wang et al., 2013）。

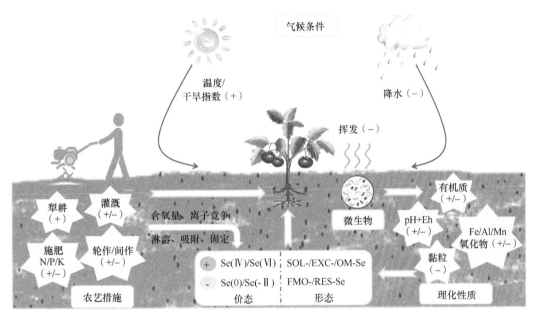

图 2-4　土壤硒生物有效性的影响因素

土壤中 Se(0) 和 Se(-Ⅱ) 均不溶于水，较难被植物吸收利用，而 Se(Ⅳ) 和 Se(Ⅵ) 均易溶于水，是作物吸收的主要硒形态（梁东丽等，2017）。同时，由于 Se(Ⅳ) 易被土壤黏土矿物、有机质等吸附，因此土壤中 Se(Ⅵ) 的移动性和有效性较 Se(Ⅳ) 更高（彭琴等，2017）。此外，不同 pH 和 Eh 条件下硒的热力学稳定价态各异，在还原性和酸性条件下，土壤硒生物有效性较低，但在氧化条件下其移动性和有效性较高。

根据植物对不同形态硒吸收利用能力的强弱，可将土壤硒依次分为水溶态（SOL-Se）、可交换态和碳酸盐结合态（EXC-Se）、铁锰氧化物结合态（FMO-Se）、有机结合态（OM-Se）和残渣态（RES-Se）五种形态（梁东丽等，2017；Qin et al.，2017）。其中，SOL-Se 移动性最强、最易被植物吸收（Schiavon and Pilon-Smitts，2017）。EXC-Se 主要指与水合氧化物结合（Ryden et al.，1987），或被黏土矿物及腐殖质（Zhang et al.，2014）表面吸附的 Se(Ⅳ)，其生物有效性远低于 SOL-Se，但可转化为有效态从而被植物吸收（Kulp and Pratt，2004）。因此，土壤中的水溶态和可交换态硒含量之和常被视为植物可利用态硒即有效态硒含量（刘娜娜，2021）。

土壤有机结合态硒主要包括富啡酸结合态硒（FA-Se）和胡敏酸结合态硒（HA-Se）（王丹，2019），其中，HA-Se 结构稳定，很难被分解利用（Coppin et al.，2006）；而 FA-Se 分子质量低，易被矿化成无机硒和小分子有机硒从而被作物吸收利用（Qin et al.，2012）。已经有研究印证了水稻（Qin et al.，2013；王潇等，2019）、茶树叶（孙月婷等，2018）和玉米（王松山，2012；Qin et al.，2013）硒含量与土壤有机结合态硒含量间的显著相关关系。因此，有机结合态硒也被看成是土壤有效硒的潜在来源，而 FA-Se 与 HA-Se 的比例（FA/HA-Se）决定了 OM-Se 在土壤生物有效硒中的作用。FMO-Se 主要是指与 Fe/Mn 相关的硒，与 RES-Se 一样均很难被植物吸收（Wang et al.，2012）。

（二）土壤性质

1. pH+Eh

在众多土壤理化性质中，pH 和 Eh 在控制土壤硒形态和生物有效性方面发挥着特别重要的作用（Antoniadis et al., 2017; Nakamaru and Altansuvd, 2014），它们直接决定着硒在土壤溶液中的存在形态。在一定的 pH 及 Eh 条件下，硒以一种或几种形态为主存在于环境中（Torres et al., 2010）（图 2-5）。

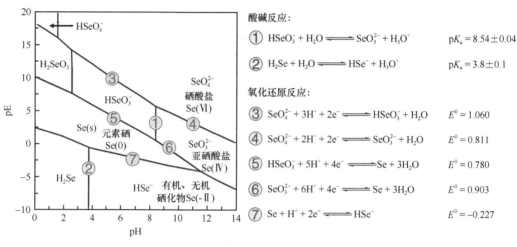

图 2-5 不同 pH-pE 条件下（25℃）土壤硒形态的组成
（改自 Séby et al., 2001; Petrović, 2021）

不同价态硒的有效性不同，而有效硒（水溶态+可交换态）通常以 Se(Ⅵ) 和 Se(Ⅳ) 为主。根据 Séby 的 pE-pH 关系图（图 2-6）可知，当 pH＞7 时，去质子化的+6 价硒是硒的主要形态，且随 pE 降低，其逐渐向低价态转化；而在相同 pE 条件下，低价态的硒含量随 pH 降低而增加，亦即 Se(Ⅵ) 被逐步还原为 Se(Ⅳ) 甚至更低价态的 Se(0) 或 Se(-Ⅱ)。因此，土壤硒有效性随着 pH 和 Eh 的下降而降低（Nakamaru and Altansuvd, 2014）。还原条件有利于土壤中的硒转化为 Se(-Ⅱ) 和 Se(0)，而氧化条件下主要以 Se(Ⅵ) 和 Se(Ⅳ) 为主。

研究证实，无论是硒酸盐或亚硒酸盐，在中国 18 种不同土壤的转化过程中，土壤 pH（4.89～8.51）与吸附或老化速率均呈负相关，亦即随土壤 pH 升高，硒的吸附量下降，老化时间也随之延长（冯璞阳等，2016; Li et al., 2015; Wang et al., 2017a）。这是由于 Se(Ⅵ) 主要通过静电作用与土壤颗粒结合，随着土壤 pH 升高，铁锰氧化物与有机质所带的正电荷数量减少，与 Se(Ⅵ) 竞争吸附的离子量增加，使吸附量下降，提高了土壤溶液中 Se(Ⅵ) 的含量（Li et al., 2015; Wang et al., 2017a）。但在酸性土壤中，硒易于被铁/铝氧化物还原吸附并与有机质发生络合作用。因此，碱性土壤中硒的有效性高于酸性土壤（Lee et al., 2011）。而与 Se(Ⅵ) 类似，伴随着 pH 降低，Se(Ⅳ) 也易于被吸附或固持在土壤固相的内层结构中（Li et al., 2016）。

土壤 pH 不仅影响土壤的吸附过程，还会影响沉淀作用和微孔扩散作用。当土壤 pH 远小于重金属的 pK_a 时，表面聚合和沉淀趋势以及微孔扩散作用均随 pH 升高而增大，从而降低重金属的生物有效性（Wang et al., 2015）。而与阳离子相反，对阴离子态的 Se(Ⅳ) 和 Se(Ⅵ) 来说，当土壤 pH 远小于硒的 pK_a 值时，沉淀作用及微孔扩散作用随 pH 升高而降低（即提高其生物有效性），且这一过程还受土壤中硒配位交换过程中的吸附作用的影响。此外，pH 还会影响土壤表面可变电荷的净电荷数。当 pH 高于土壤的零点电荷时，土壤表面的官能团发生质子分离，土壤表面可变电荷倾向于带负电，与同样带负电荷的 Se(Ⅵ) 和 Se(Ⅳ) 发生静电排斥作用（Lewis-Russ, 1991），使得硒从土壤固相中释放出来，进而提高其生物有效性。

2. 铁/铝氧化物

除了 pH 对土壤硒吸附老化产生的作用外，土壤铁/铝氧化物也是影响土壤硒有效性的主要因素（Wang et al., 2017a）。土壤中的铁/铝氧化物因具有较强的吸附能力和较大的比表面积，被认为是影响土壤硒吸附的主要因素之一（Muller et al., 2012）。Li 等（2015）发现我国 18 种土壤中的铁/铝氧化物对 Se(Ⅳ) 的吸附有重要作用，且其含量与 Se(Ⅳ) 的吸附量呈正相关；冯璞阳等（2016）也在相同的土壤中证实铁的氧化物和氢氧化物是 Se(Ⅵ) 重要的吸附剂。不同的是，Se(Ⅵ) 主要通过静电吸附作用与铁铝氧化物外层络合，而 Se(Ⅳ) 则主要和铁铝氧化物形成内层络合物（Dhillon and Dhillon, 1999）。

土壤 pH 也影响铁/铝氧化物对土壤硒的吸附及解吸作用，从而影响其有效性。pH 较低时，铁/铝氧化物表面会产生较多的正电荷，促进了其对硒的吸附（Kampf et al., 2009）；而随着土壤 pH 升高，土壤中氢氧根（OH^-）含量增加，从而与同样带负电荷的 Se(Ⅳ) 和 Se(Ⅵ) 竞争金属氧化物上的吸附位点（Fontes and Alleoni, 2006; Goh and Lim, 2004），提高了硒的生物有效性。

3. 有机质

土壤有机质具有较大的比表面积和许多官能团，具有较强的螯合能力，且腐殖质可以与土壤中的黏土矿物、铁/铝氧化物等结合形成有机-无机复合胶体，使得土壤颗粒的比表面积和表面活性增大（胡宁静等，2007），因此，土壤有机质对土壤中硒的移动性和有效性有重要的影响（Wang et al., 2012）。土壤中的硒或者与土壤有机物质发生配位作用，或者通过植物和微生物进入蛋白质或氨基酸有机分子中（Qin et al., 2012）。土壤中与矿物质结合的硒占总硒的比例不足 20%，而 40%～50% 甚至更高比例的硒与有机质相结合（Wang et al., 2012）。

有机质对硒有效性的影响具有复杂的双重效应（Dinh et al., 2017）（图 2-6），且受到有机质自身结构和性质的影响（Supriatin et al., 2015）。一方面，土壤有机质可通过土壤溶解性有机质（DOM）、小分子有机酸（LMWOA）等物质络合或溶解土壤中的硒，进而提高其有效性（Dinh et al., 2017; Supriatin et al., 2016）；另一方面，土壤有机质通过与大分子有机物螯合、沉淀物质的固定作用，降低硒的有效性（Tolu et al., 2014a; 2014b）。

在低 pH 和高有机质的条件下，小分子有机酸可以溶解和释放固持在土壤固相表面的硒，从而提高土壤硒的生物有效性（Dinh et al., 2017; Sharma et al., 2015）。而且，土壤有机质还可以通过调节土壤溶液的电荷强度促进表面电荷的释放及表面负电荷的形成，增加对硒含氧阴离子的排斥作用，从而提高硒的有效性（Dousova et al., 2015）。此

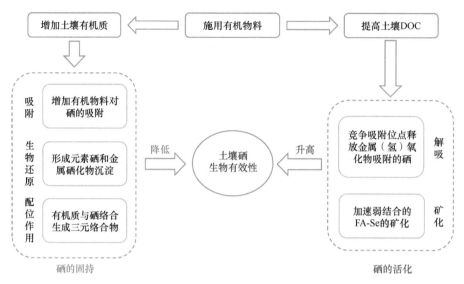

图 2-6 有机质对土壤硒生物有效性的影响（改自 Dinh et al.，2017；齐明星，2021）

外，硒、金属与有机质等三元复合体（铁/铝/锰-硒-有机质）的形成同样也可以在一定程度上提高硒的有效性及迁移能力（Fernández-Martínez and Charlet，2009）。

与此不同，土壤中的大分子腐殖物质又会与硒形成络合物（配位化合物），特别是螯合物，从而降低其移动性及有效性。王松山等（2011）发现我国 16 种不同理化性质的土壤中，硒大部分结合在有机质中。Coppin 等（2006）研究得出土壤有机质在硒的固持或固液分配中扮演了重要的角色，有机质较高的土壤中含有较多的含氧官能团，如酚羟基、羧基、苯醌、羟基醌等，这些官能团很容易与土壤中水溶态硒发生络合或者螯合反应，导致土壤有效态硒含量下降（Martin et al.，2017），特别是酸性条件下这些官能团与硒的结合能力更强（Li et al.，2017）。Bruggeman 等（2007）发现土壤腐殖酸对 Se(Ⅵ) 主要为还原作用，而对 Se(Ⅳ) 主要是持续吸附作用，且在吸附过程中将其转化为有机结合态硒。在土壤固相的有机组分中，有机结合态硒主要以 HA-Se 为主（Supriatin et al.，2015）；而在土壤溶液中，一方面，溶解性有机质（DOM）可通过竞争吸附作用降低土壤对硒的吸附，增加溶解态硒的含量；另一方面，DOM 与硒络合后产生的小分子有机酸结合态硒也是土壤有效硒的重要组分（Supriatin et al.，2016）。

除化学过程外，土壤中微生物的代谢过程也影响着土壤有机质的含量和组成，并通过影响硒的矿化过程进而促进硒的形态转化（Hageman et al.，2013）。有机结合态硒在微生物的作用下可矿化为无机硒，同时亚硒酸盐也可经微生物转化作用与土壤有机质结合（Di Tullo et al.，2016）。在好氧条件下，亚硒酸盐可被细菌用作呼吸过程中的电子终端受体。土壤中硒的氧化作用非常缓慢，但在还原条件下被还原固定的硒可重新释放进入土壤溶液中，从而提高土壤硒的有效性（Shaheen et al.，2016）。

4. 黏粒

土壤黏粒对重金属的有效性和形态有重要影响（Pierart et al.，2015）。硒与黏土矿物的强结合能力导致了黏土中硒的有效性低下（Munier-Lamy et al.，2007），含氧阴离子硒与土壤组分以静电吸附或者形成配合物的形式发生相互作用（Séby et al.，2001）。硒能够被黏土矿物铝氧八面体的正电荷吸附，且其吸附量随 pH 变化很大，黏土矿物所带负电荷随土壤 pH 下降而减少，其对硒的吸附作用也随之增大（Antoniadis et al.，2017）。

按照土壤质地不同，可以将土壤分为砂土、壤土和黏土三种类型。砂土砂粒含量高，质地疏松，透气性好，但易导致无机硒肥的淋溶损失。而黏土与砂土正好相反，质地黏重，同时透水性较差，使得更多的硒固持在土壤中。研究发现，土壤中硒的有效性与黏粒含量呈负相关（Li et al.，2015；冯璞阳等，2016）。徐文等（2010）发现，粒径小于 0.025 mm 的黏粒对土壤硒的吸附量远大于粒径大于 1 mm 的黏粒，且显著降低了土壤硒的生物有效性，这是因为细颗粒比粗颗粒具有更大的交换容量和更强的吸附能力，导致黏土对硒的固持能力更高（Gabos et al.，2014）。Wang 等（2017）在全国 18 种不同理化性质的土壤对硒酸盐老化的影响因子分析中也证实，黏粒可通过对有机质和铁/铝氧化物的间接作用影响老化过程，从而加速了硒的老化进程。

（三）农艺措施

土壤中硒的生物有效性也受农业生产管理措施如耕作、灌溉、施肥、轮作和间作等的影响（Longchamp et al.，2013；Stroud et al.，2010）。

1. 耕作方式

Lessa 等（2016）的研究表明，耕作土壤中硒的移动性大于免耕土壤，即耕作土壤中的硒更容易被植物吸收（Wang et al.，2017b）。通过犁耕可将深层土壤中硒携带到表层，从而促进了作物对土壤硒的吸收利用（Lessa et al.，2016）。Wu 等（1993）和 Van Mantgem 等（1996）的研究表明，虽然耕作土壤中总硒含量与免耕土壤中无显著差异，但耕作处理土壤中水溶态硒含量却显著高于免耕土壤。此外，种植前进行翻耕能够提高土壤中的含氧量，使得植物可吸收和利用的 Se(Ⅵ) 和 Se(Ⅳ) 含量增加，也在一定程度上提高了土壤硒的生物有效性（Kausch and Pallud，2013）。

2. 土壤水分及灌溉方式

土壤水分主要通过影响土壤氧化还原电位（Eh）、pH、溶解性有机质（DOC）、微生物活性等影响金属的生物有效性（Zhang et al.，2019；Lv et al.，2020）。土壤水分含量决定了土壤中的氧浓度，水分较少时氧浓度升高，土壤处于氧化状态，Eh 较高（Li et al.，2010；Zhou et al.，2018），而 Eh 和 pH 对土壤硒价态和形态的化学转化起主要作用（Zhou et al.，2018；Zhang et al.，2019）。

王瑞昕等（2021）采用盆栽试验研究发现：在水稻的各生育期，好氧和干湿交替较淹水灌溉在一定程度上提高了土壤 pH 和 Eh，使得土壤水溶态和可交换态硒含量增加，从而提高了土壤硒的有效性。Zhai 等（2021）也用培养试验证明，水分饱和条件较水分胁迫和适宜条件降低了不同类型土壤 Eh 值，增加了土壤 pH 和 DOC 含量，促进了土壤 Se(Ⅵ) 向 Se(Ⅳ)、Se(0) 和 Se(-Ⅱ) 的还原，以及水溶态硒和可交换态硒向残渣态硒和有机结合态硒的转化，因而加速了外源硒酸盐在三种不同类型土壤中的老化进程，促进了外源硒酸盐的固定作用。另外，水稻根系表面在淹水条件下会形成铁膜，其能促进硒在根系的累积，从而提高了硒向水稻籽粒的转运（杨旭健等，2022），因此，自然富硒土壤在干湿交替处理下，水稻籽粒硒含量和土壤水溶态硒含量均显著高于淹水处理（张青等，2018）。

灌溉方式也是影响作物硒积累的主要因素，喷灌处理比漫灌处理更利于作物对硒的吸收（Suarez et al.，2003）。灌溉措施对土壤中硒的影响包括以下几个方面：①若灌溉含硒水，则会导致硒在土壤中的积累；②灌溉引起土壤剖面中有效硒的淋溶损失增加；③灌溉还会引起土壤理化性质（如酸碱度、Eh 值等）的改变（Li et al.，2010；Zhao et al.，2007），进而影响土壤中硒形态的转化和生物有效性。Suarez 等（2003）研究发现，甘蓝和菜豆对硒的吸收量随着灌溉水中硒含量的增加而增加，Bañuelos（2002）也报道了类似的结果。而 Wu 等（1994）和 Zhao 等（2007）的研究均表明，灌溉后土壤水溶态硒含量显著降低，且植物中硒的累积量显著下降，土壤硒的淋溶损失是其产生的主要原因。此外，灌溉水中含有的硫酸盐与硒的竞争作用也会抑制作物对硒的吸收（Zhao et al.，2007；Bañuelos，2002）。

3. 轮作和间作管理

轮作和间作管理也会影响土壤中硒的生物有效性。作物种类、硒肥施用方式和种植方式等共同决定了轮作对硒生物有效性的影响。Borowska 等（2012）发现，在马铃薯-冬小麦-春大麦-玉米轮作系统中，与 5 月和 7 月相比，3 月的土壤总硒含量最高，且在轮作种植期间随着作物对硒的吸收，土壤硒含量逐渐下降。Wang 等（2017b）同样发现，在冬小麦-夏玉米轮作体系中，玉米季土壤硒含量减少，这与玉米的生物量和硒积累量高于小麦有关。

Chilimba 等（2014）发现间作系统显著影响了作物硒含量，且叶面喷施硒肥时，间作处理土壤硒含量高于单作土壤。这是由于间作条件下作物种植更为密集，从而有效地减缓了硒在土壤表面的沉积速率及硒的淋溶、固定和吸附，增加了土壤中的硒被植物根系吸收的概率（Chilimba et al.，2014）。但长期使用间作系统种植，将导致土壤中大量的硒被植物吸收及农田土壤硒含量低等问题。另外，轮作和间作管理可以改变土壤的理化性质（张娜，2014），也在一定程度上影响了土壤中硒的有效性。

4. 无机肥料

施用有机和无机肥料对土壤硒生物有效性的影响不尽相同。施肥处理对土壤中硒生物有效性的影响不仅取决于肥料的类型、用量和施用方式，还取决于作为有机硒肥的农业废弃物中的硒含量。

无机肥料如氮（N）（陈玉鹏等，2017）、硫（S）（覃思跃等，2016）和磷（P）（赵文龙等，2013a；2013b）的施用均会影响土壤中硒的形态转化及植物对硒的吸收能力。一般来说，施用这些肥料主要通过以下机制影响硒的生物有效性：①在土壤颗粒表面与硒产生的吸附竞争作用；②硒酸盐与硫酸盐竞争硫转运蛋白，亚硒酸盐与磷酸盐竞争磷转运蛋白，从而影响植物对硒的吸收；③通过改变土壤理化性质（如降低 pH、增加有机质等）影响土壤中硒形态的再分布；④施肥促进了作物生长，使得作物硒累积量随生物量下降（即生物稀释作用）（Liu et al.，2015；Reis et al.，2018）。

因此，在生物强化中可以通过合理地施用无机肥料来调控土壤硒的利用效率。

5. 有机物料

施入有机肥料（如动物粪便等）是中国农业生产中的一种长期培肥措施。Dinh（2019）发现施入牛粪和鸡粪显著降低了土壤中外源硒的生物有效性，且外源亚硒酸盐下降幅度更显著。牛粪和鸡粪对外源硒有效性的影响取决于外源硒的种类，施入鸡粪显著降低了亚硒酸盐处理硒的有效性，这是由于与牛粪相比，鸡粪降低土壤 pH 的作用更大且在培养过程中产生更多的类腐殖酸有机物，使亚硒酸盐更容易被吸附固定在土壤固相中；而施入牛粪显著降低了硒酸盐处理硒的有效性，这是由于鸡粪比牛粪释放更多的 LMWOA，更容易与土壤中吸附能力较弱的硒酸盐竞争吸附位点，且鸡粪比牛粪含氮量更高，分解过程中可能产生更多的硝酸盐，进而使硝酸盐而不是硒酸盐作为微生物生长的电子受体，这抑制了硒酸盐的还原。Sharma 等（2011）的研究表明，受施入有机物料的固定作用，有机物料还田后土壤硒生物有效性下降。但 Li 等（2017）的研究却发现，粪肥和秸秆中的小分子有机酸能将有机结合态的硒重新释放到土壤溶液中，从而提高硒在土壤中的生物有效性。秸秆在土壤中腐解释放的溶解性有机质（DOM）与硒发生相互作用，还影响着土壤微生物群落多样性、酶活性和作物根系的分布（Ghimire et al., 2017）、土壤中有机质含量及微环境的 pH 等，这些均影响土壤中硒的形态转化及其有效性（Wang et al., 2020）。

Stavridou 等（2011）发现，富硒有机物料还田可提升土壤硒的有效性，且植物中的硒多以硒代甲硫氨酸为主，生物有效性较高；而非富硒植物腐解释放的大分子有机物质则会降低有效硒的含量，微生物还原，还原的硒与胡敏酸结合，从而对硒起固定作用。Bañuelos 等（2015）的研究表明，富硒植物沙漠王羽（*Stanleya pinnata*，主要硒形态为 SeCyst）还田后，短期内土壤硒形态存在一个较快的无机化（即矿化）过程，使 SeCyst 转化为 Se(IV)，显著提高了西兰花和胡萝卜的硒累积量。Wang 等（2018）报道富硒小麦秸秆与富硒小白菜还田加速了有机物质的矿化，导致土壤水溶态硒、可交换态硒和富啡酸结合态硒的含量提高，使得种植的小白菜地上和地下部硒含量显著高于对照处理。因此，富硒秸秆还田能显著提高土壤硒的有效性，可作为缺硒地区植物补硒的一项措施。

（四）气候条件

气候条件可通过影响硒的大气沉降作用或对土壤硒的吸附作用等直接或间接影响土壤中硒的生物有效性（Jones et al., 2017）。气候因素如干旱指数、降水量、蒸散量和温度等均是影响土壤硒含量的因素（El-Ramady et al., 2015b; Jones et al., 2017）。其中，干旱指数能够通过控制土壤氧化还原条件影响硒的淋溶，从而影响硒的溶解和迁移（硒的氧化态较还原态更易溶解和移动）（Kulp and Pratt, 2004; Winkel et al., 2015）。干旱指数增加时，土壤趋于氧化状态，氧化态硒的含量增加（Jones et al., 2017），土壤的硒有效性增加。

此外，降水量和温度等也会对植物硒吸收产生重要影响（Bitterli et al., 2010）。例如，Bisbjerg（1972）和 El-Ramady 等（2015a）的研究发现，低温和高降水量可以减少植物中硒的积累。Geering 等（1968）的研究也表明，降水会影响土壤氧化还原条件，使其向缺氧状态转变，导致土壤中硒的还原。此外，高降水量还会导致土壤有效硒因淋溶而过度损失,从而减少植物中硒的积累。

目前,有关气候因子对土壤硒成因方面的研究报道较多,但有关其对土壤硒有效性影响方面的资料欠缺。

第五节　土壤硒有效性的调控

土壤硒有效性是决定作物硒吸收和硒含量的关键。因此,合理调控土壤硒有效性对于提高缺硒地区居民硒摄入水平、降低高硒/硒中毒地区居民潜在健康风险均具有重要意义。目前,相关研究主要通过使用土壤改良剂(如有机肥、秸秆、钙镁磷肥、粉煤灰、泥炭等)改变土壤理化性质,从而实现土壤中硒生物有效性的内源调控。

通过施用改良剂,提高土壤pH、增加竞争离子数量及减少吸附位点,是提高缺硒地区土壤硒生物有效性的重要途径。赵妍(2011)的研究发现,施用粉煤灰、木质素泥炭和秸秆均能显著提高强酸性高硒茶园土壤pH,但由于秸秆和木质素泥炭对土壤硒均具有较强的吸附作用,因此,施用粉煤灰对土壤有效硒含量的持续提高有较好的促进作用。杨旎等(2014)对酸性高硒茶园土壤硒有效性调控方面的研究也得到相似的结果。况琴(2019)的研究也表明,牛粪、生物炭和钢渣还田均能提高土壤pH,但施入牛粪及生物炭后,由于有机质的大量引入造成土壤中硒吸附固定,而施用钢渣提供的碱性条件导致土壤硒的有效性增加,因此应用潜力巨大。此外,刘勤(2003)和赵文龙等(2013b)发现,施用磷肥可以提高水稻和小白菜对土壤硒的吸收。诸旭东等(2015)的研究表明,将碱性磷肥施于土壤中,可通过提高土壤pH及增加竞争离子数量的方式,促进水稻籽粒对土壤中硒的吸收。马迅(2017)比较了氨基酸叶面肥、钙镁磷肥和活性硅肥后同样发现,钙镁磷肥对酸性富硒土壤中硒有效性的促进作用更显著。然而,在一些含硒原料的利用中需要关注可能引起的副作用,如在利用粉煤灰补硒的同时,应考虑F、B等元素可能带来的负面影响。

有机物料的吸附作用、大分子腐殖物质的络合作用及微生物的作用可有效降低高硒土壤中硒的生物有效性。Dhillon等(2007)和Dhiman等(2000)的研究发现,微生物活动促进了硒的还原及挥发,是有机物料还田抑制富硒土壤中生长的小麦和油菜对土壤硒的吸收利用的主要原因。Dinh(2019)对有机肥的研究结果也表明,鸡粪和牛粪还田均会显著降低土壤中外源硒的生物有效性,且对Se(Ⅳ)的影响最为显著。而王丹(2019)的研究发现秸秆源DOM还田对土壤外源硒生物有效性的影响是双重的,即在有机质高的酸性土壤中,DOM释放的FA和HON(疏水性有机物质)与硒结合,提高了硒的生物有效性;而在有机质含量低的碱性土壤中,大分子的HA与硒络合,降低了土壤硒的有效性。

第六节　结　　论

土壤既是农业生产的基础,又是人和动物体内硒的基本来源。受成土母质、气候和地形条件等地球化学环境因素的影响,全球范围内土壤硒含量差异较大,硒缺乏和硒中毒的情况同时存在且屡见不鲜。世界范围内大部分土壤硒含量为0.1~2.00 mg/kg,我国表层土壤硒含量（平均约 0.29 mg/kg）略低于世界平均水平（约 0.40 mg/kg），且仍有 51% 的土地面积存在不同程度的缺硒。土壤中的硒形态和价态随环境条件发生相互转化，而硒与土壤组分间不断发生的吸附-解吸、沉淀-溶解、（生物）氧化-还原等一系列环境化学过程是影响土壤硒形态转化的关键，有关土壤中硒的化学行为过程还有很多地方（例如，微生物介导的氧化过程中硒的生物转化及其机理、自然条件下硒在土壤各组分间的转化及分配平衡等）仍然需要进一步探究。

土壤中有效态硒的含量反映了土壤供给硒的能力，其测定包括化学浸提法及 DGT 法两种，而土壤硒生物有效性的评价则主要包括化学浸提法测定结果、参数评价法及植物指示法三种。由于土壤是一个多相非均质的复杂体系，且不同类型土壤性质差异较大，国内外迄今尚没有一个普遍适用的土壤有效硒测定及其有效性评价的方法。综合现有方法操作的难易程度、测定有效硒与植物硒含量的相关性及成本等，推荐 0.1 mol/L 磷酸盐缓冲溶液浸提法作为土壤有效硒测定方法。

土壤硒的有效性受多方因素综合影响，除土壤中硒的形态及价态分布等直接影响因素外，土壤理化性质（如 pH、Eh、质地、有机质、共存离子等）、农业活动（犁耕、灌溉、施肥等）及气候条件（干旱指数、降水量、温度等）等也可通过改变土壤硒形态/价态、促进/抑制作物生长等间接影响土壤硒的生物有效性。因此，通过各种措施调控土壤中硒的生物有效性，对于提高缺硒地区居民硒摄入水平、降低高硒/硒中毒地区居民潜在健康风险均具有重要意义。

参 考 文 献

陈玉鹏, 彭琴, 梁东丽, 等. 2017. 施氮对小麦硒 (Ⅳ) 吸收、转运和分配的影响. 环境科学, 38(2): 825-831.

冯璞阳, 李哲, 者渝芸, 等. 2016. 我国 18 种不同理化性质的土壤对硒酸盐的吸附解吸作用研究. 环境科学, 37(8): 3160-3168.

龚河阳, 李月芬, 汤洁, 等. 2015. 吉林省西部土壤硒含量、形态分布及影响因素. 吉林农业大学学报, 37(2): 177-184.

郭莉, 杨忠芳, 阮起, 等. 2012. 北京市平原区土壤中硒的含量和分布. 现代地质, 26(5): 859-864.

郭璐, 满楠, 梁东丽, 等. 2013. 小白菜对外源硒酸盐和亚硒酸盐动态吸收的差异及其机制研究. 环境科学, 34(8): 3272-3279.

何振立, 周启星, 谢正苗, 等. 1998. 污染及有益元素的土壤化学平衡. 北京: 中国环境科学出版社: 341-361.

胡宁静, 骆永明, 宋静. 2007. 长江三角洲地区典型土壤对镉的吸附及其与有机质、pH 和温度的关系. 土壤学报, 44(3): 437-443.

黄春雷, 魏迎春, 简中华, 等. 2013. 浙中典型富硒区土壤硒含量及形态特征. 地球与环境, 41: 155-159.

黄建国, 袁玲. 1997. 四川盆地主要紫色土硒的状况及其有效性研究. 土壤学报, 34(2): 152-152.

黄杰, 崔泽玮, 杨文晓, 等. 2018. 陕西泾惠渠灌区土壤-小麦体系中硒的空间分布特征. 环境科学学报, 38(2): 289-296.

姬华伟, 任蕊, 陈继平, 等. 2021. 关中不同类型土壤硒含量特征及其对玉米籽粒硒含量的影响. 西北地质, 54(4): 239-249.

江官军. 2016. 南方红壤硒的有效性调控研究. 北京: 中国地质大学硕士学位论文.

况琴. 2019. 丰城地区土壤有效硒的调控及作物富硒研究. 南昌: 南昌大学硕士学位论文.

郎春燕, 黄秀丽, 李小娇. 2013. 成都东郊稻田土壤中硒形态的分布特征. 西南农业学报, 26(2): 642-646.

李振宁. 2010. 河北省平原区土壤中硒异常源追踪及生态效应评价. 石家庄: 石家庄经济学院硕士学位论文.

梁东丽, 彭琴, 崔泽玮, 等. 2017. 土壤中硒的形态转化及其对有效性的影响研究进展. 生物技术进展, 7(5): 374-380.

廖自基. 1992. 微量元素的环境化学及生物效应. 北京: 中国环境科学出版社.

刘娜娜. 2021. 陕西塿土和广西红壤硒有效性差异及其主控因子初探. 杨凌: 西北农林科技大学硕士论文.

刘勤. 2003. 磷肥和硒施用对稻米硒、钙、锌等营养累积的影响. 广东微量元素科学, 10(6): 20-24.

罗海怡, 罗先熔, 刘攀峰, 等. 2022. 广西三江县土壤硒含量分布特征及其影响因素研究. 现代地质, 36(2): 645-654.

罗军强, 张元培, 吴颖, 等. 2016. 湖北天门中东部地区土壤全量硒及有效硒对农作物硒的影响. 资源环境与工程, 30(6): 848-851.

马迅. 2017. 不同内源调控措施对江西丰城土壤中硒有效性的影响. 南京: 南京农业大学硕士学位论文.

马玉锦. 2009. 库尔勒市绿地系统公共绿地的硒赋存形态分布. 新疆有色金属, 32(3): 47-49.

满楠. 2013. 土壤中外源硒酸盐和亚硒酸盐的老化过程. 杨凌: 西北农林科技大学硕士学位论文.

毛香菊, 刘璐, 程新涛, 等. 2021. 河南新密典型富硒区土壤 Se 元素地球化学特征及空间分布规律. 地质通报, 40(10): 1664-1670.

潘金德, 李晓春, 毛春国, 等. 2007. 瑞安市农产地土壤硒含量、形态与分布及其与土壤性质的关系. 浙江农业科学, (6): 682-684.

彭琴, 李哲, 梁东丽, 等. 2017. 不同作物对外源硒动态吸收、转运的差异及其机制. 环境科学, 38(4): 1667-1674.

齐明星. 2021. 溶解性有机质对土壤硒转化的影响初探. 杨凌: 西北农林科技大学硕士学位论文.

齐艳萍, 武瑞, 杨焕民. 2012. 硒在大庆市龙凤湿地环境-植物系统中的分配特征. 水土保持通报, 32(4): 222-224.

覃思跃, 赵文龙, 李俊, 等. 2016. 硒与硫单一及交互作用对小白菜硒生物有效性的影响. 环境科学学报, 36(4): 1500-1507.

瞿建国, 徐伯兴, 龚书椿. 1998. 上海不同地区土壤中硒的形态分布及其有效性研究. 土壤学报, 35(3): 398-403.

孙歆, 韦朝阳, 王五一. 2006. 土壤中砷的形态分析和生物有效性研究进展. 地球科学进展, 21(6): 625-632.

孙月婷, 刘琪, 凤海元. 2018. 土壤各形态硒与植物硒含量平衡模型研究. 现代农业科技, 5: 175-178.

谭建安. 1989. 中华人民共和国地方病与环境图集. 北京: 科学出版社.

仝双梅, 连国奇, 秦趣, 等. 2013. 贵州省开阳县土壤硒含量及其制约因素研究. 甘肃农业大学学报, 48(2): 105-109.

王丹. 2019. 秸秆还田对土壤硒生物有效性的影响及作用机制. 杨凌: 西北农林科技大学博士学位论文.

王涵植, 罗杰, 蔡立梅, 等. 2020. 广东省惠来县土壤硒分布特征及影响因素研究. 中国农业资源与区划, 41(6): 262-269.

王锐, 邓海, 贾中民, 等. 2020. 硒在土壤-农作物系统中的分布特征及富集土壤阈值. 环境科学, 51(12): 5571-5578.

王瑞昕, 杨静, 方正, 等. 2021. 水分管理对水稻籽粒硒积累及根际土壤细菌群落多样性的影响. 土壤学报, 58(06): 1574-1584.

王松山, 梁东丽, 魏威, 等. 2011. 基于路径分析的土壤性质与硒形态的关系. 土壤学报, 48(4): 823-830.

王松山, 吴雄平, 梁东丽, 等. 2010. 不同价态外源硒在石灰性土壤中的形态转化及其生物有效性. 环境科学学报, 30: 2499-2505.

王松山. 2012. 土壤中硒形态和价态及生物有效性研究. 杨凌: 西北农林科技大学硕士学位论文.

王潇, 张震, 朱江, 等. 2019. 青阳县富硒土壤中硒的形态与水稻富硒的相关性研究. 地球与环境, 47(3): 336-344.

王志强, 杨建峰, 魏丽馨, 等. 2022. 石嘴山地区碱性土壤硒地球化学特征及生物有效性研究. 物探与化探, 46(1): 229-237.

吴少尉, 池泉, 陈文武, 等. 2004. 土壤中硒的形态连续浸提方法的研究. 土壤, 36: 92-95.
谢薇. 2019. 天津近郊与远郊菜地中硒与重金属差异性研究. 安徽农业科学, 47(20): 88-93.
徐强. 2016. 黑龙江省土壤硒的形态、分布及其影响因素研究. 哈尔滨: 东北农业大学硕士学位论文.
徐文, 唐文浩, 邝春兰, 等. 2010. 海南省土壤中硒含量及影响因素分析. 安徽农业科学, 38(6): 3026-3027.
徐争启, 倪师军, 张成江, 等. 2011. 四川省万源市土壤硒形态特征及影响因素分析. 安徽农业科学, 39(3): 1455-1458.
严佳. 2014. 磷-硒交互作用对茶园土壤硒吸附行为及有效性的影响. 南京: 南京农业大学硕士学位论文.
杨兰芳. 2000. 土壤中的硒. 湖北民族学院学报(自然科学版), 18(1): 43-46.
杨旎, 宗良纲, 严佳, 等. 2014. 改良剂与生物有机肥配施方式对强酸性高硒茶园土壤硒有效性的影响. 土壤. 46(6): 1069-1075.
杨旭健, 田宇豪, 沈宏. 2022. 酸性土壤条件下铁膜调控技术对水稻根表铁膜和籽粒硒累积的影响. 中国生态农业学报(中英文), 30(7): 1174-1185.
喻大松. 2015. 陕西紫阳和青海平安富硒环境中硒分布特征及其对人体健康的影响. 杨凌: 西北农林科技大学硕士学位论文.
张春来, 杨慧, 黄芬, 等. 2021. 广西马山县岩溶区土壤硒含量分布及影响因素研究. 物探与化探, 45(6): 1497-1503.
张华, 史红文, 邱喜阳, 等. 2009. 镉-硫-硒复合作用下硒在土壤-蔬菜系统中的迁移分配研究. 湖南科技大学学报(自然科学版), 24(2): 112-115.
张娜. 2014. 两种水栽方式及其水旱轮作对设施土壤性质影响的差异研究. 扬州: 扬州大学硕士学位论文.
张青, 王煌平, 孔庆波, 等. 2018. 干湿交替灌溉对富硒土壤硒形态及水稻硒积累的影响. 水土保持学报, 32(01): 327-331+338.
张亚峰, 苗国文, 马强, 等. 2019. 青海东部碱性土壤中硒的形态特征. 物探与化探, 43(05): 1138-1144.
张艳玲, 潘根兴, 李正文, 等. 2002. 土壤-植物系统中硒的迁移转化及低硒地区食物链中硒的调节. 生态环境学报, 11(4): 388-391.
张忠, 周丽, 张勤. 1997. 地球化学样品中硒的循序提取技术. 岩矿测试, 16: 255-261.
赵文龙, 胡斌, 王嘉薇, 等. 2013a. 磷与四价硒的共存对小白菜磷、硒吸收及转运的影响. 环境科学学报, 33(7): 2020-2026.
赵文龙, 梁东丽, 石美, 等. 2013b. 磷酸盐与硒酸盐相互作用对小白菜磷和硒吸收的影响. 农业环境科学学报, 32(12): 2331-2338.
赵妍. 2011. 强酸性高硒茶园土壤硒有效性调控及其机理研究. 南京: 南京农业大学硕士学位论文.
周菲, 彭琴, 王敏, 等. 2022. 土壤-植物体系中硒生物有效性评价研究进展. 科学通报, 67(6): 461-472.
朱青, 郭熙, 韩逸, 等. 2020. 南方丘陵区土壤硒空间分异特征及其影响因素——以丰城市为例. 土壤学报, 57(4): 834-843.
朱永官, 陈怀满, 董元华, 等. 2018. 环境土壤学(第三版), 北京: 科学出版社: 130-166.
诸旭东, 宗良纲, 胡秋辉, 等. 2015. 一种土壤内源调控生产富硒大米的方法: 中国, CN201510249162.3.
Abrams M M, Burau R G. 1989. Fractionation of selenium and detection of selenomethionine in a soil extract. Communitionin Soil Science and Plant Analysis, 20: 221-237.
Adriano D C. 2001. Trace elements in terrestrial environments: Biogeochemistry, bioavaibility, and risks of metals. New York: Springer.
Akiho H, Ito S, Matsuda H. 2010. Effect of oxidizing agents on selenate formation in a wet FGD. Fuel, 89(9): 2490-2495.
Akiho H, Yamamoto T, Tochihara Y, et al. 2012. Speciation and oxidation reaction analysis of selenium in aqueous solution using X-ray absorption spectroscopy for management of trace element in FGD liquor. Fuel, 102: 156-161.
Antoniadis V, Levizou E, Shaheen S M, et al. 2017. Trace elements in the soil-plant interface: Phytoavailability,

translocation, and phytoremediation—A review. Earth-Science Reviews, 171: 621-645.

Bañuelos G S, Arroyo I, Pickering I J, et al. 2015. Selenium biofortification of broccoli and carrots grown in soil amended with Se-enriched hyperaccumulator *Stanleya pinnata*. Food Chemistry, 166(1): 603-608.

Bañuelos G S. 2002. Irrigation of broccoli and canola with boron- and selenium-laden effluent. Journal of Environmental Quality, 31(6): 1802-1808.

Bar-Yosef B, Meek D. 1987. Selenium sorption by kaolinite and montmorillonite. Soil Sci, 144: 11-19.

Bassil J, Naveau A, Bueno M, et al. 2018. Leaching behavior of selenium from the karst infillings of the hydrogeological experimental site of poitiers. Chemical Geology, 483: 141-150.

Bawrylak-Nowak B. 2013. Comparative effects of selenite and selenate on growth and selenium accumulation in lettuce plants under hydroponic conditions. Plant Growth Regulation, 70: 149-157.

Berrow M L, Ure A M. 1989. Geological materials and soils. In: Ihnat M. Occurrence and Distribution of Selenium. Boca Raton: CRC Press: 213-242.

Bisbjerg B. 1972. Risø report no. 200: studies on selenium in plants and soils. Copenhagen: Danish Atomic Energy.

Bitterli C, Bañuelos G S, Schulin R. 2010. Use of transfer factors to characterize uptake of selenium by plants. Journal of Geochemical Exploration, 107(2): 206-216.

Bodnar M, Szczyglowska M, Konieczka P, et al. 2016. Methods of selenium supplementation: Bioavailability and determination of selenium compounds. Critical Reviews in Food Science and Nutrition, 56(1): 36-55.

Bolan N, Kunhikrishnan A, Thangarajan R, et al. 2014. Remediation of heavy metal(loid)s contaminated soils-to mobilize or to immobilize? Journal of Hazardous Materials, 266: 141-166.

Borowska K, Koper J, Milanowski M. 2012. Seasonal changes of selenium and selected oxidoreductases in soil under different fertilization and crop rotation. Ecological Chemistry and Engineering A, 19: 719-730.

Broadley M R, White P J, Bryson R J, et al. 2006. Biofortification of UK food crops with selenium. Proceedings of the Nutrition Society, 65: 169-181.

Bruggeman C, Maes A, Vancluysen J. 2007. The interaction of dissolved Boom Clay and Gorleben humic substances with selenium oxyanions (selenite and selenate). Applied Geochemistry, 22(7): 1371-1379.

Bujdos M, Mulova A, Kubova J, et al. 2005. Selenium fractionation and speciation in rocks, soils, waters and plants in polluted surface mine environment. Environmental Geology, 47(3): 353-360.

Catrouillet C, Davranche M, Dia A, et al. 2016. Does As(III) interact with Fe(II), Fe(III) and organic matter through ternary complexes? Journal of Colloid and Interface Science, 470: 153-161.

Chen Q L, Shi W M, Wang X C. 2010. Selenium speciation and distribution characteristics in the rhizosphere soil of rice (*Oryza sativa* L.) seedlings. Communications in Soil Science and Plant Analysys, 41: 1411-1425.

Chilimba A D C, Young S D, Joy E J M. 2014. Agronomic biofortification of maize, soybean and groundnut with selenium in intercropping and sole cropping systems. African Journal of Agricultural Research, 9: 3620-3626.

Combs G F. 2001. Selenium in global food systems. British Journal of Nutrition, 85(5): 517-547.

Coppin F, Chabroullet C, Martin-Garin A, et al. 2006. Methodological approach to assess the effect of soil ageing on selenium behaviour: first results concerning mobility and solid fractionation of selenium. Biology and Fertility of Soils, 42(5): 379-386.

Cutter G A. 1985. Determination of selenium speciation in biogenic particles and sediments. Analytical Chemistry, 57: 2951-2955.

Darcheville O, Février L, Haichar F Z, et al. 2008. Aqueous, solid and gaseous partitioning of selenium in an oxic sandy soil under different microbiological states. Journal of Environmental Radioactivity, 99(6): 981-992.

Davison W, Zhang H. 2012. Progress in understanding the use of diffusive gradients in thin films (DGT)—back to basics. Environmental Chemistry, 9: 1-13.

De Temmerman L, Waegeneers N, Thiry C, et al. 2014. Selenium content of Belgian cultivated soils and its uptake by field crops and vegetables. Science of the Total Environment, 468: 77-82.

Dhillon K S, Dhillon S K. 1999. Adsorption-desorption reactions of selenium in some soils of India. Geoderma, 93(1-2): 19-31.

Dhillon K S, Dhillon S K. 2003. Distribution and Management of Seleniferous Soils, Advances in Agronomy. New York: Academic Press: 119-184.

Dhillon S K, Dhillon K S. 2000. Selenium adsorption in soils as influenced by different anions. J Plant Nutr Soil Sci, 163: 577-582.

Dhillon S K, Hundal B K, Dhillon K S. 2007. Bioavailability of selenium to forage crops in a sandy loam soil amended with Se-rich plant materials. Chemosphere, 66: 1734-1743.

Dhiman S D, Nandal D P, Hari O M. 2000. Productivity of rice (*Oryza sativa*)-wheat (*Triticum aestivum*) cropping system as affected by its residue management and fertility levels. Indian Journal of Agronomy, 45(1): 1-5.

Di Tullo P, Pannier F, Thiry Y, et al. 2016. Field study of time-dependent selenium partitioning in soils using isotopically enriched stable selenite tracer. Science of the Total Environment, 562: 280-288.

Dinh Q T, Cui Z W, Huang J, et al. 2018. Selenium distribution in the Chinese environment and its relationship with human health: a review. Environment International, 112: 294-309.

Dinh Q T, Li Z, Tran T A T, et al. 2017. Role of organic acids on the bioavailability of selenium in soil: A review. Chemosphere, 184: 618-635.

Dinh Q T, Wang M, Tran T A T, et al. 2019. Bioavailability of selenium in soil-plant system and a regulatory approach. Critical Reviews in Environmental Science and Technology, 49: 443-517.

Dinh Q T, Zhou F, Wang M, et al. 2021. Assessing the potential availability of selenium in the soil-plant system with manure application using diffusive gradients in thin-films technique (DGT) and DOM-Se fractions extracted by selective extractions. Science of the Total Environment, 763: 143047.

Dinh Q T. 2019. 有机肥施用对土壤-植物体系中硒生物有效性的影响及其机理. 杨凌: 西北农林科技大学博士学位论文.

Dousova B, Buzek F, Herzogova L, et al. 2015. Effect of organic matter on arsenic(V) and antimony(V) adsorption in soils. European Journal of Soil Science, 66(1): 74-82.

Du Y, Luo K, Ni R, et al. 2018. Selenium and hazardous elements distribution in plant-soil-water system and human health risk assessment of Lower Cambrian, Southern Shaanxi, China. Environmental Geochemistry and Health, 40(5): 2049-2069.

Dungan R S, Frankenberger W T. 1999. Microbial transformations of selenium and the bioremediation of seleniferous environments. Bioremediation Journal, 3(3): 171-188.

El-Ramady H R, Abdalla N, Alshaal T, et al. 2015a. Giant reed for selenium phytoremediation under changing climate. Environmental Chemistry Letters, 13(4): 359-380.

El-Ramady H, Abdalla N, Alshaal T, et al. 2015b. Selenium in soils under climate change, implication for human health. Environmental Chemistry Letters, 13(1): 1-19.

Elrashidi M A, Adriano D C, Workman S M, et al. 1987. Chemical equilibria of selenium in soils: A theoretical development. Soil Science, 144(2): 274-280.

Emmanuelle B, Virginie M, Fabienne S. 2012. Selenium exposure in subjects living in areas with high selenium concentrated drinking water: Results of a French integrated exposure assessment survey. Environment International, 40(1): 155-161.

Eswayah A S, Smith T J, Gardiner P H E. 2016. Microbial transformations of selenium species of relevance to bioremediation. Applied and Environmental Microbiology, 82(16): 4848-4859.

Eurola M, Alfthan G, Aro A, et al. 2003. Results of the finish selenium monitoring program 2000-2001, MTT Agrifood Research Finland, Jokioinen.

Fan J, Zeng Y, Sun J. 2018. The transformation and migration of selenium in soil under different eh conditions. Journal of Soils and Sediments, 18(9): 2935-2947.

Feinberg A, Stenke A, Peter T, et al. 2020. Constraining atmospheric selenium emissions using observations, global modeling, and Bayesian inference. Environmental Science and Technology, 54(12): 7146-7155.

Feng C X, Chi G X, Liu J J, et al. 2012. Geochemical constraints on the origin and environment of lower cambrian, selenium-rich siliceous sedimentary rocks in the Ziyang area, Daba region, central China. International Geology Review, 54 (7): 765-778.

Fernández-Martínez A, Charlet L. 2009. Selenium environmental cycling and bioavailability: a structural chemist point of view. Reviews in Environmental Science and Bio-technology, 8: 81-110.

Fontes M P F, Alleoni L R. 2006. Electrochemical attributes and availability of nutrients, toxic elements, and heavy metals in tropical soils. Scientia Agricola, 63: 589-608.

Fordyce F M. 2013. Selenium Deficiency and Toxicity in the Environment. Essentials of Medical Geology: Revised Edition: 375-416.

Fordyce F, Guangdi Z, Green K, et al. 2000. Soil, grain and water chemistry in relation to human selenium-responsive diseases in Enshi District, China. Applied Geochemistry, 15(1): 117-132.

Fordyce F. 2007. Selenium geochemistry and health. Ambio, 36(1): 94-97.

Fujita M, Ike M, Hashimoto R, et al. 2005. Characterizing kinetics of transport and transformation of selenium in water-sediment microcosm free from selenium contamination using a simple mathematical model. Chemosphere, 58(6): 705-714.

Gabos M B, Alleoni L R F, Abreu C A. 2014. Background levels of selenium in some selected Brazilian tropical soils. Journal of Geochemical Exploration, 145: 35-39.

Geering H R, Cary E E, Jones L H P, et al. 1968. Solubility and redox criteria for the possible forms of selenium in soils. Soil Science Society of America Journal, 32(1): 35-40.

Geoffroy N, Demopoulos G P. 2011. The elimination of selenium(IV) from aqueous solution by precipitation with sodium sulfide. Journal of Hazardous Materials, 185(1): 148-154.

Gerla P J, Sharif M U, Korom S F. 2011. Geochemical process controlling the spatial distribution of selenium in soil and water, west central South Dakota. Environment Earth Science, 62: 1551-1560.

Ghimire R, Lamichhane S, Acharya B S, et al. 2017. Tillage, crop residue, and nutrient management effects on soil organic carbon in rice-based cropping systems: A review. Journal of Integrative Agriculture, 16(1): 1-15.

Goh K H, Lim T T. 2004. Geochemistry of inorganic arsenic and selenium in a tropical soil: effect of reaction time, Ph, and competitive anions on arsenic and selenium adsorption. Chemosphere, 55: 849-859.

Goldberg S. 2011. Chemical equilibrium and reaction modeling of arsenic and selenium in soils. In: Selim H M. Dynamics and bioavailability of heavy metals in the rootzone. Boca Raton: CRC Press: 65-92.

Hagarová M, Žemberyová D, Bajčan. 2005. Sequential and single step extraction procedures used for fractionation of selenium in soil samples. Chemical Papers- Slovak Academy of Sciences, 59: 93-98.

Hageman S P W, van der Weijden R D, Weijma J, et al. 2013. Microbiological selenate to selenite conversion for selenium removal. Water Research, 47(7): 2118-2128.

Hamelink J, Landrum P F, Bergman H, et al. 1994. Bioavailability: Physical, Chemical, and Biological Interactions. Boca Raton: CRC Press.

Han F X, Su Y B, Sridhar B M, et al. 2004. Distribution, transformation and bioavailability of trivalent and hexvalent chromium in contaminated soil. Plant and Soil, 265: 243-252.

Harper M P, Davison W, Tych W. 2000. DIFS—a modelling and simulation tool for DGT induced trace metal remobilisation in sediments and soils. Environmental Modelling and Software, 15(1): 55-66.

Haug A, Graham R D, Christophersen O A, et al. 2007. How to use the world's scarce selenium resources efficiently to increase the selenium concentration in food. Microbial Ecology in Health & Disease, 19(4): 209-228.

He Y, Xiang Y, Zhou Y, et al. 2018. Selenium contamination, consequences and remediation techniques in water and soils: A review. Environmental Research, 164: 288-301.

Hibbs B J, Lee M, Ridgway R. 2017. Land use modification and changing redox conditions releases selenium and sulfur from historic marsh sediments. Journal of Contemporary Water Research & Education, 161: 48-65.

Huang S S, Hua M, Feng J S, et al. 2009. Assessment of selenium pollution in agricultural soils in the Xuzhou District, Northwest Jiangsu, China. Journal of Environmental Sciences, 21: 481-487.

Jones D L, Winkel L H E. 2016. Global predictions of selenium distributions in soils. In: Banuelos G S. Global Advances in Selenium Research From Theory to Application. London: Taylor and Francis Group.

Jones G D, Droz B, Greve P, et al. 2017. Selenium deficiency risk predicted to increase under future climate change. Proceedings of the National Academy of Sciences, 114(11): 2848-2853.

Jordan N, Ritter A, Foerstendorf H, et al. 2013. Adsorptionmechanismof Se(Ⅳ) onto maghemite. Geochimica et Cosmochimica Acta, 103(2): 63-75.

Jung B, Safan A, Batchelor B, et al. 2016. Spectroscopic study of Se(Ⅳ) removal from water by reductive precipitation using sulfide. Chemosphere, 163: 351-358.

Jung Y, Alam S U, Sun Y, et al. 2014. Removal and recovery of phosphate from water using sorption. Critical Reviews in Environment Scienceand Technology, 44(8): 847-907.

Kabala C, Singh B R. 2001. Fractionation and mobility of copper, lead, and zinc in soil profiles in the vicinity of a copper smelter. Journal of Environmental Quality, 30: 485-492.

Kämpf N, Curi N, Marques J J. 2009. Óxidos de alumínio, silício, manganês e titânio. In: Alleoni L R F, Melo V F. Química e mineralogia do solo. 1rd edn UFV Viçosa: 573-610.

Kang M L, Bin M, Fabrizio B, et al. 2013. Interaction of aqueous Se(Ⅳ)/Se(Ⅳ) with FeSe/FeSe$_2$: Implication to Se redoxprocess. Journal of Hazardous Materials, 248-249: 20-28.

Kang Y, Yamada H, Kyuma K, et al. 1991. Selenium in soil humic acid. Soil Science and Plant Nutrition, 37(2): 241-248.

Kausch M F, Pallud C E. 2013. Modeling the impact of soil aggregate size on selenium immobilization. Biogeosciences, 10(3): 1323-1336.

Keskinen R, Räty M, Yli-Halla M. 2011. Selenium fractions in selenate-fertilized field soils of Finland. Nutrient Cycling in Agroecosystems, 91: 17-29.

Keskinen R, Turakainen M, Hartikainen H. 2010. Plant availability of soil selenate additions and selenium distribution within wheat and ryegrass. Plant and Soil, 333: 301-313.

Kulp T R, Pratt L M. 2004. Speciation and weathering of selenium in upper cretaceous chalk and shale from South Dakota and Wyoming, USA. Geochimica et Cosmochimica Acta, 68(18): 3687-3701.

Lanno R, Wells J, Conder J, et al. 2004. The bioavailability of chemicals in soil for earthworms. Ecotoxicology and Environmental Safety, 57: 39-47.

Lee S, Woodard H J, Doolittle J J. 2011. Selenium uptake response among selected wheat Triticum aestivum varieties and relationship with soil selenium fractions. Soil Science and Plant Nutrition, 57(6): 823-832.

Leenheer J A. 1981. Comprehensive approach to preparative isolation and fractionation of dissolved organic carbon from natural waters and wastewaters. Environmental Science and Technology, 15: 578-587.

Lessa J H L, Araujo A M, Silva G N T, et al. 2016. Adsorption-desorption reactions of selenium(Ⅳ) in tropical cultivated and uncultivated soils under Cerrado biome. Chemosphere, 164: 271-277.

Lewis-Russ A. 1991. Measurement of surface charge of inorganic geologic materials: techniques and their consequences. In: Sparks D L. Advances in Agronomy. Cambridge, MA: Academic Press: 199-243.

Li H F, Lombi E, Stroud J L, et al. 2010. Selenium speciation in soil and rice: influence of water management and Se fertilization. Journal of Agricultural and Food Chemistry, 58(22): 11837-11843.

Li J, Peng Q, Liang D L, et al. 2016. Effects of aging on the fraction distribution and bioavailability of selenium in three different soils. Chemosphere, 144: 2351-2359.

Li M, Yang B, Xu K, et al. 2020. Distribution of Se in the rocks, soil, water and crops in Enshi County, China. Applied Geochemistry, 122: 104707.

Li Z, Liang D L, Peng Q, et al. 2017. Interaction between selenium and soil organic matter and its impact on soil selenium bioavailability: A review. Geoderma, 295: 69-79.

Li Z, Man N, Wang S S, et al. 2015. Selenite adsorption and desorption in main Chinese soils with their characteristics and physicochemical properties. Journal of Soils and Sediments, 15(5): 1150-1158.

Lin Z Q, Terry N. 2003. Selenium removal by constructed wetlands: quantitative importance of biological volatilization in the treatment of selenium-laden agricultural drainage water. Environmental Science and Technology, 37(3): 606-615.

Liu N N, Wang M, Zhou F, et al. 2021b. Selenium bioavailability in soil-wheat system and its dominant influential factors: a field study in Shaanxi province, China. Science of the Total Environment, 770: 144664.

Liu X W, Zhao Z Q, Duan B H, et al. 2015. Effect of applied sulphur on the uptake by wheat of selenium applied as selenite. Plant and Soil, 386(1-2): 35-45.

Liu Y, Tian X, Liu R, et al. 2021a. Key driving factors of selenium-enriched soil in the low-Se geological belt: A case study in Red Beds of Sichuan Basin, China. Catena, 196: 104926.

Loganathan P, Vigneswaran S, Kandasamy J, et al. 2014. Removal and recovery of phosphate from water using sorption. Critical Reviews in Environment Science and Technology, 44(8): 847-907.

Long J, Luo K L. 2017. Trace element distribution and enrichment patterns of Ediacaran-early Cambrian, Ziyang selenosis area, Central China: constraints for the origin of selenium. Journal of Geochemical Exploration, 172: 211-230.

Longchamp M, Angeli N, Castrec-Rouelle M. 2013. Selenium uptake in *Zea mays* supplied with selenate or selenite under hydroponic conditions. Plant and Soil, 362(1-2): 107-117.

Lopez-Hernandez D, Siegert G, Rodriguez J V. 1986. Competitive adsorption of phosphate with malate and oxalate by tropical soils. Soil Science Society of America Journal, 50(6): 1460-1462.

Luo K L, Xu L R, Tan J A, et al. 2004. Selenium source in the selenosis area of the Daba region, South Qinling mountain, China. Environmental Geology, 45(3): 426-432.

Lv H, Chen W, Zhu Y, et al. 2020. Efficiency and risks of selenite combined with different water conditions in reducing uptake of arsenic and cadmium in paddy rice. Environmental Pollution, 262: 114283.

Ma B, Nie Z, Liu C L, et al. 2014. Kinetics of $FeSe_2$ oxidation by ferric iron and its reactivity compared with FeS_2. Science China: Chemistry, 57(9): 1300-1309.

Mäkelä-Kurtto R, Sippola J. 2002. Monitoring of Finnish arable land: changes in soil quality between 1987 and 1998. Agricultural and Food Science, 11: 273-284.

Maksimovic Z J, Djujic I, Jovic V, et al. 1992. Selenium deficiency in Yugoslavia. Biological Trace Element Research, 33: 187-195.

Mao J D, Xing B S. 1999. Fractionation and distribution of selenium in soils. Communications in Soil Science and Plant Analysis, 30(17-18): 2437-2447.

Martin D P, Seiter J M, Lafferty B J, et al. 2017. Exploring the ability of cations to facilitate binding between inorganic oxyanions and humic acid. Chemosphere, 166: 192-196.

Masscheleyn P H, Delaune R D, Patrick W H. 1991. Arsenic and selenium chemistry as affected by sediment redox potential and pH. Journal of Environment Quality, 20(3): 522-527.

Mccarty L S, Mackay D. 1993. Enhancing ecotoxicological modeling and assessment. Body residues and modes of toxic action. Environmental Science and Technology, 27(9): 1718-1728.

Menzies N W, Donn M J, Kopittke P M. 2007. Evaluation of extractants for estimation of the phytoavailable trace metals in soils. Environmental Pollution, 145: 121-130.

Minich W B. 2022. Selenium metabolism and biosynthesis of selenoproteins in the human body. Biochemistry (Moscow), 87(1): S168-S177.

Muller J, Abdelouas A, Ribet S, et al. 2012. Sorption of selenite in a multicomponent system using the "dialysis membrane": Method. Applied Geochemistry, 27(12): 2524-2532.

Munier-Lamy C, Deneux-Mustin S, Mustin C, et al. 2007. Selenium bioavailability and uptake as affected by four different plants in a loamy clay soil with particular attention to mycorrhizae inoculated ryegrass. Journal of Environmental Radioactivity, 97: 148-158.

Nakamaru Y M, Altansuvd J. 2014. Speciation and bioavailability of selenium and antimony in non-flooded and wetland soils: A review. Chemosphere, 111: 366-371.

Nakamaru Y M, Sekine K. 2008. Sorption behaviour of selenium and antimony in soils as a function of phosphate ion concentration. Soil Science and Plant Nutrition, 54: 332-341.

Nazemi L, Nazmara S, Eshraghyan M R, et al. 2012. Selenium status in soil, water and essential crops of Iran. Iranian Journal of Environmental Health Science and Engineering, 9(1): 1-8.

Neumann P M, De Souza M P, Pickering I J, et al. 2003. Rapid microalgal metabolism of selenate to volatile dimethylselenide. Plant, Cell and Environment, 26(6): 897-905.

Ni R, Luo K, Tian X, et al. 2016. Distribution and geological sources of selenium in environmental materials in Taoyuan County, Hunan Province, China. Environmental Geochemistry and Health, 38(3): 927-938.

Ohta Y, Suzuki K T. 2008. Methylation and demethylation of intermediates selenide and methylselenol in the metabolism of selenium. Toxicology and Applied Pharmacology, 226(2): 169-177.

Oldfield J E. 2002. Se World atlas. Grimbergen, Belgium: Se-Tellurium Development Association (STDA).

Olegario J T, Yee N, Miller M, et al. 2010. Reduction of Se(IV) to Se(-II) by zerovalent iron nanoparticle suspensions. Journal of Nanoparticle Research, 12(6): 2057-2068.

Park J H, Lamb D, Paneerselvam P, et al. 2011. Role of organic amendments on enhanced bioremediation of heavy metal(loid) contaminated soils. Journal of Hazardous Materials, 185(2-3): 549-574.

Peng Q, Li J, Wang D, et al. 2019. Effects of ageing on bioavailability of selenium in soils assessed by diffusive gradients in thin-films and sequential extraction. Plant and Soil, 436(1): 159-171.

Peng Q, Wang D, Wang M, et al. 2020. Prediction of selenium uptake by pak choi in several agricultural soils based on diffusive gradients in thin-films technique and single extraction. Environmental Pollution, 256: 113414.

Peng Q, Wang M K, Cui Z W, et al. 2017. Assessment of bioavailability of selenium in different plant-soil systems by diffusive gradients in thin-films (DGT). Environmental Pollution, 225: 637-643.

Perez-Sirvent C, Martinez-Sanchez M J, Garcia-Lorenzo M L, et al. 2010. Selenium content in soils from Murcia Region (Se, Spain). Journal of Geochemical Exploration, 107: 100-109.

Petrović M. 2021. Selenium: widespread yet scarce, essential yet toxic. Chem Texts, 7(2): 1-17.

Phan T H V. 2017. Mechanism of Arsenic release in ecosystems of Southeast Asia delta: Mekong Deltas Vietnam. Geochemistry. Université Grenoble Alpes, 2017.

Pierart A, Shahid M, Sejalon-Delmas N, et al. 2015. Antimony bioavailability: knowledge and research perspectives for sustainable agricultures. Journal of Hazardous Materials, 289: 219-234.

Premarathna H L, McLaughlin M J, Kirby J K, et al. 2010. Potential availability of fertilizer selenium in field capacity and submerged soils. Soil Science Society of America Journal, 74(5): 1589-1596.

Qin H B, Zhu J M, Liang L, et al. 2013. The bioavailability of selenium and risk assessment for human selenium poisoning in high-Se areas, China. Environment International, 52: 66-74.

Qin H B, Zhu J M, Lin Z Q, et al. 2017. Selenium speciation in seleniferous agricultural soils under different cropping systems using sequential extraction and X-ray absorption spectroscopy. Environmental Pollution, 225: 361-369.

Qin H B, Zhu J M, Su H. 2012. Selenium fractions in organic matter from Se-rich soils and weathered stone coal in selenosis areas of China. Chemosphere, 86(6): 626-633.

Reilly C. 1996. Se in Food and Health. London: Blackie.

Reis H P G, Barcelos J P D Q, Junior E F, et al. 2018. Agronomic biofortification of upland rice with selenium and nitrogen and its relation to grain quality. Journal of Cereal Science, 79: 508-515.

Richard J P. 2003. Time-delay systems: an overview of some recent advances and open problems. Automatica, 39(10): 1667-1694.

Rosenfield I, Beath O A. 1964. Selenium, geobotany, biochemistry, toxicity and nutrition. Academic, 92(5): 414-415.

Ryden J C, Syers J K, Tillman R W. 1987. Inorganic anion sorption and interactions with phosphate sorption by hydrous ferric oxide gel. Journal of Soil Science, 38(2): 211-217.

Sarathchandra S U, Watkinson J H. 1981. Oxidation of elemental selenium to selenite by *Bacillus megaterium*. Science, 211(4482): 600-601.

Schiavon M, Pilon-Smits E A H. 2017. Selenium biofortification and phytoremediation phytotechnologies: A review. Journal of Environment Quality, 46(1): 10-19.

Scott M J, Morgan J J. 1996. Reactions at oxide surfaces. 2. Oxidation of Se(IV) by synthetic birnessite. Environmental Science and Technology, 30(6): 1990-1996.

Séby F, Potin-Gautier M, Giffaut E, et al. 1998. Assessing the speciation and the biogeochemical processes affecting the mobility of selenium from a geological repository of radioactive wastes to the biosphere. Analusis, 26: 193-198.

Séby F, Potin-Gautier M, Giffaut E, et al. 2001. A critical review of thermodynamic data for selenium species at 25 degrees C. Chemical Geology, 171: 173-194.

Shaheen N, Irfan N M, Khan I N, et al. 2016. Presence of heavy metals in fruits and vegetables: health risk implications in Bangladesh. Chemosphere, 152: 431-438.

Shaheen S M, Frohne T, White J R, et al. 2017. Redoxinduced mobilization of copper, selenium, and zinc in deltaic soils originating from Mississippi (U.S.A.) and Nile (Egypt) River Deltas: A better understanding of biogeochemical processes for safe environmental management. Journal of Environmental Management, 186: 131-140.

Shand C A, Eriksson J, Dahlin A S, et al. 2012. Selenium concentrations in national inventory soils from Scotland and Sweden and their relationship with geochemical factor. Journal of Geochemical Exploration, 121(121): 4-14.

Shao Y, Cai C, Zhang H, et al. 2018. Controlling factors of soil selenium distribution in a watershed in Se-enriched and longevity region of South China. Environmental Science and Pollution Research, 25(20): 20048-20056.

Sharma N, Prakash R, Srivastava A, et al. 2009. Profile of selenium in soil and crops in seleniferous area of Punjab, India by neutron activation analysis. Journal of Radioanalytical and Nuclear Chemistry, 281(1): 59-62.

Sharma S, Bansal A, Dogra R, et al. 2011. Effect of organic amendments on uptake of selenium and biochemical grain composition of wheat and rape grown on seleniferous soils in Northwestern India. Journal of Plant Nutrition and Soil Science, 174(2): 269-275.

Sharma V K, McDonald T J, Sohn M, et al. 2015. Biogeochemistry of selenium. A review. Environmental Chemistry Letters, 13(1): 49-58.

Skalnaya M G, Jaiswal S K, Prakash R, et al. 2017. The level of toxic elements in edible crops from seleniferous area (Punjab, India). Biological Trace Element Research, 184(3): 1-6.

Skalny A V, Burtseva T I, Salnikova E V, et al. 2019. Geographic variation of environmental, food, and human hair selenium content in an industrial region of Russia. Environmental Research, 171: 293-301.

Spallholz J E, Boylan L M, Robertson J D, et al. 2008. Selenium and arsenic content of agricultural soils from Bangladesh and Nepal. Toxicological & Environmental Chemistry, 90(2): 203-210.

Stavridou E. 2011. The effects of cropping systems on selenium and glucosinolate concentrations invegetables. Denmark: Aarhus University.

Stroud J L, Li H F, Lopez-Bellido F J, et al. 2010. Impact of sulphur fertilisation on crop response to selenium fertilisation. Plant and Soil, 332(1-2): 31-40.

Suarez D L, Grieve C M, Poss J A. 2003. Irrigation method affects selenium accumulation in forage *Brassica* species. Journal of Plant Nutrition, 26(1): 191-201.

Sun G X, Meharg A A, Li G, et al. 2016. Distribution of soil selenium in China is potentially controlled by deposition and volatilization? Scientific Reports, 6: 20953.

Sun Q, Chen J, Ding S, et al. 2014. Comparison of diffusive gradients in thin film technique with traditional methods for evaluation of zinc bioavailability in soils. Environmental Monitoringand Assessment, 186: 6553-6564.

Sun W, Huang B, Zhao Y, et al. 2009. Spatial variability of soil selenium as affected by geologic and pedogenic processes and its effect on ecosystem and human health. Geochemical Journal, 43(4): 217-225.

Supriatin S, Weng L P, Comans R N J. 2015. Selenium speciation and extractability in Dutch agricultural soils. The Science of the Total Environment, 532: 368-382.

Supriatin S, Weng L P, Comans R N J. 2016. Selenium-rich dissolved organic matter determines selenium uptake in wheat grown on low-selenium arable land soils. Plant and Soil, 408(1-2): 73-94.

Suzuki K T, Ohta Y, Suzuki N. 2006. Availability and metabolism of 77Se-methylseleninic acid compared simultaneously with those of three related selenocompounds. Toxicology and Applied Pharmacology, 217(1): 51-62.

Swaine D J. 1955. The Trace Element Content of Soils. Commonwealth Agricultural Bureaux, Farnham Royal, Bucks, England.

Tam S C, Chow A, Hadley D. 1995. Effects of organic-component on the immobilization of selenium on iron oxyhydroxide. Science of the Total Environment, 164(1): 1-7.

Tamás M, Mándoki Z, Csapó J. 2010. The role of selenium content of wheat in the human nutrition A literature review. Acta Universitatis Sapientiae Alimentaria, 3: 5-34.

Tan J A, Zhu W Y, Wang W Y, et al. 2002. Selenium in soil and endemic diseases in China. Science of the Total Environment, 284(1): 227-235.

Tan J A. 1989. The Atlas of Endemic Diseases and Their Environments in the People's Republic of China. Beijing: Science Press.

Tao Z, Chu T, Du J, et al. 2000. Effect of fulvic acids on sorption of U(Ⅵ), Zn, Yb, I and Se(Ⅳ) onto oxides of aluminum, iron and silicon. Applied Geochemistry, 15: 133-139.

Terry N, Zayed A M, de Souza M P, et al. 2000. Selenium in higher plants. Annual Review of Plant Physiology and Plant Molecular Biology, 51: 401-432.

Thiry C, Ruttens A, Temmerman L D, et al. 2012. Current knowledge in species-related bioavailability of selenium in food. Food Chemistry, 130(4): 767-784.

Thurman E M, Malcolm R L. 1981. Preparative isolation of aquatic humic substances. Environmental Science and Technology, 15: 463-466.

Tian H, Ma Z Z, Chen X L, et al. 2016. Geochemical characteristics of selenium and its correlation to other elements and minerals in selenium-enriched rocks in Ziyang county, Shaanxi province, China. Journal of Earth Science, 27: 763-776.

Tian X L, Luo K L. 2017. Distribution and enrichment patterns of selenium in the Ediacaran and early Cambrian strata in the Yangtze Gorges area, South China. Science China-Earth Sciences, 60 (7): 1268-1282.

Tokunaga T K, Lipton D S, Benson S M, et al. 1991. Soil selenium fractionation, depth profiles and time trends in a vegetated site at Kesterson Reservoir. Water Air and Soil Pollution, 57-58: 31-41.

Tolu J, Di Tulllo P, Le Hécho I, et al. 2014a. A new methodology involving stable isotope tracer to compare simultaneously short- and long-term selenium mobility in soils. Analytical and Bioanalytical Chemistry, 406: 1221-1231.

Tolu J, Le Hécho I, Bueno M, et al. 2011. Selenium speciation analysis at trace level in soils. Analytica Chimica Acta, 684: 126-133.

Tolu J, Thiry Y, Bueno M, et al. 2014b. Distribution and speciation of ambient selenium in contrasted soils, from mineral to organic rich. Science of the Total Environment, 479: 93-101.

Torma A E, Habashi F. 1972. Oxidation of copper (II) selenide by Thiobacillus ferrooxidans. Canadian Journal of Microbiology, 18(11): 1780-1781.

Torres J, Pintos V, Dominguez S, et al. 2010. Selenite and selenate speciation in natural waters: interaction with divalent metal ions. Journal of Solution Chemistry, 39(1): 1-10.

Van Mantgem P J, Wu L, Banuelos G S. 1996. Bioextraction of selenium by forage and selected field legume species in selenium-laden soils under minimal field management conditions. Ecotoxicology and Environmental Safety, 34(3): 228-238.

Wang D, Dinh Q T, Thu T T A, et al. 2018. Effect of selenium-enriched organic material amendment on selenium fraction transformation and bioavailability in soil. Chemosphere, 199: 417-426.

Wang D, Peng Q, Yang W X, et al. 2020. DOM derivations determine the distribution and bioavailability of DOM-Se in selenate applied soil and mechanisms. Environmental Pollution, 259: 113899.

Wang D, Xue M Y, Wang Y K, et al. 2019a. Effects of straw amendment on selenium aging in soils: Mechanism and influential factors. Science of the Total Environment, 657: 871-881.

Wang D, Zhou F, Yang W X, et al. 2017a. Selenate redistribution during aging in different Chinese soils and the dominant influential factors. Chemosphere, 182: 284-292.

Wang J, Li H, Li Y, et al. 2013. Speciation, distribution, and bioavailability of soil selenium in the Tibetan Plateau Kashin-Beck disease area—a case study in Songpan County, Sichuan Province, China. Biological Trace Element Research, 156: 367-375.

Wang J, Zhang C B, Jin Z X. 2009. The distribution and phytoavailability of heavy metal fractions in rhizosphere soils of *Paulowniu fortunei* (seem) Hems near a Pb/Zn smelter in Guangdong, PR China. Geoderma, 148(3-4): 299-306.

Wang M K, Cui Z W, Xue M Y, et al. 2019b. Assessing the uptake of selenium from naturally enriched soils by maize (*Zea mays* L.) using diffusive gradients in thin-films technique (DGT) and traditional extractions. Science of the Total Environment, 689: 1-9.

Wang Q, Yu Y, Li J, et al. 2017b. Effects of different forms of selenium fertilizers on Se accumulation, distribution, and residual effect in winter wheat-summer maize rotation system. Journal of Agricultural and Food Chemistry, 65(6): 1116-1123.

Wang S S, Liang D L, Wang D, et al. 2012. Selenium fractionation and speciation in agriculture soils and accumulation in corn (*Zea mays* L.) under field conditions in Shaanxi Province, China. Science of the Total Environment, 427: 159-164.

Wang Y N, Zeng X B, Lu Y H, et al. 2015. Effect of aging on the bioavailability and fractionation of arsenic in soils derived from five parent materials in a red soil region of Southern China. Environmental Pollution, 207: 79-87.

Wang Y, Zhang X, Zhang X, et al. 2017c. Characterization of spectral responses of dissolved organic matter DOM for atrazine binding during the sorption process onto black soil. Chemosphere, 180: 531-539.

Wen H J, Carignan J, Qiu Y H, et al. 2006. Selenium speciation in kerogen from two Chinese selenium deposits: environmental implications. Environmental Science and Technology, 40: 1126-1132.

Weng L, Vega F A, Supriatin S, et al. 2011. Speciation of Se and DOC in soil solution and their relation to Se bioavailability. Environmental Science and Technology, 45: 262-267.

Wijnja H, Schulthess C P. 2000. Interaction of carbonate and organic anions with sulfate and selenate adsorption on an aluminum oxide. Soil Science Society of America Journal, 64: 898-908.

Wilber C G. 1980. Toxicology of selenium: a review. Clinical Toxicology, 17(2): 171-230.

Williams P N, Zhang H, Davison W, et al. 2011. Organic matter-solid phase interactions are critical for predicting arsenic release and plant uptake in Bangladesh paddy soils. Environmental Science and Technology, 45: 6080-6087.

Winkel L H E, Vriens B, Jones G D, et al. 2015. Selenium cycling across soil-plant-atmosphere interfaces: a critical review. Nutrients, 7(6): 4199-4239.

Wu L, Emberg A, Biggar J A. 1994. Effects of elevated selenium concentration on selenium accumulation and nitrogen-fixation symbiotic activity of Melilotus-Indica L. Ecotoxicology and Environmental Safety, 27(1): 50-63.

Wu L, Enberg A, Tanji K K. 1993. Natural establishment and selenium accumulation of herbaceous plant-species in soils with elevated concentrations of selenium and salinity under irrigation and tillage practices. Ecotoxicology and Environmental Safety, 25(2): 127-140.

Wu X, Lag J. 1988. Selenium in Norwegian farmland soils. Acta Agriclture Scandinavica, 38(3): 271-276.

Xu Y, Li Y, Li H, et al. 2018. Effects of topography and soil properties on soil selenium distribution and bioavailability (phosphate extraction): a case study in Yongjia County, China. Science of the Total Environment, 633: 240-248.

Yadav S K, Singh I, Sharma A, et al. 2005. Selenium status in soils of northern districts of India. Journal of Environmental Management, 88(4): 770-774.

Yamada H, Kamada A, Usuki M, et al. 2009. Total selenium content of agricultural soils in Japan. Soil Scienceand Plant Nutrition, 55: 616-622.

Yu T, Yang Z F, Hou Q Y, et al. 2016. Topsoil selenium distribution in relation to geochemical factors in main agricultural areas of China. In: Banuelos G S. Global Advances in Selenium Research From Theory to Application. London: Taylor and Francis Group.

Yuan L, Yin X, Zhu Y, et al. 2012. Selenium in plants and soils, and selenosis in Enshi, China: implications for selenium biofortification. In: Yin X, Yuan L. Phytoremediation and Biofortification. Netherlands: Springer: 7-31.

Zhai H, Kleawsampanjai P, Wang M, et al. 2021. Effects of soil moisture on aging of exogenous selenate in three different soils and mechanisms. Geoderma, 390: 114966.

Zhang M, Tang S H, Huang X, et al. 2014. Selenium uptake, dynamic changes in selenium content and its influence on photosynthesis and chlorophyll fluorescence in rice (*Oryza sativa* L.). Environmental and Experimental Botany, 107: 39-45.

Zhang Q, Chen H, Huang D, et al. 2019. Water managements limit heavy metal accumulation in rice: Dual effects of iron-plaque formation and microbial communities. Science of the Total Environment, 687: 790-799.

Zhang X P, Deng W, Yang X M. 2002. The background concentrations of 13 soil trace elements and their relationships to parent materials and vegetation in Xizang (Tibet), China. Journal of Asian Earth Sciences, 21(2): 167-174.

Zhang Y Q, Moore J N. 1996. Selenium fraction ation and speciation in a wetland system. Environ Sci Technol, 30(8): 2613-2619.

Zhao F J, Lopez-Bellido F J, Gray C W, et al. 2007. Effects of soil compaction and irrigation on the concentrations of selenium and arsenic in wheat grains. Science of the Total Environment, 372(2-3): 433-439.

Zhou F, Li Y N, Ma Y Z, et al. 2021. Selenium bioaccessibility in native seleniferous soil and associated plants: comparison between in vitro assays and chemical extraction methods. Science of the Total Environment, 762: 143119.

Zhou X, Li Y, Lai F. 2018. Effects of different water management on absorption and accumulation of selenium in rice. Saudi Journal of Biological Sciences, 25(6): 1178-1182.

Zhu J, Wang N, Li S, et al. 2008. Distribution and transport of selenium in Yutangba, China: Impact of human activities. Science of the Total Environment, 392(2-3): 252-261.

第三章 土壤中硒的生物转化

硒是生物体不可或缺的微量元素，但是当其浓度过高时也会对生物体造成损害。在自然界中硒主要以四种形态存在：还原态的硒化物[Se(-II)]、单质硒[Se(0)]、氧化态的亚硒酸盐[Se(IV)]和硒酸盐[Se(VI)]。在厌氧的还原环境中，硒主要以Se(0)和Se(-II)形式存在于土壤中（Zhang and Moore，1996；Martens and Suarez，1997）。好氧土壤中以Se(IV)和Se(VI)为主，其中，碱性土壤中硒易被化学或生物氧化，主要以Se(VI)形式存在，而中等氧化还原电位环境中则以Se(IV)为主。Se(IV)和Se(VI)均可溶于水，前者具有更高的吸附性，因此移动性较差，同时其也是毒性最强的形态，能影响细胞呼吸和酶活性、损伤细胞抗氧化系统等（Bebien et al.，2001），而Se(VI)因与土壤颗粒的亲和力差，较易发生迁移，进而被生物利用，因此Se(VI)较其他形态硒具有更强的生物可利用性。此外，土壤中还存在有机硒，如甲基硒（MeSe）、二甲基硒（DMeSe）、三甲基硒（TMeSe）及硒代甲硫氨酸（SeMet）等（Masscheleyn et al.，1991），但是这种低分子质量的有机硒在土壤中容易分解（Martens and Suarez，1999）。总体来说，土壤性质和硒的形态影响着其流动性和生物可利用性。

土壤中硒的形态转化主要依靠微生物调节（Shrift，1964），包括吸收、转运、还原、氧化、甲基化、氨基酸化和蛋白化等过程。微生物调控硒的生物地球化学循环的同时，也影响着硒的毒性，其可以将氧化态Se(IV)和Se(VI)异化还原为Se(0)，或同化还原合成硒蛋白，或者甲基化为具有高挥发性的DMeSe等。由于甲基硒的毒性比无机硒[Se(IV)和Se(VI)]低很多（无机硒毒性是甲基硒的500～700倍）（Dungan and Frankenberger，1999），而Se(0)性质稳定，难以被生物利用，故利用微生物将硒转化为低毒性和低生物利用度的甲基硒或元素硒已成为修复硒污染土壤的有效手段。同时，由于微生物还原产生的单质硒等生物纳米颗粒具有优越的光电子特性及抗生物活性，因此其在生物传感器、医药、光催化等领域也具有广阔的应用前景（范书伶等，2020；王丹等，2017）。微生物参与调控硒的生物地球化学循环的途径和机制是目前硒生物转化的研究重点。此外，土壤动物也是参与土壤中硒生物地球化学循环过程的重要一环，认识硒对土壤动物的影响及其在硒转化中发挥的作用，对于全面理解硒的生物地球化学循环具有重要意义。

第一节 微生物对硒的吸收转运

硒的吸收与转运是微生物硒代谢的起始。许多微生物均能将硒吸收并转运，例如，乳酸菌包括粪肠球菌（*Enterococcus faecalis*）、类布氏乳杆菌（*Lactobacillus parabuchneri*）、副干酪乳杆菌（*Lactobacillus Paracasei*）、植物乳杆菌（*Lactobacillus plantarum*），在含Se(IV)的培养基中能吸收硒并在体内积累（Morschbacher et al.，2018）。硒可以通过多种途径进入细胞。第一，硒与硫具有很高的物理化学相似性，因此硒的代谢路径也与硫相近，即Se(VI)和Se(IV)能通过硫酸盐渗透酶系统进入细胞（Sirko et al.，1990；

Birringer et al., 2002), 同时硫的浓度水平也影响着细胞对硒的吸收及细胞内硒的代谢产物, 进而影响了硒对细胞的毒性效应 (Mapelli et al., 2012)。第二, Se(Ⅳ)还可能利用亚硝酸盐转运系统进入细胞 (Antonioli et al., 2007; Heider and Bock, 1993)。此外, 不同的微生物对于硒的吸收转运还具有许多不同的机制。例如, 类球红细菌 (*Rhodobacter sphaeroides*) 中, Se(Ⅳ) 由多元醇转运体介导进入细胞质中 (Bebien et al., 2001); 而雷尔氏菌 (*Ralstonia metallidurans* CH34) 中主要以 DedA 蛋白作为转运体或辅助体参与 Se(Ⅳ) 的吸收 (Ledgham et al., 2005); 酵母中 Jen1p 转运体能高效催化细胞吸收 Se(Ⅳ) 并促进其在胞内累积, 由于其与乳酸的转运机制相似, 因此 Se(Ⅳ) 和乳酸盐存在竞争性抑制 (McDermott et al., 2010); 在大肠杆菌 (*Escherichia coli*) 中, Se(Ⅳ) 的吸收过程则主要由低亲和磷酸转运系统 (PIT) 的关键蛋白 PitA 介导, 且磷酸盐能通过竞争作用抑制大肠杆菌对 Se(Ⅳ) 的吸收 (Zhu et al., 2020)。总之, 硒在不同微生物中的吸收转运途径多样, 目前仍有许多途径有待发现。

此外, 硒在细胞表面的吸附也是其进一步被微生物转化的途径之一。氧化环境中硒以带负电的阴离子形式存在, 而枯草芽孢杆菌 (*Bacillus subtilis*) 能通过细胞包膜巯基位点将 Se(Ⅳ) 吸附在带负电的细胞表面, 从而进一步将 Se(Ⅳ) 转化为有机硒化合物 (Yu et al., 2018)。巯基位点存在于多种细菌的细胞膜中, 因此巯基控制的 Se(Ⅳ) 吸附可能是氧化环境中微生物利用 Se(Ⅳ) 的重要途径。

第二节 硒的微生物还原

硒是生命所必需的微量元素, 主要以硒代半胱氨酸 (SeCys) 和硒代甲硫氨酸 (SeMet) 的形式存在于硒蛋白中, 硒的过量摄入或不足均会导致多种疾病。我国低硒地区分布广泛, 约占国土总面积的 72% (朱燕云等, 2018), 因此近年来富硒农产品受到研究者的广泛关注, 农田环境中具有亚硒酸盐 [Se(Ⅳ)] 和硒酸盐 [Se(Ⅵ)] 还原能力的菌株逐渐得到分离培养。例如, 水稻土中分离得到的 *Chitinophaga* sp. 和 *Comamonas testosteroni* 能将 Se(Ⅳ) 还原, 提高土壤中硒的生物利用率, 促进水稻对硒的吸收 (Huang et al., 2021a); 茶园富硒土中分离得到能将 Se(Ⅳ) 好氧还原的杆菌 *Lysinibacillus xylanilyticus* 和 *Lysinibacillus macrolides* (Zhang et al., 2019); 另外, 在一些极端环境中也分离出了硒还原菌, 如 *Stenotrophomonas bentonitica* 能在厌氧和碱性条件下将 Se(Ⅳ) 还原 (Ruiz-Fresneda et al., 2019); 活性污泥中也分离得到了具有 Se(Ⅳ) 还原能力的菌株 (Nguyen et al., 2019)。Peng 等 (2016) 研究发现具有硒同化能力的细菌在环境中广泛存在, 而细菌、真菌及放线菌等多种微生物均能通过异化还原作用将 Se(Ⅳ) 和 Se(Ⅵ) 还原产生硒纳米颗粒 (SeNP), 从而大大降低硒的毒性和生物可利用性。因此, 微生物硒还原作用在生产含硒产品和消除环境污染方面具有广阔的应用前景。

一、硒的同化还原

硒的同化还原是合成硒蛋白所必需的过程。硒进入细胞后被进一步同化为硒代半胱氨酸 (SeCys) 和硒代甲硫氨酸 (SeMet), 其中 SeCys 是人体第 21 种氨基酸, 存在于许多酶中, 是所有活细胞所必需的 (Stadtman, 1996)。硫氧还蛋白还原酶 (TrxR)、谷胱甘肽过氧化物酶（GSH-Px）和甲状腺激素脱碘酶（DIO）是表征较为完善的硒蛋白，参与了细胞内信号转导、氧化还原稳态和甲状腺激素代谢的氧化还原调节（Papp et al.，2007）。在不同微生物体内，硒以不同的形式存在，例如，富硒酵母中硒主要以 SeMet 形式存在，而在罗伊乳杆菌中则主要以 SeCys 的形式掺入硒蛋白中（Galano et al.，2013）。

硒在细胞内的代谢途径尚不明晰。研究表明，进入细胞的硒可以被还原为 Se(-Ⅱ)，进而通过特定途径并入 SeCys，而 SeMet 则是由游离氨基酸 SeCys 通过与甲硫氨酸相同的途径合成；另外，Se(Ⅳ) 也能与谷胱甘肽反应生成硒二谷胱甘肽（GS-Se-GS），硒二谷胱甘肽及其还原物谷胱甘肽硒醇（GS-SeH）是硒代谢过程中的关键中间体（Turner et al.，1998）。硒磷酸是合成硒蛋白和硒修饰 tRNA 的关键硒供体，其中硒磷酸合成酶由 ATP 和硒化物合成，是微生物同化还原硒的关键基因（Peng et al.，2016）。研究表明，谷胱甘肽、硒半胱氨酸裂解酶、半胱氨酸脱硫酶和硒结合蛋白可能参与了硒传递进入硒磷酸合成酶的过程；而大肠杆菌中硫氧还蛋白系统也参与了该过程，从而实现了特定硒蛋白的生物合成，并且避免了游离硒化物对生物体的毒性（Tobe and Mihara，2018）。此外，丝氨酸参与了嗜热乳酸链球菌（*Streptococcus thermophilus*）将 Se(Ⅳ) 转化为 SeCys 的过程（Castaneda-Ovando et al.，2019）。总体来说，不同微生物对硒的同化还原过程还缺乏深入系统的研究，硒蛋白是硒实现生理功能的关键生物分子，其形成路径和机制值得进一步探究。

二、硒酸盐的异化还原

微生物异化还原是硒还原的主要途径（Zehrt and Oremland，1987；Macy et al.，1989；Switzer et al.，1998；Stolz and Oremland，1999；Narasingarao and Häggblom，2006；Rauschenbach et al.，2011），具有此功能的微生物种类繁多，且生理生化特性差异极大。在原核微生物中，通过还原硒的氧化物产生能量的过程极为普遍，其中代表性微生物有嗜泉古菌、嗜热细菌、革兰氏阳性菌、变形菌及光营养型的微生物紫细菌等。土壤中 Se(Ⅵ) 的厌氧还原过程中，脱氯单胞菌属（*Dechloromonas*）、红环菌科（Rhodocyclaceae）和丛毛单胞菌科（Comamonadaceae）的丰度显著增加，表明这些微生物协同参与了 Se(Ⅵ) 的还原（Navarro et al.，2015）；另外，在废石中也发现脱硫菌（*Desulfosporosinus*）和发酵黏土弯菌（*Pelosinus fermentans*）参与了乳酸诱导的 Se(Ⅵ) 还原过程并促进了硒的生物固定（Aoyagi et al.，2021）。硒同位素分馏研究表明，微生物还原 Se(Ⅵ) 和 Se(Ⅳ) 的速率与机制存在明显差异，前者主要由与菌株生理特异性相关的酶（硒酸盐还原酶 SerA）主导，因此有团队根据 SerA 设计了引物以用于定量检测环境中的硒酸盐还原菌（Wen et al.，2016）；而后者则主要是胞外还原过程（Schilling et al.，2020），通过胞外聚合物中硫醇、醛和酚等基团，以及亚硝酸盐还原酶、亚硫酸盐还原酶、硫氧还蛋白和硫氧还蛋白还原酶、谷胱甘肽和谷胱甘肽还原酶等多种非酶促和酶促还原途径完成。

此外，不同微生物对于硒酸盐 [Se(Ⅵ)] 的还原过程及机制也存在巨大差异。例如，在没有外部电子供体的条件下，腐败希瓦氏菌（*Shewanella putrefaciens* 200R）通过外层络合将 Se(Ⅳ) 吸附，并在细胞表面将其还原为 Se(0)（Kenward et al.，2006）。*Shigella fergusonii* 和 *Pantoea vagans* 还原 Se(Ⅵ) 则包括两个阶段：微生物首先将 Se(Ⅵ) 还原为

Se(Ⅳ)，然后进一步还原生成 Se(0)，其中第二个还原阶段是限速步骤，故导致毒性更强的 Se(Ⅳ) 积累，而二者共同培养则可显著降低 Se(Ⅳ) 的累积（Ji and Wang，2019）。因此，深入认识 Se(Ⅵ) 的微生物还原过程及机制，对于调控硒的转化、控制其环境效应具有重要意义。

（一）硒酸盐还原与烷烃氧化

微生物介导的 Se(Ⅵ) 还原过程能与烷烃氧化过程耦合发生。首先，Se(Ⅵ) 还原可与甲烷厌氧氧化过程相耦合，在甲烷厌氧氧化菌的作用下被还原为 SeNP（Luo et al.，2018）。其次，Se(Ⅵ) 还原可与甲烷好氧氧化过程耦合发生。研究表明，好氧条件下 Se(Ⅵ) 的还原机制主要有两种：第一，甲烷氧化菌（如甲基单胞菌 *Methylomonas*）氧化甲烷的同时直接偶联 Se(Ⅵ) 还原，在添加硝酸盐时二者的还原过程同时发生，且硝酸盐部分抑制了 Se(Ⅵ) 的还原；第二，甲烷氧化菌的代谢产物也能作为电子供体，协同 Se(Ⅵ) 还原（Lai et al.，2016）。同时，甲烷氧化古菌和细菌的共同作用也能促进 Se(Ⅵ) 的还原，其中甲烷八叠球菌（*Methanosarcina*）和甲基孢囊菌（*Methylocystis*）是主要的甲烷好氧氧化菌，与节杆菌（*Arthrobacter*）和贪噬菌（*Variovorax*）协同作用促进 Se(Ⅵ) 还原为 Se(Ⅳ)，进而生成 Se(0)（Shi et al.，2020b）。该过程受供氧速率的影响，过高会导致介导 Se(Ⅵ) 还原的基因 *narG* 下调，从而抑制 Se(Ⅵ) 的还原速率（Wang et al.，2021）。进一步研究发现，生物反应器群落中丰度最高的甲基孢囊菌（*Methylocystis*）本身缺乏 Se(Ⅵ) 还原的遗传潜力，而假黄单胞菌（*Pseudoxanthomonas*）中的硝酸盐还原酶可以利用甲烷氧化菌释放的发酵副产物作为电子供体还原 Se(Ⅵ)（Shi et al.，2021）。此外，乙烷和丙烷也能作为膜生物反应器中 Se(Ⅵ) 还原的电子供体，生成产物主要为 Se(0)，其中分枝杆菌（*Mycobacterium*）和红球菌（*Rhodococcus*）起主导作用；同时，与烷烃氧化、反硝化和聚羟基脂肪酸循环相关的基因也在生物膜中富集（Lai et al.，2020）。该发现将硒与烷烃的生物地球化学循环联系起来，拓展了对微生物介导 Se(Ⅳ) 还原机制的认识。

（二）硒酸盐与硝酸盐还原

硝酸盐在热力学上是一种极好的电子受体，因此当其浓度较高时能够抑制 Se(Ⅳ) 的还原。研究发现位于细胞周质空间的硒酸盐还原酶（SerABC）（Schröder et al.，1997）和硝酸盐还原酶（Steinberg et al.，1992；Sabaty et al.，2001；Gates et al.，2011）均可介导 Se(Ⅵ) 的还原，其中 SerABC 通过细胞色素接受电子继而催化还原 Se(Ⅵ) 生成 Se(Ⅳ)，Se(Ⅳ) 在细胞质中进一步还原形成 Se 纳米球，最终分泌到周围介质中（Butler et al.，2012）。Subedi 等（2017）在矿山附近的天然湿地环境中发现土著微生物可以同时还原 Se(Ⅵ) 和硝酸盐，利用纯硒酸盐培养基发现其优势菌群主要为假单胞菌属（*Pseudomonas*）、赖氨酸芽孢杆菌属（*Lysinibacillus*）和索氏菌属（*Thauera*），且其开放阅读框与已知的硒酸还原酶亚基 SerA 和 SerB 同源。培养基同时含有硒酸盐和硝酸盐时，罕见类群微小杆菌（*Exiguobacterium*）、泰氏菌属（*Tissierella*）和梭菌属（*Clostridium*）成为优势菌群。另外，在矿山环境中发现肠杆菌目（Enterobacterales）和梭菌目（Clostridiales）能介导硒酸盐的还原，推测硒酸盐还原酶可能是硒蛋白合成酶（Nkansah-Boadu et al.，2021）。

（三）硒酸盐与硫酸盐还原

硫酸盐还原菌还原转化硒的反应路径如图 3-1 所示（Hockin and Gadd，2003）。首先，Se(Ⅵ) 和 Se(Ⅳ) 能通过硫酸盐渗透酶系统进入细胞（Sirko et al.，1990；Birringer et al.，2002）；然后，硫酸盐还原菌能分别在硫酸盐过量和有限的条件下将 Se(Ⅵ) 异化还原为 Se(0) 和 Se(-Ⅱ)，其中前者主要通过酶解作用还原，而后者则是异化硫酸盐还原路径（DSR）发挥了关键作用（Stolz and Oremland，1999；Hockin and Gadd，2006）；同时，普通脱硫弧菌（*Desulfovibrio vulgaris*）、巨大脱硫弧菌（*Desulfovibrio gigas*）等硫酸盐还原菌均能将 Se(Ⅳ) 还原为二甲基硒（DMeSe）和二甲基二硒（DMeDSe）（Michalke et al.，2000），而脱硫脱硫弧菌（*Desulfovibrio desulfuricans*）则将 Se(Ⅵ) 和 Se(Ⅳ) 还原为 Se(0) 并在细胞内积累（Tomei et al.，1995）；此外，Se(Ⅳ) 还能与硫酸盐还原菌产生的硫化物通过非生物反应生成 Se(0) 和 S(0)，并在细胞外沉淀。近期研究表明，当膜生物反应器中硫酸盐浓度过高时，会与钼酸盐竞争运输，使得编码依赖钼的硒酸盐还原酶的基因丰度大大降低，导致硒酸盐还原酶活性降低，进而抑制 Se(Ⅵ) 的还原过程（Shi et al.，2020a）。因此，不同条件下 Se(Ⅵ) 与硫酸盐还原的内在机制还需进一步揭示。

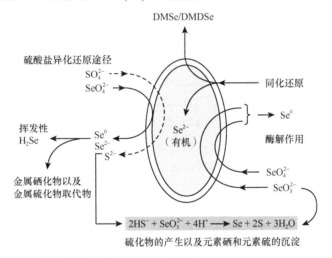

图 3-1　硫酸盐还原菌对硒的转化（改自 Hockin and Gadd，2003）

三、亚硒酸盐的异化还原

亚硒酸盐 [Se(Ⅳ)] 是环境中生物毒性极高的物种，其能在人体内通过内源性蛋白辅助形成 SeNP，而后者对糖酵解酶的隔离抑制了 ATP 的生成，导致线粒体结构和功能被破坏；同时，SeNP 的表面效应能产生氧化应激，进一步引起细胞毒性（Bao et al.，2015）。环境中许多微生物都能通过将 Se(Ⅳ) 还原为不溶的 Se(0) 来降低毒性，一定程度上表现出对 Se(Ⅳ) 的抗性，因此该微生物代谢过程及机制越来越受到重视。Se(Ⅳ) 通常在微生物的细胞质或周质中被还原，该过程与周质钼氧化还原酶（SerT）（Tan et al.，2018）、谷胱甘肽（GSH）（Kessi and Hanselmann，2004）、谷胱甘肽还原酶（GR）、富马酸还原酶（FccA）（Li et al.，2014a）、亚硝酸盐还原酶（Basaglia et al.，2007）及硫氧还蛋白还原酶（TrxR）（Bjornstedt et al.，1995）等生物分子相关。

（一）亚硒酸盐还原菌

许多微生物均具有 Se(Ⅳ) 还原能力，然而不同微生物的还原机理存在较大差异。朱永官团队从云南德宏的水稻土中分离得到了具有 Se(Ⅳ) 和硝酸盐还原功能的兼性厌氧菌芽孢杆菌（*Bacillus oryziterrae*），其能将 Se(Ⅳ) 还原为 SeNP，并在菌株内鉴定出了具有 SeNP 高亲和性的蛋白质（Bao et al., 2016）。假单胞菌（*Pseudomonas putida* KT2440）等能将 Se(Ⅳ) 还原为 SeNP，但不能还原 Se(Ⅵ)（Avendano et al., 2016；Staicu et al., 2015）。许多乳酸菌（包括 *Lactobacillus brevis* CRL 2051、*Lactobacillus plantarum* CRL 2030 等）均能将 Se(Ⅳ) 还原为球形 SeNP 和硒代半胱氨酸（SeCys）。肠球菌 *Enterococcus casseliflavus* 47 和 82 能将 Se(Ⅳ) 还原产生二甲基二硒（DMeDSe）（Martinez et al., 2020）。嗜麦芽糖寡养单胞菌（*Stenotrophomonas maltophilia* SeITE02）能将 Se(Ⅳ) 还原为球形 SeNP，其中主要在细胞质蛋白中检测到 Se(Ⅳ) 还原活性，进一步通过蛋白质组学分析表明醇脱氢酶同源物参与了该过程（Lampis et al., 2017）；而 *Stenotrophomonas bentonitica* 还原 Se(Ⅳ) 形成的无定形 SeNP 会转变为三方晶系纳米颗粒，同样地，蛋白质在其合成和转化过程中也发挥了重要作用；此外，其还原产物还包括挥发性的 DMeDSe 和二甲基硫硒化合物（DMeSeS），这些产物均具有较低的迁移率和化学毒性。因此，微生物还原作用在降低硒的迁移性和毒性方面具有重要意义（Ruiz-Fresneda et al., 2020）。

金属污染土壤中常见的 Se(Ⅳ) 还原菌有枯草芽孢杆菌（*Bacillus subtilis*）、微小杆菌（*Exiguobacterium* sp.）、地衣芽孢杆菌（*Bacillus licheniformis*）和类产碱假单胞菌（*Pseudomonas pseudoalcaligenes*）等，然而其各自的还原机制还有待揭示。以蕈状芽孢杆菌（*Bacillus mycoides* SeITE01）为例，其通过生物过程形成 SeNP 的机制如下（图 3-2）：①利用小分子硫醇包括芽孢硫醇（BSH）的活性或通过硫氧还蛋白/硫氧还蛋白还原酶（Trx/TrxRed）系统将 Se(Ⅳ) 还原为 SeNP 并在胞质沉淀；②利用膜还原酶的活性在细胞内将 Se(Ⅳ) 还原为 SeNP；③通过胞溶作用将生成的 SeNP 释放至胞外；④膜还原酶还可能在胞外催化 Se(Ⅳ) 还原并形成 SeNP 沉淀；⑤从细胞中释放的含巯基的多肽和其他化合物可直接与 Se(Ⅳ) 反应生成 SeNP；⑥ *Bacillus mycoides* SeITE01 培养基上清液中加入 NADH 后 Se(Ⅳ) 被还原为 SeNP，这可能是由于胞外蛋白具有介导 Se(Ⅳ) 还原沉淀的能力；⑦新生成的 SeNP 由于具有较高的比表面积而存在不稳定性，因此倾向于通过 Ostwald 成熟机制进一步生长从而增加尺寸，以达到稳定的低能态（Lampis et al., 2014）。

兼性厌氧的阴沟肠杆菌（*Enterobacter cloacae* SLD1a-1）能同时还原 Se(Ⅳ) 和 Se(Ⅵ) 形成 SeNP，且前者的还原速率明显高于后者，二者还原速率均与温度有关。值得注意的是，Se(Ⅵ) 和 Se(Ⅳ) 的还原是由不同的酶系统催化的，其中在低氧环境中 Se(Ⅵ) 还原酶的活性由富马酸硝酸盐还原调节基因 *FNR* 调控，Se(Ⅵ) 的还原活性则依赖于还原酶的跨膜运输（Ma et al., 2007；Yee et al., 2007）。Xia 等研究发现兼性厌氧菌类希瓦氏菌 *Alishewanella* sp. WH16-1 能在好氧条件下同时将 Se(Ⅳ) 和 Cr(Ⅵ) 还原为 Se(0) 和 Cr(Ⅲ)，并从中分离和鉴定出了亚硒酸盐还原酶 CsrF，其能在体内将二者还原，并能在体外将硫酸盐和三价铁还原，在重金属生物修复方面表现出巨大潜力（Xia et al., 2018）。

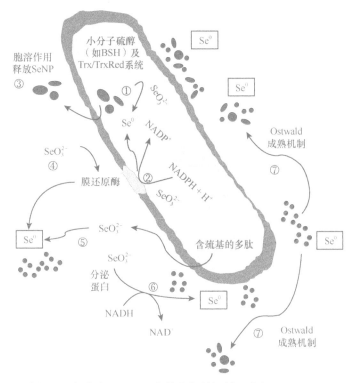

图 3-2　蕈状芽孢杆菌中 SeNP 形成的生物学机制（改自 Lampis et al.，2014）

（二）亚硒酸盐的异化还原作用

许多微生物均能进行 Se(Ⅳ) 的异化还原，通常认为该过程是通过细胞周质的亚硝酸盐还原酶（DeMoll-Decker and Macy，1993）、氢化酶Ⅰ（Yanke et al.，1995）或非酶途径（Tomei et al.，1992）介导的，但其还原机制仍未完全探明。DeMoll-Decker 和 Macy（1993）在 *Thauera selenatis* 中发现其细胞周质中存在一种亚硝酸盐异化还原酶可以催化 Se(Ⅳ) 还原，而缺乏这种酶活性的突变体既不能还原亚硝酸盐也不能还原 Se(Ⅳ)。根瘤菌（*Rhizobium sullae* HCNT1）体内的亚硝酸盐还原酶使之具有 Se(Ⅳ) 还原能力，并提高了对 Se(Ⅳ) 的抗性（Basaglia et al.，2007）。Yanke 等（1995）研究表明巴氏芽孢梭菌（*Clostridium pasteurianum*）中氢化酶Ⅰ也可以还原 Se(Ⅳ)，且 CO_2 可能通过增强氢化酶Ⅰ的活性和细胞色素 b 的水平促进 Se(Ⅳ) 还原过程，而硝酸盐由于具有较高的还原电位且能抑制硒酸盐和亚硒酸盐还原酶活性而抑制硒还原（Bao et al.，2013）。

此外，异化金属还原菌奥奈达希瓦氏菌（*Shewanella oneidensis* MR-1）也能厌氧还原 Se(Ⅳ)。首先，奥奈达希瓦氏菌 MR-1 氧化乳酸提供电子产生 NADH，进而通过 NADH 脱氢酶和氢醌库将电子传递至 c 型细胞色素 CymA，然后 CymA 将电子传递至各种还原酶，从而促进微生物厌氧呼吸和 Se(Ⅳ) 还原过程，呼吸产生的能量有助于细胞生存以及抵抗 Se(Ⅳ) 的毒性，而细胞周质中的富马酸还原酶 FccA 作为 Se(Ⅳ) 的末端还原酶能进一步清除已经进入周质的 Se(Ⅳ)，阻止其进入细胞质，最终在细胞周质和表面产生红色球形 SeNP。然而，MR-1 体内的硝酸盐还原酶 NapA 和 NapB、亚硝酸盐还原酶 NrfA 以及厌氧呼吸蛋白 Mtr 均无 Se(Ⅳ) 还原活性，同时微生物分泌的黄素单核苷酸（FMN）和核黄素（RF）对 Se(Ⅳ) 还原过程也并无贡献（图 3-3）（Li et al.，2014a）。

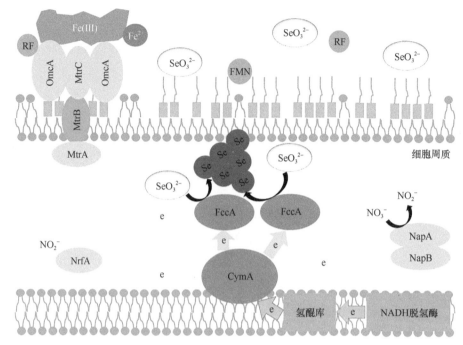

图 3-3 *Shewanella oneidensis* MR-1 体内亚硒酸盐还原和厌氧呼吸途径（改自 Li et al.，2014a）

（三）亚硒酸盐还原与甲烷氧化

甲烷厌氧氧化能与 Se(Ⅳ) 还原过程相耦合，研究表明甲烷氧化菌模式菌株荚膜甲基球菌（*Methylococcus capsulatus*）和甲基弯菌（*Methylosinus trichosporium* OB3b）均能将 Se(Ⅳ) 还原为红色球形 SeNP 和挥发性的甲基硒，其中 SeNP 主要形成于细胞壁上，但二者都不能还原 Se(Ⅵ)（Eswayah et al.，2017）。在甲烷厌氧氧化耦合 Se(Ⅳ) 还原过程中，硝酸根作为电子受体会与 Se(Ⅳ) 竞争电子，从而抑制 Se(Ⅳ) 的还原（Bai et al.，2019）。此外，好氧甲烷氧化菌也能在甲烷存在的条件下将 Se(Ⅳ) 转化为 SeNP 和挥发性的甲基硒及硒化物，具体反应路径如图 3-4 所示。首先，谷胱甘肽（GSH）与 Se(Ⅳ) 反应产生

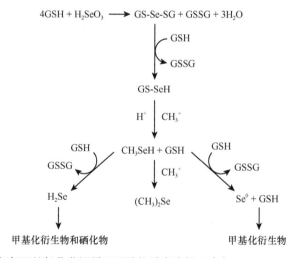

图 3-4 好氧甲烷氧化菌还原亚硒酸盐反应路径（改自 Eswayah et al.，2019）

硒二谷胱甘肽（GS-Se-GS）及二谷胱甘肽（GSSG），然后 GS-Se-GS 与 GSH 反应生成谷胱甘肽硒醇（GS-SeH），后者进一步与甲烷反应产生甲基硒醇（CH_3SeH），其是甲烷介导的、使硒向所有颗粒态和甲基化产物转化的中间体（Eswayah et al.，2019）。

（四）亚硒酸盐还原与硫代谢

Se(Ⅳ) 的微生物还原与含硫蛋白和嗜硫微生物也密切相关。Harrison 等（1984）研究发现巴氏芽孢梭菌（*Clostridium pasteurianum*）体内的亚硫酸盐异化还原酶可以将 Se(Ⅳ) 还原，即 Se(Ⅳ) 与亚硫酸盐存在还原竞争。芽孢杆菌（*Bacillus* sp. Y3）还原 Se(Ⅳ) 过程中，硫氧还蛋白还原酶（TrxR）、硫酸盐和能量代谢相关蛋白含量均显著上调，TrxR 转录水平也有所增加，且 TrxR 过表达的大肠杆菌（*Escherichia coli*）对 Se(Ⅳ) 耐受性提高，Se(Ⅳ) 还原效率增加，证明 TrxR 是重要的 Se(Ⅳ) 还原酶（Yasir et al.，2020）。Wells 等（2019）从硒还原芽孢杆菌 MLS10 中鉴定出亚硒酸盐异化还原酶 Srr，利用 LC-MS/MS 表征发现该蛋白质为聚硫还原酶（PsrA）催化亚基的同源物。Huang 等（2021b）利用超耐亚硒酸盐的雷氏普罗威登斯菌 *Providencia rettgeri* HF16-A 研究发现其能在 24 h 内将 5 mmol/L Se(Ⅳ) 完全还原为 SeNP，以 NADPH 和 NADH 为电子供体进行体外试验发现其胞质蛋白上能发生硒还原过程，并基于蛋白质组学首次在生理条件下发现亚硫酸盐还原酶介导了 Se(Ⅳ) 的还原。此外，嗜硫代谢菌［包括施氏假单胞菌（*Pseudomonas stutzeri* NT-I）、假单胞菌 *Pseudomonas* sp. RB、嗜麦芽糖寡养单胞菌（*Stenotrophomonas malophilia* TI-1）、人苍白杆菌（*Ochrobactrum anthropi* TI-2 和 TI-3）］也能将 Se(Ⅳ) 还原为挥发性有机硒（Kuroda et al.，2019）。

（五）亚硒酸盐还原与谷胱甘肽

谷胱甘肽（GSH）在 Se(Ⅳ) 还原过程中也发挥了重要作用。Newton 和 Fahey（1989）发现 GSH 在 α、β、γ 变形杆菌中可以以毫摩尔级水平存在，同时，Kessi 和 Hanselmann（2004）报道了 GSH 与 Se(Ⅳ) 的高反应性，并证实 GSH 参与了 Se(Ⅳ) 还原并合成硒代谷胱甘肽的过程。而链霉菌（*Streptomyces* sp. ES2-5）能通过 GSH 介导 Se(Ⅳ) 还原并在细胞内合成 SeNP，随后通过菌丝溶解或分裂释放到胞外（Tan et al.，2016）。Bebien 等（2002）则发现在鼠伤寒沙门菌（*Salmonella enterica serovar* Typhimurium）和 *E. coli* 培养物中添加 Se(Ⅳ) 能诱导谷胱甘肽还原酶（GR）的合成。这些发现都支持了微生物体内 Se(Ⅳ) 还原与 GSH 的关联性。该类硒还原微生物在硒污染的环境中数目较多，且在缺乏硒酸盐的环境中依然可以生存。但是，利用暗沟肠杆菌 Z0206 的研究表明，富马酸还原酶在 Se(Ⅳ) 还原中发挥着主导作用，而 GSH 介导的反应并非 Se(Ⅳ) 还原的主要途径（Song et al.，2017）。因此，对于 GSH 介导 Se(Ⅳ) 还原的机理仍有待系统研究。

（六）亚硒酸盐还原的影响因素

许多因素均影响着 Se(Ⅳ) 的微生物还原过程。首先，电子穿梭体能显著促进 Se(Ⅳ) 的微生物还原。例如，利用奥奈达希瓦氏菌 MR-1（*Shewanella oneidensis* MR-1）研究发现蒽醌-2,6-二磺酸盐（AQDS）和核黄素能显著加速 Se(Ⅳ) 还原，同时，电子穿梭体将还原过程从细胞内转移到胞外，诱导 SeNP 在胞外形成（Xia et al.，2016）；而磺化卟啉通过氧化还原转化降低了还原反应的活化能，同时增加了胞外聚合物（EPS）中具有氧化还原活性的物质，从而显著提高了奥奈达希瓦氏菌 MR-1 还原 Se(Ⅳ) 的速率（Zhao et al.，2020）。醌类等具有氧化还原活性的物质也能显著促进大肠杆菌（*Escherichia coli*）和厌氧性黏菌（*Anaeromyxobacter dehalogenans*）等微生物对 Se(Ⅳ) 的还原作用（He and Yao，2011；Wang et al.，2011）。此外，EPS 中硫醇、醛和酚基团能将 Se(Ⅳ) 还原并生成 Se(0)，该过程主要受溶解氧的影响，溶解氧通过调节 EPS 的氧化还原状态进而影响 Se(Ⅳ) 的还原过程，这也是微生物耐 Se(Ⅳ) 胁迫的生存机制之一（Zhang et al.，2020）。

溶液组成也影响着 Se(Ⅳ) 的微生物还原。研究表明，溶液中磷酸盐浓度较低时，大肠杆菌将吸收的 Se(Ⅳ) 还原产生大量 Se(0)，而当磷酸盐浓度升高时，胞内 Se 浓度降低且还原产物由 Se(0) 转变为毒性更大的有机硒，可能是由于 PitA 既是 Se(Ⅳ) 转运蛋白，又是低亲和磷酸转运系统关键蛋白，使得大肠杆菌对磷酸盐和 Se(Ⅳ) 存在竞争吸收，磷酸盐由此抑制了 Se(Ⅳ) 的还原和生物矿化（Zhu et al.，2020）。

（七）亚硒酸盐的真菌还原

目前，细菌介导硒还原转化的过程及机制得到了较为广泛的研究，此外，真菌对于富硒土壤中硒的转化也十分重要，真菌的活性影响着硒的形态和分布（Holland and Carter，1983）。对 6 种耐金属子囊菌（*Ascomycete*）的研究表明，溶液态硒的去除效率主要取决于真菌的种类、硒的形态及浓度。6 种子囊菌均能将 Se(Ⅵ) 和 Se(Ⅳ) 还原形成直径为 50～300 nm 的无定型或准晶型球状 SeNP，并且可以通过挥发作用去除 15%～20% 的硒，而硒浓度的增加则会导致 *Ascomycete* 生长速率降低（Rosenfeld et al.，2017）。在好氧条件下两种子囊菌真菌（*Paraconiothyrium sporulosum* 和 *Stagonospora* sp.）能将 Se(Ⅳ)/Se(Ⅵ) 还原为 SeNP 和 Se(-Ⅱ)，同时偶联真菌源+2 价锰氧化形成锰氧化物（Rosenfeld et al.，2020）。从环境中分离的真菌伯克霍尔德菌（*Burkholderia fungorum*）也能还原 Se(Ⅳ) 并在细胞内外生成球形 SeNP，推测过程如下：首先通过电子供体介导的胞质酶激活过程将胞内 Se(Ⅳ) 还原形成 SeNP，继而随着分泌过程或细胞裂解将其释放到细胞外（Khoei et al.，2017）。此外，酵母细胞中硒的代谢过程非常复杂，通常通过一系列的转化过程使硒的氧化程度降低，进而形成硒化物，再根据需要用于硒蛋白的合成，或转化为甲基硒以从体内去除。例如，酿酒酵母（*Saccharomyces cerevisiae*）能将无机硒和硒糖、硒-甲基硒代半胱氨酸等有机硒化合物代谢为硒代甲硫氨酸（SeMet）（Ogra et al.，2018）；球状茎点霉菌（*Phoma glomerata*）、出芽短梗霉菌（*Aureobasidium pullulans*）、哈茨木霉菌（*Trichoderma harzianum*）等真菌也能将 Se(Ⅳ) 还原为 SeNP 并使其分布于细胞内外（Liang et al.，2019）。值得注意的是，纳米颗粒的形成受胞外聚合物的影响，其表面被不同的分子所覆盖，同时茎点霉真菌对硒的吸收转化作用也改变了矿石中硒的分布与浓度（Liang et al.，2020）。

四、硒的生物成矿

硒的生物成矿是指硒氧阴离子被微生物还原形成稳定的红色纳米 Se(0) 的过程，并且能在微生物还原作用下持续生长形成 SeNP。值得注意的是，SeNP 的合成场所尚不清晰，其可能是在胞内酶的作用下形成，也可能是 Se(0) 被外排至胞外后在 EPS 的作用下形成，同时相应的外排路径也存在争议（图 3-5）。Losi 和 Frankenberger（1997）利用阴沟肠杆菌（*Enterobacter cloacae* SLD1a-1）研究表明 Se(Ⅵ) 和 Se(Ⅳ) 能被细胞膜上的还原酶还原，并将 SeNP 快速外排至胞外。Se(0) 可以通过囊泡外排（Piacenza et al.，2018；Wang et al.，2018），然而一些 SeNP 粒径远大于囊泡直径，因此可能通过细胞裂解外排或存在其他外排路径（Kushwaha et al.，2021）。SeNP 是微生物还原的重要产物，也是微生物修复硒污染环境以及回收硒的重要途径，例如，Tugarova 等（2020）便利用普遍存在且具有 Se(Ⅳ) 还原功能的植物促生根瘤菌固氮螺菌（*Azospirillum brasilense*）建立了合成并回收胞外 SeNP 的方法。

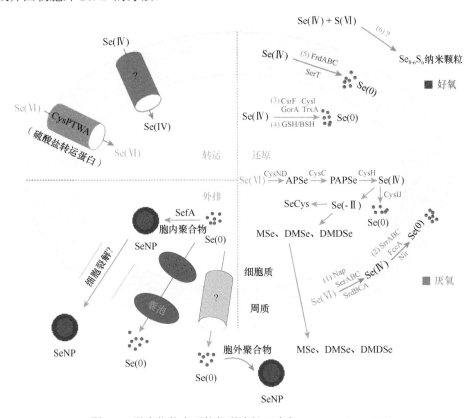

图 3-5　微生物体内硒的代谢途径（改自 Wang et al.，2022）

(1) 革兰氏阴性菌和革兰氏阳性菌可分别通过硝酸还原酶和硫酸盐呼吸还原酶在细胞周质将 Se(Ⅵ) 厌氧还原为 Se(Ⅳ)。(2) 亚硒酸盐呼吸还原酶 SrrABC、富甲酸还原酶 FccA 和亚硝酸盐还原酶 Nir 可将细胞周质中的 Se(Ⅳ) 厌氧还原为 SeNP。(3) Se(Ⅳ) 可被细胞质中的多种还原酶好氧还原为 Se(0)。(4) Se(Ⅳ) 也可被 GSH 和 BSH 等硫醇好氧还原为 Se(0)。(5) 富马酸还原酶和亚硒酸还原酶 SerT 可在周质中将 Se(Ⅳ) 好氧还原为 Se(0)。(6) 在硫酸盐和 Se(Ⅳ) 存在的情况下，可在细胞外通过未知的途径形成 $Se_{8-n}S_n$ 纳米颗粒混合物。

SeNP 作为一种重要的纳米材料，其形状和稳定性影响着颗粒的迁移和归趋。研究表明，胞外聚合物形成了 SeNP 表面的有机覆盖涂层，控制着颗粒表面电荷，从而决定了其胶体性质（Jain et al.，2015），而富含带电氨基酸的蛋白质是控制 SeNP 形成和稳定性的主要因素（Xu et al.，2018）。例如，大肠杆菌（*Escherichia coli* K-12）产生的球状 SeNP 性质稳定，但是沙福芽孢杆菌（*Bacillus safensis* JG-B5T）产生的肽酶能切割 SeNP 的蛋白涂层，导致生成的球状 SeNP 转化为热力学稳定的三方晶系纳米棒，而厌氧污泥产生的 SeNP 由于蛋白质包覆较少，也会逐渐转化为棒状颗粒，证明 SeNP 的来源和环境影响着其稳定性（Fischer et al.，2021）。

第三节　硒的微生物氧化

硒的氧化溶解是其生物地球化学循环的重要过程，还原态硒的氧化过程存在于不同的地质环境及土壤中，氧化增溶作用大大提高了其生物利用性，而目前关于硒的微生物氧化过程还未受到足够的关注。

1923 年，Lipman 和 Waksman 首次报道了微生物可以将 Se(0) 氧化为硒酸以获得能量。土壤环境中 Se(0) 的氧化大多是由微生物引发的，并以相对较慢的速率发生，产物包括 Se(Ⅵ) 和 Se(Ⅳ)，同时该过程也受到环境中氧化剂产生的非生物氧化作用和环境因素的影响。另外，自养和异养条件下均存在微生物的氧化作用，且无机碳源比葡萄糖作为碳源时氧化作用更明显，这也反映了有机质可能抑制硒的氧化（Losi and Frankenberger，1998）。从土壤中分离的巨大芽孢杆菌（*Bacillus megaterium*）能将 Se(0) 氧化为 Se(Ⅳ)，而 Se(Ⅵ) 的产量不及 Se(Ⅳ) 的 1%，虽然该菌株氧化硒的总量不足硒添加量的 1.5%，但此过程的发现完善了硒的生物地球化学循环（Sarathchandra and Watkinson，1981）。Dowdle 和 Oremland（1998）利用 ^{75}Se(0) 为示踪剂研究了土壤泥浆中 Se(0) 的微生物氧化过程，结果表明化学异养微生物和化学自养硫杆菌参与了 Se(0) 的氧化，主要产物为 Se(Ⅳ)，仅有少量 Se(Ⅵ) 产生，而纤毛菌（*Leptothrix* MnB1）和土壤富集培养液中主要产物为 Se(Ⅵ)，表明土壤中 Se(Ⅳ) 的微生物氧化在一定程度上受其在土壤颗粒表面吸附程度的制约。此外，分离自硒矿区的革兰氏阴性菌弧形茎菌（*Caulobacter vibrioides* T5M6）能将 Se(0) 和硒矿［包含 Se(-Ⅱ)、Se(0)、Se(Ⅳ) 和 Se(Ⅵ)］氧化溶解为 Se(Ⅳ)（Wang et al.，2016）；在通气条件下，小麦根际土壤的微生物也能驱动 Se(Ⅳ) 氧化为 Se(Ⅵ)（Liu et al.，2014）。

由于硒具有与硫类似的外部电子层结构，在特定的环境条件下微生物可能通过不同的硫氧化菌，利用类似氧化硫的途径来氧化硒以获得能量，如氧化亚铁硫杆菌（*Thiobacillus ferrooxidans*）等（Torma and Habashi，1972）。同时，芽孢杆菌（*Bacillus* sp. JG17）也能通过释放胞外活性硫代谢物（包括亚硫酸盐、硫化物和硫代硫酸盐）在中性氧化条件下将 Se(0) 氧化溶解（Goff et al.，2019）。遗憾的是，近年少有关于硒氧化微生物的进一步研究，其相关机理仍需深入探究。

第四节　硒的微生物甲基化

据报道，每年有 7000～18 000 t 硒进入大气，其中 60%～80% 是海洋生物产生的气态硒，其他自然来源包括火山活动、植物挥发、岩石和土壤的风化等，此外，南极海域冻土中也检测到了硒的挥发（Ye et al.，2021）。据估算，硒在大气中的停留时间只有 45 d，很少发生半球间的长距离迁移。大气中硒的去除主要通过沉降作用（特别是湿沉降），硒的大气沉降量为（35～100）×10^8 g/a（Blazina et al.，2014），沉降量主要取决于与硒源之间的距离，陆地降水中硒的平均浓度为 0.3～1.1 μg/L。自然环境中，甲基化硒是主要的挥发性硒化合物，此外还有硒化氢（H_2Se）和甲硒醇（MeSeH）。

一、硒的甲基化过程

甲基化硒主要以挥发性的二甲基硒（DMeSe）、二甲基二硒（DMeDSe）和二甲基硫硒化合物（DMeSeS）形态存在。普遍认为 DMeSe 是最常见的形态（Stork et al.，1999），但在不同环境中的主要存在形态也有差异，例如，瑞士的泥炭湿地中挥发性硒的形态主要是 DMeDSe（Vriens et al.，2014）。硒的微生物甲基化过程广泛存在于细菌和真菌（包括酵母）中，此外，一些动物、植物、原生动物（包括四膜虫 *Tetrahymena thermophila*）和藻类也可以将无机硒转化为挥发性的甲基硒（Chasteen and Bentley，2003；Vriens et al.，2016）。具有硒甲基化功能的微生物如表 3-1 所示。

表 3-1　能够将硒甲基化为挥发性产物的微生物（改自 Chasteen and Bentley，2003）

微生物	底物	产物		
		DMeSe	DMeDSe	DMeSeS
藻类				
小球藻属（*Chorella* sp.）	Se(IV)	+	+	+
蓝绿藻簇（*Cyanophyte-dominated* Mat）	Se(IV)	+	+	+
莱茵衣藻（*Chlamydomonas reinhardtii*）	Se(IV, VI)	+	+	
细菌				
气单胞菌（*Aeromonas* sp. VS6）	Se(VI)	+	+	+
弗氏柠檬酸杆菌（*Citrobacter freundii* KS8）	Se(VI)	+	+	+
噬胶梭菌（*Clostridium collagenovorans*）	Se(IV)	+	+	
棒状杆菌（*Corynebacterium* sp.）	Se(VI)	+	+	
巨大脱硫弧菌（*Desulfovibrio gigas*）	Se(IV)	+	+	
普通脱硫弧菌（*Desulfovibrio vulgaris*）	Se(IV)	+	+	
阴沟肠杆菌（*Enterobacter cloacae*）	Se(IV)	+		
甲酸甲烷杆菌（*Methanobacterium formicicum*）	Se(IV)	+	+	
巴氏甲烷八叠球菌（*Methanosarcina barkeri*）	Se(IV)	+	+	
铜绿假单胞菌（*Pseudomonas aeruginosa* VS7）	Se(VI)	+	+	+
荧光假单胞菌（*Pseudomonas fluorescens* K27）	Se(VI)	+	+	+
假单胞菌（*Pseudomonas* sp. VW1）	Se(VI)	+	+	+
类球红细菌（*Rhodobacter sphaeroides*）	Se(IV, VI, 0)	+	+	+
聚磷菌（*Rhodocyclus tenuis*）	Se(IV, VI, 0)	+	+	
深红红螺菌（*Rhodospirillum rubrum* S1）	Se(VI, 0)	+	+	
真菌				
镰状支顶孢菌（*Acremonium falciforme*）	Se(IV)	+	+	
直立顶孢霉菌（*Acremonium strictum*）	Se(IV, VI)	+	+	
链格孢菌（*Alternaria alternate*）	Se(IV, VI)	+		
棒曲霉（*Aspergillus clavatus*）	Se(VI)	+		

续表

微生物	底物	产物		
		DMeSe	DMeDSe	DMeSeS
头孢霉（*Cephalosporium* sp.）	Se(Ⅳ, Ⅵ)	+		
镰刀霉（*Fusarium* sp.）	Se(Ⅳ, Ⅵ)	+		
被孢霉（*Mortierella* sp.）	Se(Ⅳ, Ⅵ)	+		
橘青霉（*Penicillium citrinum*）	Se(Ⅳ)	+	+	
青霉菌（*Penicillium* sp.）	Se(Ⅳ)	+		
球状茎点霉（*Phoma glomerata*）	Se(Ⅳ)	+	+	
须壳孢属（*Pyrenochaeta* sp.）	Se(Ⅳ, Ⅵ)	+		
短帚霉（*Scopulariopsis brevicaulis*）	Se(Ⅳ)	+		

注：DMeSe，二甲基硒；DMeDSe，二甲基二硒；DMeSeS，二甲基硫硒化合物；Se(Ⅳ)，亚硒酸盐；Se(Ⅵ)，硒酸盐；Se(0)，元素硒；+表示检出。

与砷不同，甲基化硒均是完全的还原态[Se(-Ⅱ)]。对砷来说，原核生物通常通过甲基化转移酶将砷甲基化（Zhu et al.，2014），而硒的甲基化机理比砷更加复杂。硒通过与砷类似的 Challenger 机理进行甲基化（Chasteen and Bentley，2003），首先 Se(Ⅵ) 被还原为 Se(Ⅳ)，然后再进行甲基化并同时氧化为 Se(Ⅵ)，具体过程如下：硒酸盐→亚硒酸盐→甲基硒酸盐→一甲基亚硒酸盐→二甲基硒酸盐→二甲基亚硒酸盐→二甲基硒。甲基亚硒酸盐还可以转化为硒甲烷或硒甲醇，然后转化为 DMeDSe（Reamer and Zoller，1980）。硒代半胱氨酸（SeCys）也可以在微生物作用下转化为 DMeSe 和 DMeDSe，但是具体机理尚不明确（Chasteen and Bentley，2003）。迄今为止，硒的甲基化过程尚未得到系统研究（Bañuelos et al.，2014；雷磊等，2011；郑世学等，2013）。

二、土壤中硒的甲基化

土壤中硒的甲基化是一种将硒从土壤中去除的过程（Dungan and Frankenberger，1999；Ranjard et al.，2002），甲基化硒的挥发会抑制硒进入食物链，导致硒的严重缺乏。随着全球变暖，硒的甲基化和挥发通量增加，其挥发效率分别是砷和硫的 40 倍和 100 倍，该过程在硒的生物地球化学循环中起着至关重要的作用（Vriens et al.，2014）。

土壤和沉积物中能发生硒甲基化的微生物主要是细菌和真菌（Challenger，1945）。从加拿大湖泊沉积物中分离的三种细菌（包括气单胞菌 *Aeromonas* sp.、黄杆菌 *Flavobacterium* sp. 和假单胞菌 *Pseudomonas* sp.）以及未知的真菌均可以将 Se(Ⅵ)/Se(Ⅳ)、SeCys、硒脲和硒代甲硫氨酸（SeMet）转化为挥发性 DMeSe 和 DMeDSe，这是首次发现纯培养微生物可以生成 DMeDSe（Chau et al.，1976）。Doran 和 Alexander（1977）发现土壤中的棒状杆菌（*Corynebacterium*）可以将 Se(Ⅵ)、Se(Ⅳ)、Se(0) 和几种硒氨基酸转化为 DMeSe。此外，产甲烷古菌（包括甲酸甲烷杆菌 *Methanobacterium formicicum*、巴氏甲烷八叠球菌 *Methanosarcina barkeri*）和硫酸盐还原菌（包括普通脱硫弧菌 *Desulfovibrio vulgaris*、巨大脱硫弧菌 *Desulfovibrio gigas*）均能将 Se(Ⅳ) 还原为 DMeSe 和 DMeDSe（Michalke et al.，2000）。聚磷菌（*Rhodocyclus tenuis*）和深红红螺菌（*Rhodospirillum rubrum*）在光照培养条件下可以将 Se(Ⅵ) 转化为 DMeSe 和 DMeDSe（McCarthy et al.，1993），同时聚磷菌也可以将 Se(Ⅳ) 转化为 DMeSe。Moreno-Martin 等（2021）利用顶空固相微萃取-串联气相色谱质谱（HS-SPME-GC-MS）检测发现大肠杆菌能够将 Se(Ⅳ) 和壳聚糖修饰的硒纳米颗粒（Ch-SeNP）转化为 DMeSe 和 DMeDSe；金黄色葡萄球菌（*Staphylococcus aureus*）则将 Se(Ⅳ) 转化为 DMeSe 和 DMeDSe，而仅将 Ch-SeNP 转化为 DMeDSe，表明硒转化形成的挥发性有机硒与细菌种类及硒的初始化学形态有关（Moreno-Martin et al.，2021）。

细菌可以通过硫嘌呤甲基转移酶（bTPMT）实现硒的甲基化。研究表明，*E. coli* 表达假单胞杆菌（*Pseudomonas syringe*）的硫嘌呤甲基转移酶基因后，可以将 Se(Ⅵ)、Se(Ⅳ) 和 SeCys 转化为 DMeSe 和 DMeDSe，该酶可能是人为或自然硒污染场地修复的关键酶（Ranjard et al.，2003；Ranjard et al.，2002）。托拉氏假单胞菌（*Pseudomonas tolaasii*）则通过杂化酶反应将硒高效快速地转化为 DMeDSe 和 DMeSeS，其中硫氨基酸的状态决定着硒的甲基化过程，因此，硫可能是大气中硒通量的主要控制因素（Liu et al.，2021）。

硒稳定同位素分馏效应可以用于揭示甲基化硒的来源和形成过程（Schilling et al.，2011）。微生物对硒的甲基化贡献较大，因此，土壤中硒的挥发过程会被多种影响微生物生命活动的因素所影响，如温度、土壤湿度、通气状况、微量元素及土壤污染状况等（Stork et al.，1999）。目前大部分硒的甲基化研究都是在实验室培养条件下进行的，对于不同生态系统的土壤中硒挥发情况，特别是低硒土壤中硒的挥发和参与的微生物种类尚不清晰。

三、植物根际硒的甲基化

除了微生物外，植物也可以将硒甲基化进而挥发，不同植物作用下硒挥发的速率差异明显（Pilon-Smits et al.，1999；Terry et al.，1992）。值得注意的是，植物对硒挥发的贡献在很大程度上依赖于根际微生物。1994 年，Zayed 和 Terry 首次发现细菌参与了植物硒挥发过程，并且证明根际微生物是影响根系硒挥发的重要因素（Zayed and Terry，1994）。此外，植物硒挥发的能力还受根部硒浓度和化学形态的影响。研究表明，添加 Se(Ⅵ) 和 Se(Ⅳ) 为底物时，硒挥发速率与土壤硒浓度和植物组织硒浓度呈线性相关（Terry et al.，1992）。田间条件下，在低硒土壤中加入硒代甲硫氨酸（SeMet），硒挥发速率增加了 14 倍，而添加还原态硒只增加了 4 倍（Wang et al.，1991）；将 SeMet 和硒代半胱氨酸（SeCys）分别添加到土壤中，发现 50%～80% 的 SeMet 被挥发，而 SeCys 处理的土壤中则几乎没有硒的挥发（Martens and Suarez，1997）。但是，目前关于不同植物根际介导硒挥发的功能微生物仍缺乏研究。

第五节　土壤动物对硒的转化

硒污染土壤通过食物链对陆地生态系统构成严重威胁，其对土壤动物也具有毒性效应，土壤动物及其肠道微生物对硒的吸收代谢是硒生物地球化学过程中重要的一环。

一、蚯蚓对硒的转化

蚯蚓是土壤中重要的消费者、分解者和调节者，同时因其高蛋白质含量和生物活性成分常被用作动物饲料，因此，阐明蚯蚓在富硒土壤等基质生长过程中对硒的吸收代谢，对于深入认识硒在土壤中完整的生物转化过程以及生产富硒饲料均具有重要意义。研究表明，硒暴露会抑制蚯蚓生长，且不同种类的蚯蚓对于硒暴露的反应不同（Xiao et al., 2018）。赤子爱胜蚓是研究硒毒性常用的模式土壤动物，研究表明高浓度 Se(Ⅳ) 影响了赤子爱胜蚓的三羧酸循环过程和代谢平衡，且 Se(Ⅳ) 浓度越高，代谢紊乱程度越大（Shao et al., 2019），因此其通常将 Se(Ⅳ) 富集于尾部以减轻毒性（Yue et al., 2021）。同时，硒暴露提高了蚓粪中细菌的多样性，从蚓粪中分离的蜡样芽孢杆菌（Bacillus cereus）和毛刺气单胞菌（Aeromonas encheleia）均能快速高效地还原 Se(Ⅳ)，表明蚯蚓肠道菌群具有缓冲 Se(Ⅳ) 毒性的作用（Gan et al., 2021）。当添加 40 mg/kg Se(Ⅳ) 至基质进行赤子爱胜蚓培养时，Se(Ⅳ) 对蚯蚓生物量和繁殖均无不良影响，赤子爱胜蚓在体内将 Se(Ⅳ) 大部分转化为有机硒，代谢产物中硒代甲硫氨酸（SeMet）和硒代胱氨酸（SeCys2）的比例占 75% 以上。胃肠消化是蚯蚓最有效的硒释放途径，体外模拟胃肠消化试验，结果表明蚯蚓体内硒的生物可利用率高达 90%，表明富硒蚯蚓可以作为一种较好的富硒饲料（Yue et al., 2019）。

二、线虫对硒的转化

秀丽隐杆线虫也是常用于研究硒的毒性和代谢途径的模式土壤动物。不同形态硒对秀丽隐杆线虫的毒性效应差异显著，研究发现硒的毒性与其生物可利用性呈反比，如有机硒的生物可利用性高于无机硒，而毒性则小于后者。有机硒进入体内后会进一步发生代谢，利用高效液相色谱-质谱分析表明，硒代甲硫氨酸（SeMet）的代谢产物为硒-腺苷甲硫氨酸（AdoSeMet）和硒腺苷半胱氨酸（AdoSeHcy），而甲基硒代半胱氨酸（MeSeCys）的代谢产物为甲基硒代谷胱甘肽（MeSeGSH）和 γ-谷氨酰-MeSeCys（Rohn et al., 2018）。在个体水平上，硒诱导的氧化应激导致了秀丽隐杆线虫运动和产卵所需的胆碱能神经元的下降和退化（Estevez et al., 2012）；在种群水平上，Se(Ⅳ) 影响着线虫的存活率和种群增长率（Li et al., 2014c）。秀丽隐杆线虫中转录因子 HF-1 在耐硒方面发挥重要作用，并且 CYSL-1 是关键的硒信号传感器。此外，线虫对硒的抗性与参与硫化物和亚硫酸盐胁迫响应起关键作用的酶（硫化物醌氧化还原酶和亚硫酸盐氧化酶）不相关，而是受硫双加氧酶 ETHE-1 和硫转移酶 MPST-7 的影响（Romanelli-Credrez et al., 2020）。

值得注意的是，Se 也是一种必需的微量元素，参与了许多生物过程，使机体保持最佳免疫功能。例如，Se(Ⅳ) 能通过调节 SKN-1 依赖的信号通路，增强秀丽隐杆线虫先天免疫关键基因的表达，从而使之对铜绿假单胞菌 PA14 菌株产生免疫反应（Li et al., 2014b）。此外，微量 Se(Ⅳ) 能通过转录因子 DAF-16 和线虫唯一的硒蛋白——硫氧还蛋白还原酶-1（TrxR-1）介导秀丽隐杆线虫的抗氧化作用（Li et al., 2014d），进一步研究表明 TrxR-1 能介导硒有机形态 N-γ-(L-谷酰基)-L-硒代甲硫氨酸（Glu-SeMet）的反应，以增强线虫的抗逆性，改善衰老，提高其在氧化和热胁迫条件下的存活率（Chang et al.,

2017）。同时，硒蛋白 TrxR-1 能与 GSR-1 谷胱甘肽还原酶共同作用，促进线虫蜕皮过程中旧角质层的清除（Stenvall et al.，2011）。

当硒与重金属共存时，对土壤动物的作用复杂。汞和硒共暴露对秀丽隐杆线虫的毒性效应复杂，研究表明二者对线虫生长的影响并不取决于体内汞/硒含量比，而是取决于化学物种形态（Wyatt et al.，2016）。此外，Se(Ⅳ)的抗氧化特性使之能显著降低秀丽隐杆线虫细胞内的活性氧（reactive oxygen species，ROS）水平，从而使线虫免受铅[Pb(Ⅱ)]诱导的神经毒性作用（Li et al.，2013）。Se(Ⅳ)还能通过硫醇包覆的多肽诱导 CdSe 纳米颗粒形成，从而降低镉（Cd）对线虫的毒性，具体机制如下：首先线虫体内的 Cd 与谷胱甘肽、植物螯合肽结合，然后进一步与硒代半胱氨酸（SeCys）相互作用形成复合物，最终产生 CdSe/CdS 纳米颗粒，二者主要分布于线虫的咽和肠道内并不断从体内排出，有利于线虫的生存（Li et al.，2019；Li et al.，2021）。

第六节 结 论

微生物对于土壤中硒元素的迁移转化发挥着至关重要的作用，主要包括还原、氧化、甲基化等过程，深入认识硒在土壤中的转化机制对于了解其赋存形态及毒性效应具有重要意义。值得注意的是，微生物硒还原过程是一种解毒策略，还是产生质子驱动 ATP 合成和细胞生长目前还不清楚，二者之间基本的生理差异仍未得到全面的认识；此外，具有 Se(0) 还原能力的菌株报道较少，其还原潜力和相关机理有待深入研究。因此，需要不断发现并研究新的硒代谢微生物，深入研究其代谢机理，全面揭示微生物对硒吸收、转化、外排的作用机制。

利用先进的微生物生态学工具和高通量筛选技术，有助于在群落水平上了解硒的代谢机理，揭示自然环境中硒的生物转化及矿化机制，并筛选具有硒代谢功能活性的微生物，进一步分离出具有生物技术应用前景的微生物。这些功能微生物不仅可以用于污染场地的生物修复，以及粮食作物和牲畜饲料的营养生物强化，同时，硒生物矿化菌还可用于制造硒量子点和硒纳米颗粒等材料，这些材料由于具有稳定、无毒、生物相容性高等独特的物化性质，已被广泛应用于医学、催化、电子等多个领域。

硒的生物地球化学循环与碳、氮、硫等元素的生物地球化学过程密切相关，然而关于其复杂的相关性及内在机理研究尚少，微生物功能基因分析和宏基因组技术将为深入揭示硒与其他元素的相互作用，以及识别关键影响因素提供强有力的支撑。

参 考 文 献

范书伶, 王平, 张珩琳, 等. 2020. 环境中硒的迁移、微生物转化及纳米硒应用研究进展. 科学通报, 65(26): 2853-2862.

雷磊, 朱建明, 秦海波, 等. 2011. 硒的微生物地球化学研究进展. 地球与环境, 39(1): 97-104.

谭建安. 1989. 中华人民共和国地方病与环境图集. 北京: 科学出版社.

王丹, 夏险, 王革娇, 等. 2017. 微生物对硒的还原及其产物的应用研究进展——纪念硒发现 200 周年. 微生物学通报, 44(7): 1728-1735.

郑世学, 粟静, 王瑞, 等. 2013. 硒是双刃剑？——谈微生物中的硒代谢. 华中农业大学学报, 32(5): 1-8.

朱燕云, 吴文良, 赵桂慎, 等. 2018. 硒在动植物及微生物体中的转化规律研究进展. 农业资源与环境学报, 35(3): 189-198.

Antonioli P, Lampis S, Chesini I, et al. 2007. Stenotrophomonas maltophilia SeITE02, a new bacterial strain suitable for bioremediation of selenite-contaminated environmental matrices. Applied and Environmental Microbiology, 73(21): 6854-6863.

Aoyagi T, Mori Y, Nanao M, et al. 2021. Effective Se reduction by lactate-stimulated indigenous microbial communities in excavated waste rocks. Journal of Hazardous Materials, 403: 123908.

Avendano R, Chaves N, Fuentes P, et al. 2016. Production of selenium nanoparticles in *Pseudomonas putida* KT2440. Scientific Reports, 6: 37155.

Bai Y N, Wang X N, Lu Y Z, et al. 2019. Microbial selenite reduction coupled to anaerobic oxidation of methane. Science of the Total Environment, 669: 168-174.

Bañuelos Lin Z Q, Yin X B. 2014. Selenium in the environment and human health. London: Taylor & Francis Group.

Bao P, Chen Z, Tai R Z, et al. 2015. Selenite-induced toxicity in cancer cells is mediated by metabolic generation of endogenous selenium nanoparticles. Journal of Proteome Research, 14(2): 1127-1136.

Bao P, Huang H, Hu Z Y, et al. 2013. Impact of temperature, CO_2 fixation and nitrate reduction on selenium reduction, by a paddy soil Clostridium strain. Journal of Applied Microbiology, 114(3): 703-712.

Bao P, Xiao K Q, Wang H J, et al. 2016. Characterization and potential applications of a selenium nanoparticle producing and nitrate reducing bacterium *Bacillus oryziterrae* sp. nov. Scientific Reports, 6: 34054.

Basaglia M, Toffanin A, Baldan E, et al. 2007. Selenite-reducing capacity of the copper-containing nitrite reductase of *Rhizobium sullae*. FEMS Microbiology Letters, 269(1): 124-130.

Bebien M, Chauvin J P, Adriano J M, et al. 2001. Effect of selenite on growth and protein synthesis in the phototrophic bacterium Rhodobacter sphaeroides. Applied and Environmental Microbiology, 67(10): 4440-4447.

Bebien M, Lagniel G, Garin J, et al. 2002. Involvement of superoxide dismutases in the response of *Escherichia coli* to selenium oxides. Journal of Bacteriology, 184: 1556-1564.

Birringer M, Pilawa S, Flohe L. 2002. Trends in selenium biochemistry. Natural Product Reports, 19(6): 693-718.

Bjornstedt M, Kumar S, Holmgren A. 1995. Selenite and selenodiglutathione: reactions with thioredoxin systems. Methods in Enzymology, 252: 209-219.

Blazina T, Sun Y, Voegelin A, et al. 2014. Terrestrial selenium distribution in China is potentially linked to monsoonal climate. Nature Communications, 5: 4717.

Butler C S, Debieux C M, Dridge E J, et al. 2012. Biomineralization of selenium by the selenite-respiring bacterium *Thauera selenatis*. Biochemical Society Transactions, 40(6): 1239-1243.

Carey A M, Lombi E, Donner E, et al. 2012. A review of recent developments in the speciation and location of arsenic and selenium in rice grain. Analytical and Bioanalytical Chemistry, 402: 3275-3286.

Castaneda-Ovando A, Segovia-Cruz J A, Flores-Aguilar J F, et al. 2019. Serine-enriched minimal medium enhances conversion of selenium into selenocysteine by *Streptococcus thermophilus*. Journal of Dairy Science, 102(8): 6781-6789.

Challenger F. 1945. Biological methylation. Chemical Reviews, 36(3): 315-361.

Chang C H, Ho C T, Liao V H. 2017. N-gamma-(L-Glutamyl)-L-selenomethionine enhances stress resistance and ameliorates aging indicators via the selenoprotein TRXR-1 in *Caenorhabditis elegans*. Molecular Nutrition & Food Research, 61(8): 1600954.

Chasteen T G, Bentley R. 2003. Biomethylation of selenium and tellurium: microorganisms and plants. Chemical Reviews, 103(1): 1-25.

Chau Y K, Wong P T S, Silverberg B A, et al. 1976. Methylation of selenium in the aquatic environment. Science, 192(4244): 1130-1131.

Combs G F. 2001. Selenium in global food systems. British Journal of Nutrition, 85: 517-547.

DeMoll-Decker H, Macy J M. 1993. The periplasmic nitrite reductase of *Thauera selenatis* may catalyze the reduction of selenite to elemental selenium. Archives of Microbiology, 160: 241-247.

Doran J W, Alexander M. 1977. Microbial transformations of selenium. Applied and Environmental Microbiology: 33(1): 31-37.

Dowdle P R, Oremland R S. 1998. Microbial oxidation of elemental selenium in soil slurries and bacterial cultures. Environmental Science & Technology, 32(23): 3749-3755.

Dungan, R S, Frankenberger W T. 1999. Microbial transformations of selenium and the bioremediation of seleniferous environments. Bioremediation Journal, 3: 171-188.

Estevez A O, Mueller C L, Morgan K L, et al. 2012. Selenium induces cholinergic motor neuron degeneration in *Caenorhabditis elegans*. Neurotoxicology, 33(5): 1021-1032.

Eswayah A S, Hondow N, Scheinost A C, et al. 2019. Methyl selenol as a precursor in selenite reduction to Se/S species by methane-oxidizing bacteria. Applied and Environmental Microbiology, 85(22): e01379-01319.

Eswayah A S, Smith T J, Scheinost A C, et al. 2017. Microbial transformations of selenite by methane-oxidizing bacteria. Applied Microbiology and Biotechnology, 101(17): 6713-6724.

Fernández-Martínez A, Charlet L. 2009. Selenium environmental cycling and bioavailability: a structural chemist point of view. Reviews in Environmental Science and Bio/Technology, 8(1): 81-110.

Fischer S, Jain R, Krause T, et al. 2021. Impact of the microbial origin and active microenvironment on the shape of biogenic elemental selenium nanomaterials. Environmental Science & Technology, 55(13): 9161-9171.

Fordyce F M. 2005. Selenium deficiency and toxicity in the environment. In: Selinus O, et al. Essentials of Medical Geology. The Amsterdam: Elsevier Academic Press: 373-415.

Galano E, Mangiapane E, Bianga J, et al. 2013. Privileged incorporation of selenium as selenocysteine in *Lactobacillus reuteri* proteins demonstrated by selenium-specific imaging and proteomics. Molecular & Cellular Proteomics, 12(8): 2196-2204.

Gan X, Huang J C, Zhang M, et al. 2021. Remediation of selenium-contaminated soil through combined use of earthworm *Eisenia fetida* and organic materials. Journal of Hazardous Materials, 405: 124212.

Gates A J, Butler C S, Richardson D J, et al. 2011. Electrocatalytic reduction of nitrate and selenate by NapAB. Biochemical Society Transactions, 39: 236-242.

Goff J, Terry L, Mal J, et al. 2019. Role of extracellular reactive sulfur metabolites on microbial Se(0) dissolution. Geobiology, 17(3): 320-329.

Greenwood N N, Earnshow A. 1997. Chemistry of the Elements. Amsterdam, Netherlands: Elsevier.

Harrison G, Curle C, Laishley E J. 1984. Purification and characterization of an inducible dissimilatory type sulfite reductase from *Clostridium pasteurianum*. Archives of Microbiology, 138: 72-78.

He Q, Yao K. 2011. Impact of alternative electron acceptors on selenium(Ⅳ) reduction by Anaeromyxobacter dehalogenans. Bioresource Technology, 102(3): 3578-3580.

Heider J, Bock A. 1993. Selenium metabolism in micro-organisms. Advances in Microbial Physiology, 35: 71-109.

Hockin S, Gadd G M. 2003. Linked redox precipitation of sulfur and selenium under anaerobic conditions by sulfate-reducing bacterial biofilms. Applied and Environmental Microbiology, 69: 7063-7072.

Hockin S, Gadd G M. 2006. Removal of selenate from sulfate-containing media by sulfate-reducing bacterial biofilms. Environmental Microbiology, 8: 816-826.

Holland H L, Carter I M. 1983. An investigation of the biotransformation of organic selenides by fungi. Bioorganic Chemistry, 12(1): 1-7.

Huang C, Wang H, Shi X, et al. 2021a. Two new selenite reducing bacterial isolates from paddy soil and the potential Se biofortification of paddy rice. Ecotoxicology, 30(7): 1465-1475.

Huang S, Wang Y, Tang C, et al. 2021b. Speeding up selenite bioremediation using the highly selenite-tolerant strain *Providencia rettgeri* HF16-A novel mechanism of selenite reduction based on proteomic analysis. Journal of Hazardous Materials, 406: 124690.

Jain R, Jordan N, Weiss S, et al. 2015. Extracellular polymeric substances govern the surface charge of biogenic elemental selenium nanoparticles. Environmental Science & Technology, 49(3): 1713-1720.

Ji Y, Wang Y T. 2019. Selenium reduction by a defined co-culture of *Shigella fergusonii* strain TB42616 and *Pantoea vagans* strain EWB32213-2. Bioprocess and Biosystems Engineering, 42(8): 1343-1351.

Kenward P A, Fowle D A, Yee N. 2006. Microbial selenate sorption and reduction in nutrient limited systems. Environmental Science & Technology, 40(12): 3782-3786.

Kessi J, Hanselmann K W. 2004. Similarities between the abiotic reduction of selenite with glutathione and the dissimilatory reaction mediated by *Rhodospirillum rubrum* and *Escherichia coli*. Journal of Biological Chemistry, 279(49): 50662-50669.

Khoei N S, Lampis S, Zonaro E, et al. 2017. Insights into selenite reduction and biogenesis of elemental selenium nanoparticles by two environmental isolates of *Burkholderia fungorum*. New Biotechnology, 34: 1-11.

Kuroda M, Suda S, Sato M, et al. 2019. Biosynthesis of bismuth selenide nanoparticles using chalcogen-metabolizing bacteria. Applied Microbiology and Biotechnology, 103(21-22): 8853-8861.

Kushwaha A, Goswami L, Lee J, et al. 2021. Selenium in soil-microbe-plant systems: Sources, distribution, toxicity, tolerance, and detoxification. Critical Reviews in Environmental Science and Technology, 52(13): 1-42.

Lai C Y, Song Y, Wu M, et al. 2020. Microbial selenate reduction in membrane biofilm reactors using ethane and propane as electron donors. Water Research, 183: 116008.

Lai C Y, Wen L L, Shi L D, et al. 2016. Selenate and nitrate bioreductions using methane as the electron donor in a membrane biofilm reactor. Environmental Science & Technology, 50(18): 10179-10186.

Lampis S, Vallini G. 2021. Microbial reduction of selenium oxyanions: energy-yielding and detoxification reactions. In: Lens P N L, Pakshirajan K. Environmental Technologies to Treat Selenium Pollution: Principles and Engineering. London: IWA Publishing: 101-143.

Lampis S, Zonaro E, Bertolini C, et al. 2014. Delayed formation of zero-valent selenium nanoparticles by *Bacillus mycoides* SeITE01 as a consequence of selenite reduction under aerobic conditions. Microbial Cell Factories, 13: 35.

Lampis S, Zonaro E, Bertolini C, et al. 2017. Selenite biotransformation and detoxification by *Stenotrophomonas maltophilia* SeITE02: Novel clues on the route to bacterial biogenesis of selenium nanoparticles. Journal of Hazardous Materials, 324(Pt A): 3-14.

Ledgham F, Quest B, Vallaeys T, et al. 2005. A probable link between the DedA protein and resistance to selenite. Research in Microbiology, 156(3): 367-374.

Li D B, Cheng Y Y, Wu C, et al. 2014a. Selenite reduction by *Shewanella oneidensis* MR-1 is mediated by fumarate reductase in periplasm. Scientific Reports, 4: 3735.

Li L L, Cui Y H, Lu L Y, et al. 2019. Selenium stimulates cadmium detoxification in *Caenorhabditis elegans* through thiols-mediated nanoparticles formation and secretion. Environmental Science & Technology, 53(5): 2344-2352.

Li L L, Wu Q Z, Chen J J, et al. 2021. Mechanical insights into thiol-mediated synergetic biotransformation of cadmium and selenium in nematodes. Environmental Science & Technology, 55(11): 7531-7540.

Li W H, Chang C H, Huang C W, et al. 2014b. Selenite enhances immune response against *Pseudomonas aeruginosa* PA14 via SKN-1 in *Caenorhabditis elegans*. PLoS One, 9(8): e105810.

Li W H, Ju Y R, Liao C M, et al. 2014c. Assessment of selenium toxicity on the life cycle of *Caenorhabditis elegans*. Ecotoxicology, 23(7): 1245-1253.

Li W H, Shi Y C, Chang C H, et al. 2014d. Selenite protects *Caenorhabditis elegans* from oxidative stress via DAF-16 and TRXR-1. Molecular Nutrition & Food Research, 58(4): 863-874.

Li W H, Shi Y C, Tseng I L, et al. 2013. Protective efficacy of selenite against lead-induced neurotoxicity in *Caenorhabditis elegans*. PLoS One, 8(4): e62387.

Liang X, Perez M A M, Nwoko K C, et al. 2019. Fungal formation of selenium and tellurium nanoparticles. Applied Microbiology and Biotechnology, 103(17): 7241-7259.

Liang X, Perez M A M, Zhang S, et al. 2020. Fungal transformation of selenium and tellurium located in a volcanogenic sulfide deposit. Environmental Microbiology, 22(6): 2346-2364.

Lipman J G, Waksman S A. 1923. The oxidation of selenium by a new group of autotrophic micro-organisms. Science, 57: 60-60.

Liu X, Zhao Z, Duan B, et al. 2014. Effect of applied sulphur on the uptake by wheat of selenium applied as selenite. Plant Soil, 386(1-2): 35-45.

Liu Y, Hedwig S, Schaffer A, et al. 2021. Sulfur amino acids status controls selenium methylation in *Pseudomonas tolaasii*: identification of a novel metabolite from promiscuous enzyme reactions. Applied and Environmental Microbiology, 87(12): e0010421.

Losi M E, Frankenberger W T. 1998. Microbial oxidation and solubilization of precipitated elemental selenium in soil. Journal of Environmental Quality, 27: 836-843.

Losi M E. Frankenberger W T J. 1997. Reduction of selenium oxyanions by *Enterobacter cloacae* SLD1a-1: isolation and growth of the bacterium and its expulsion of selenium particles. Applied and Environmental Microbiology, 63(8): 3079-3084.

Luo J H, Chen H, Hu S, et al. 2018. Microbial selenate reduction driven by a denitrifying anaerobic methane oxidation biofilm. Environmental Science & Technology, 52(7): 4006-4012.

Lussier C. Geochemistry of selenium release from the elk river valley coal mines. 2001. PhD thesis. The University of British Columbia.

Ma J, Kobayashi D Y, Yee N. 2007. Chemical kinetic and molecular genetic study of selenium oxyanion reduction by *Enterobacter cloacae* SLD1a-1. Environmental Science & Technology, 41(22): 7795-7801.

Macy J M, Michel T A, Kirsch D G. 1989. Selenate reduction by a *Pseudomonas* species: a new mode of anaerobic respiration. FEMS Microbiology Letters, 52: 195-198.

Mapelli V, Hillestrom P R, Patil K, et al. 2012. The interplay between sulphur and selenium metabolism influences the intracellular redox balance in Saccharomyces cerevisiae. FEMS Yeast Research, 12(1): 20-32.

Martens D A, Suarez D L. 1997. Mineralization of selenium-containing amino acids in two California soils. Soil Science Society of America Journal, 61(6): 1685-1694.

Martens D A, Suarez D L. 1999. Transformations of volatile methylated selenium in soil. Soil Biology and Biochemistry, 31(10): 1355: 1361.

Martinez F G, Moreno-Martin G, Pescuma M, et al. 2020. Biotransformation of selenium by lactic acid bacteria: formation of seleno-nanoparticles and seleno-amino acids. Frontiers in Bioengineering and Biotechnology, 8: 506.

Masscheleyn P H, Delaune R D, Patrick W H. 1991. Arsenic and selenium chemistry as affected by sediment redox potential and pH. Journal of Environmental Quality, 20(3): 522-527.

McCarthy S, Chasteen T, Marshall M, et al. 1993. Phototrophic bacteria produce volatile, methylated sulfur and selenium compounds. FEMS Microbiology Letters, 112(1): 93-97.

McDermott J R, Rosen B P, Liu Z. 2010. Jen1p: a high affinity selenite transporter in yeast. Molecular Biology of the Cell, 21(22): 3934-3941.

Michalke K, Wickenheiser E B, Mehring M, et al. 2000. Production of volatile derivatives of metal(loid)s by microflora involved in anaerobic digestion of sewage sludge. Applied and Environmental Microbiology, 66(7): 2791-2796.

Moreno-Martin G, Sanz-Landaluze J, Leon-Gonzalez M E, et al. 2021. *In vivo* quantification of volatile organoselenium compounds released by bacteria exposed to selenium with HS-SPME-GC-MS. Effect of selenite and selenium nanoparticles. Talanta, 224: 121907.

Morschbacher A P, Dullius A, Dullius C H, et al. 2018. Assessment of selenium bioaccumulation in lactic acid bacteria. Journal of Dairy Science, 101(12): 10626-10635.

Narasingarao P, Häggblom M M. 2006. Sedimenticola selenatireducens, gen. nov., sp. nov., an anaerobic selenate-respiring bacterium isolated from estuarine sediment. Systematic Applied and Microbiology, 29: 382-388.

Navarro R R, Aoyagi T, Kimura M, et al. 2015. High-resolution dynamics of microbial communities during dissimilatory selenate reduction in anoxic soil. Environmental Science & Technology, 49(13): 7684-7691.

Newton G L, Fahey R C. 1989. Glutathione in prokaryotes. In glutathione: Metabolism and physiological functions. Boca Raton: CRC Press: 69-77.

Nguyen V K, Nguyen T H, Ha M G, et al. 2019. Kinetics of microbial selenite reduction by novel bacteria isolated from activated sludge. Journal of Environmental Management, 236: 746-754.

Nkansah-Boadu F, Hatam I, Baldwin S A. 2021. Microbial consortia capable of reducing selenate in the presence of nitrate enriched from coalmining-impacted environments. Applied Microbiology and Biotechnology, 105(3): 1287-1300.

Ogra Y, Shimizu M, Takahashi K, et al. 2018. Biotransformation of organic selenium compounds in budding yeast, *Saccharomyces cerevisiae*. Metallomics, 10(9): 1257-1263.

Papp L V, Lu J, Holmgren A, et al. 2007. From selenium to selenoproteins: synthesis, identity, and their role in human health. Antioxidants & Redox Signaling, 9(7): 775-806.

Peng T, Lin J, Xu Y Z, et al. 2016. Comparative genomics reveals new evolutionary and ecological patterns of selenium utilization in bacteria. ISME Journal, 10(8): 2048-2059.

Piacenza E, Presentato A, Ambrosi E, et al. 2018. Physical-chemical properties of biogenic selenium nanostructures produced by *Stenotrophomonas maltophilia* SeITE02 and *Ochrobactrum* sp. MPV1. Frontiers in Microbiology, 9: 3178.

Pilon-Smits E A H, De Souza M P, Hong G, et al. 1999. Selenium volatilization and accumulation by twenty aquatic plant species. Journal of Environmental Quality, 28(3): 1011-1018.

Ranjard L, Nazaret S, Cournoyer B. 2003. Freshwater bacteria can methylate selenium through the thiopurine methyltransferase pathway. Applied and Environmental Microbiology, 69(7): 3784-3790.

Ranjard L, Prigent-Combaret C, Nazaret S, et al. 2002. Methylation of inorganic and organic selenium by the bacterial thiopurine methyltransferase. Journal of Bacteriology, 184(11): 3146-3149.

Rauschenbach I, Yee N, Häggblom M M, et al. 2011. Energy metabolism and multiple respiratory pathways revealed by genome sequencing of *Desulfurispirillum indicum* strain S5. Environmental Microbiology, 13: 1611-1621.

Reamer D C, Zoller W H. 1980. Selenium biomethylation products from soil and sewage sludge. Science, 208: 500-502.

Rohn I, Marschall T A, Kroepfl N, et al. 2018. Selenium species-dependent toxicity, bioavailability and metabolic transformations in *Caenorhabditis elegans*. Metallomics, 10(6): 818-827.

Romanelli-Credrez L, Doitsidou M, Alkema M J, et al. 2020. HIF-1 has a central role in *Caenorhabditis elegans* organismal response to selenium. Frontiers in Genetics, 11: 63.

Rosenfeld C E, Kenyon J A, James B R, et al. 2017. Selenium (Ⅳ, Ⅵ) reduction and tolerance by fungi in an oxic environment. Geobiology, 15(3): 441-452.

Rosenfeld C E, Sabuda M C, Hinkle M A G, et al. 2020. A fungal-mediated cryptic selenium cycle linked to manganese biogeochemistry. Environmental Science & Technology, 54(6): 3570-3580.

Ruiz-Fresneda M A, Eswayah A S, Romero-González M, et al. 2020. Chemical and structural characterization of Se^{Ⅳ} biotransformations by *Stenotrophomonas bentonitica* into Se^0 nanostructures and volatiles Se species. Environmental Science: Nano, 7(7): 2140-2155.

Ruiz-Fresneda M A, Gomez-Bolivar J, Delgado-Martin J, et al. 2019. The bioreduction of selenite under anaerobic and alkaline conditions analogous to those expected for a deep geological repository system. Molecules, 24(21): 3868.

Sabaty M, Avazeri C, Pignol D, et al. 2001. Characterization of the reduction of selenate and tellurite by nitrate reductases. Applied and Environmental Microbiology, 67(11): 5122-5126.

Sarathchandra S U, Watkinson J H. 1981. Oxidation of elemental selenium to selenite by *Bacillus megaterium*. Science, 211(4482): 600-601.

Schilling K, Basu A, Wanner C, et al. 2020. Mass-dependent selenium isotopic fractionation during microbial reduction of seleno-oxyanions by phylogenetically diverse bacteria. Geochimica et Cosmochimica Acta, 276: 274-288.

Schilling K, Johnson T M, Wilcke W. 2011. Isotope fractionation of selenium during fungal biomethylation by *Alternaria alternata*. Environmental Science & Technology, 45(7): 2670-2676.

Schröder I, Rech S, Krafft T, et al. 1997. Purification and characterization of the selenate reductase from *Thauera selenatis*. Journal Biological Chemistry, 272: 23765-23768.

Schwarz K, Foltz C M. 1957. Selenium as an integral part of factor 3 against dietary necrotic liver degeneration. Journal of American Chemical Society, 79(12): 3292-3293.

Shao X, He J, Liang R, et al. 2019. Mortality, growth and metabolic responses by (1) H-NMR-based metabolomics of earthworms to sodium selenite exposure in soils. Ecotoxicology and Environmental Safety, 181: 69-77.

Shi L D, Lv P L, McIlroy S J, et al. 2021. Methane-dependent selenate reduction by a bacterial consortium. The ISME Journal, 15: 3683-3692.

Shi L, Lv P, Niu Z, et al. 2020a. Why does sulfate inhibit selenate reduction molybdenum deprivation from Mo-dependent selenate reductase. Water Research, 178: 115832.

Shi L, Lv P, Wang M, et al. 2020b. A mixed consortium of methanotrophic archaea and bacteria boosts methane-dependent selenate reduction. Science of the Total Environment, 732: 139310.

Shrift A. 1964. A Selenium cycle in nature? Nature, 201: 1304-1305.

Sirko A, Hryniewicz M, Hulanicka D, et al. 1990. Sulfate and thiosulfate transport in *Escherichia coli* K-12: nucleotide sequence and expression of the *cysTWAM* gene cluster. Journal of Bacteriology, 172(6): 3351-3357.

Song D, Li X, Cheng Y, et al. 2017. Aerobic biogenesis of selenium nanoparticles by *Enterobacter cloacae* Z0206 as a consequence of fumarate reductase mediated selenite reduction. Scientific Reports, 7(1): 3239.

Stadtman T C. 1996. Selenocysteine. Annual Review of Biochemistry, 65: 83-100.

Staicu L C, Ackerson C J, Cornelis P, et al. 2015. *Pseudomonas moraviensis* subsp. stanleyae, a bacterial endophyte of hyperaccumulator *Stanleya pinnata*, is capable of efficient selenite reduction to elemental selenium under aerobic conditions. Journal of Applied Microbiology, 119(2): 400-410.

Steinberg N A, Blum J S, Hochstein L, et al. 1992. Nitrate is a preferred electron acceptor for growth of freshwater selenite-respiring bacteria. Applied Environmental and Microbiology, 58: 426-428.

Stenvall J, Fierro-Gonzalez J C, Swoboda P, et al. 2011. Selenoprotein TRXR-1 and GSR-1 are essential for removal of old cuticle during molting in *Caenorhabditis elegans*. Proceedings of the National Academy of Sciences of the United States of America, 108(3): 1064-1069.

Stolz J F, Basu P, Santini J M, et al. 2006. Arsenic and selenium in microbial metabolism. Annual Review of Microbiology, 60: 107-130.

Stolz J F, Oremland R S. 1999. Bacterial respiration of arsenic and selenium. FEMS Microbiololgy Reviews, 23: 615-627.

Stork A, Jury W A, Frankenberger W T. 1999. Accelerated volatilization rates of selenium from different soils. Biological Trace Element Research, 69(3): 217-234.

Subedi G, Taylor J, Hatam I, et al. 2017. Simultaneous selenate reduction and denitrification by a consortium of enriched mine site bacteria. Chemosphere, 183: 536-545.

Switzer B J, Burns B A, Buzzelli J, et al. 1998. *Bacillus arsenicoselenatis*, sp. nov., and *Bacillus selenitireducens*, sp. nov.: two haloal-kaliphiles from Mono Lake, California that respire oxyanions of selenium and arsenic. Archives of Microbiology, 171: 19-30.

Takai K, Hirayama H, Sakihama Y, et al. 2002. Isolation and metabolic characteristics of previously uncultured members of the order Aquificales in a subsurface gold mine. Applied Environmental and Microbiology, 68: 3046-3054.

Tan Y, Wang Y, Wang Y, et al. 2018. Novel mechanisms of selenate and selenite reduction in the obligate aerobic bacterium *Comamonas testosteroni* S44. Journal of Hazardous Materials, 359: 129-138.

Tan Y, Yao R, Wang R, et al. 2016. Reduction of selenite to Se(0) nanoparticles by filamentous bacterium *Streptomyces* sp. ES2-5 isolated from a selenium mining soil. Microbial Cell Factories, 15(1): 157.

Terry N, Carlson C, Raab T K, et al. 1992. Rates of selenium volatilization among crop species. Journal of Environmental Quality, 21(3): 341-344.

Tobe R, Mihara H. 2018. Delivery of selenium to selenophosphate synthetase for selenoprotein biosynthesis. Biochimica et Biophysica Acta - General Subjects, 1862(11): 2433-2440.

Tomei F A, Barton L L, Lemanski C L, et al. 1992. Reduction of selenate and selenite to elemental selenium by *Wolinella succinogenes*. Canadian Journal of Microbiology, 38: 1328-1333.

Tomei F A, Barton L L, Lemanski C L, et al. 1995. Transformation of selenate and selenite to elemental selenium by *Desulfovibrio desulfuricans*. Journal of Industrial Microbiology, 14: 329-336.

Torma A E, Habashi F. 1972. Oxidation of copper (II) selenide by *Thiobacillus ferrooxidans*. Canadian Journal of Microbiology, 18(11): 1780-1781.

Tugarova A V, Mamchenkova P V, Khanadeev V A, et al. 2020. Selenite reduction by the rhizobacterium *Azospirillum brasilense*, synthesis of extracellular selenium nanoparticles and their characterisation. New Biotechnology, 58: 17-24.

Turner R J, Weiner J H, Taylor D E. 1998. Selenium metabolism in *Escherichia coli*. BioMetals, 11: 223-227.

Vriens B, Behra R, Voegelin A, et al. 2016. Selenium uptake and methylation by the microalga *Chlamydomonas reinhardtii*. Environmental Science & Technology, 50(2): 711-720.

Vriens B, Lenz M, Charlet L, et al. 2014. Natural wetland emissions of methylated trace elements. Nature Communications, 5: 3035.

Wang D, Rensing C, Zheng S. 2022. Microbial reduction and resistance to selenium: mechanisms, applications and prospects. Journal of Hazardous Materials, 421: 126684.

Wang X, Liu G, Zhou J, et al. 2011. Quinone-mediated reduction of selenite and tellurite by *Escherichia coli*. Bioresource Technology, 102(3): 3268-3271.

Wang Y, Lai C, Wu M, et al. 2021. Roles of oxygen in methane-dependent selenate reduction in a membrane biofilm reactor stimulation or suppression. Water Research, 198: 117150.

Wang Y, Qin Y, Kot W, et al. 2016. Genome sequence of selenium-solubilizing bacterium *Caulobacter vibrioides* T5M6. Genome Announcements, 4(1): e01721-15.

Wang Y, Shu X, Zhou Q, et al. 2018. Selenite reduction and the biogenesis of selenium nanoparticles by *Alcaligenes faecalis* Se03 isolated from the gut of *Monochamus alternatus* (Coleoptera: Cerambycidae). International Journal of Molecular Sciences, 19(9): 2799.

Wang Z J, Gao Y X. 2001. Biogeochemical cycling of selenium in Chinese environments. Applied Geochemistry, 16: 1345-1351.

Wang Z J, Zhao L H, Zhang L, et al. 1991. Effect of the chemical forms of selenium on its volatilization from soils in Chinese low-selenium-belt. Journal of Environmental Sciences, 3(2): 113-119.

Wells M, McGarry J, Gaye M M. et al. 2019. Respiratory selenite reductase from *Bacillus selenitireducens* strain MLS10. Journal of Bacteriology, 201(7): e00614-18.

Wen L L, Lai C Y, Yang Q, et al. 2016. Quantitative detection of selenate-reducing bacteria by real-time PCR targeting the selenate reductase gene. Enzyme and Microbial Technology, 85: 19-24.

Winkel L H E, Johnson C A, Lenz M, et al. 2012. Environmental selenium research: from microscopic processes to global understanding. Environmental Science & Technology, 46: 571-579.

Wyatt L H, Diringer S E, Rogers L A, et al. 2016. Antagonistic growth effects of mercury and selenium in *Caenorhabditis elegans* are chemical-species-dependent and do not depend on internal Hg/Se ratios. Environmental Science & Technology, 50(6): 3256-3264.

Xia X, Wu S, Li N, et al. 2018. Novel bacterial selenite reductase CsrF responsible for Se(Ⅳ) and Cr(Ⅳ) reduction that produces nanoparticles in *Alishewanella* sp. WH16-1. Journal of Hazardous Materials, 342: 499-509.

Xia Z C, Cheng Y Y, Kong W Q, et al. 2016. Electron shuttles alter selenite reduction pathway and redistribute formed Se(0) nanoparticles. Process Biochemistry, 51(3): 408-413.

Xiao K, Song M, Liu J, et al. 2018. Differences in the bioaccumulation of selenium by two earthworm species (*Pheretima guillemi* and *Eisenia fetida*). Chemosphere, 202: 560-566.

Xu D, Yang L, Wang Y, et al. 2018. Proteins enriched in charged amino acids control the formation and stabilization of selenium nanoparticles in *Comamonas testosteroni* S44. Scientific Reports, 8(1): 4766.

Yanke L J, Bryant R D, Laishley, E J. 1995. Hydrogenase-I of *Clostridium pasteurianum* functions as a novel selenite reductase. Anaerobe, 1(1): 61-67.

Yasir M, Zhang Y, Xu Z, et al. 2020. NAD(P)H-dependent thioredoxin-disulfide reductase TrxR is essential for tellurite and selenite reduction and resistance in *Bacillus* sp. Y3. FEMS Microbiology Ecology, 96(9): fiaa126.

Ye W, Yuan L, Zhu R, et al. 2021. Selenium volatilization from tundra soils in maritime Antarctica. Environment International, 146: 106189.

Yee N, Ma J, Dalia A, et al. 2007. Se(Ⅳ) reduction and the precipitation of Se(0) by the facultative bacterium *Enterobacter cloacae* SLD1a-1 are regulated by FNR. Applied and Environmental Microbiology, 73(6): 1914-1920.

Yu Q, Boyanov M I, Liu J, et al. 2018. Adsorption of Selenite onto *Bacillus subtilis*: The Overlooked Role of Cell Envelope Sulfhydryl Sites in the Microbial Conversion of Se(Ⅳ). Environmental Science & Technology, 52(18): 10400-10407.

Yue S, Huang C, Wang R, et al. 2021. Selenium toxicity, bioaccumulation, and distribution in earthworms (*Eisenia fetida*) exposed to different substrates. Ecotoxicology and Environmental Safety, 217: 112250.

Yue S, Zhang H, Zhen H, et al. 2019. Selenium accumulation, speciation and bioaccessibility in selenium-enriched earthworm (*Eisenia fetida*). Microchemical Journal, 145: 1-8.

Zayed A M, Terry N. 1994. Selenium volatilization in roots and shoots: effects of shoot removal and sulfate level. Journal of Plant Physiology, 143(1): 8-14.

Zehrt J P, Oremland R S. 1987. Reduction of selenate to selenide by sulfate-respiring bacteria: experiments with cell suspensions and estuarine sediments. Applied and Environmental Microbiology, 53: 1365-1369.

Zhang J, Wang Y, Shao Z, et al. 2019. Two selenium tolerant Lysinibacillus sp. strains are capable of reducing selenite to elemental Se efficiently under aerobic conditions. Journal of Environmental Sciences (China), 77: 238-249.

Zhang X, Fan W Y, Yao M C, et al. 2020. Redox state of microbial extracellular polymeric substances regulates reduction of selenite to elemental selenium accompanying with enhancing microbial detoxification in aquatic environments. Water Research, 172: 115538.

Zhang Y, Moore J N. 1996. Selenium fractionation and speciation in a wetland system. Environmental Science & Technology, 30(8): 2613-2619.

Zhao R, Guo J, Song Y, et al. 2020. Mediated electron transfer efficiencies of Se(IV) bioreduction facilitated by meso-tetrakis (4-sulfonatophenyl) porphyrin. International Biodeterioration & Biodegradation, 147: 104838.

Zhu T T, Tian L J, Yu H Q. 2020. Phosphate-suppressed selenite biotransformation by *Escherichia coli*. Environmental Science & Technology, 54(17): 10713-10721.

Zhu Y G, Yoshinaga M, Zhao F J, et al. 2014. Earth abides arsenic biotransformations. Annual Review of Earth and Planetary Sciences, 42: 443-467.

第四章 植物硒吸收和同化的分子机制

硒（Se）作为人体、动物和微生物的必需微量营养元素，是机体中多种含硒蛋白质如谷胱甘肽过氧化物酶、硫氧还蛋白还原酶等的重要组分，其与人体健康如抗氧化、抗癌、提高免疫能力等方面息息相关。人体硒摄入不足会造成免疫力下降，长期严重缺硒则会导致患克山病、大骨节病的概率大大增加。据估计，全球范围内缺硒人口约有15%（5亿～11亿）（Tan et al.，2016），而我国是世界上缺硒最严重的国家之一（孙国新等，2017），国土面积中72%为缺硒区域，并且超1.05亿人口处于硒营养缺乏状态（Wang et al.，2017）。另外，过量摄入硒则可能增加患2型糖尿病的风险（Rayman，2012；Kohler et al.，2018）。基于硒元素的膳食摄入不足及其毒性范围考量，联合国粮农组织/世界卫生组织建议成年人每日硒摄入量为26～34 μg，最高安全剂量为400 μg/d（FAO and WHO，2002）。

人体摄入的硒主要来源于植源性食物，对保障人体健康具有重要作用（Tan et al.，2018；Harthill，2011；Rayman，2020）。稻米作为我国60%以上人口的主食，是硒膳食摄入的主要来源，稻米硒含量与人体硒水平密切相关（Sun et al.，2010）。相比于美国和印度等国家，我国市售普通大米的硒含量普遍较低（0.01～0.03 mg/kg），无法满足人们对硒的日常摄入需求，因此，亟须通过农艺措施或生物强化的方法提高水稻籽粒硒含量（Williams et al.，2009；Sun et al.，2010；Zhao and McGrath，2009；Zhu et al.，2009）。提高作物可食部位硒含量，需要对植物硒吸收、转运和代谢的机理进行深入解析。

硒作为一种类金属元素，存在4种不同价态的形式，包括-2价的硒化物（Se^{2-}）、0价的硒元素（Se）、+4价的亚硒酸盐（SeO_3^{2-}）和+6价的硒酸盐（SeO_4^{2-}）。硒与硫位于元素周期表的同一主族，两者具有类似的化学性质，在自然界中硒存在于金属硫化物矿石中并可部分取代硫原子（Boyd，2011）。在自然界中，硒主要以无机硒和有机硒的形式存在（Boyd，2011；Ellis and Salt，2003）。其中，常见的无机硒包括元素硒、硒化物、硒酸盐和亚硒酸盐。除了元素硒是固体且不溶于水，其他形式的硒大都易溶于水，从而进入湿地、地表水与地下水（Lenz and Lens，2009）。有机硒主要以挥发性甲基硒化物、三甲基硒和硒代氨基酸等形式存在（Pyrzynska，2002）。在硒的生物地球化学循环中，硒在岩石、沉积物和土壤中经过风化、淋溶和迁移等过程进入水体，并通过植物吸收后进入食物链，最终进入动物或人体体内（Tan et al.，2016）。

目前，虽然没有证据表明硒是高等植物必需的矿质营养元素，但硒对植物的生长发育和抗逆性均具有一定的影响，被认为是植物的有益元素（Pilon-Smits et al.，2009）。在黑麦、生菜、马铃薯和大豆等作物中的研究表明，适量添加硒可以促进植物的生长（Hartikainen，2005）。此外，硒在抵抗氧化胁迫、抑制有毒有害重金属吸收和积累、提高抗病性和抗虫性等方面都发挥作用。另一方面，过量的硒会对植物的细胞膜造成损伤，破坏细胞内水分平衡，抑制植物的呼吸作用和光合作用（Aggarwal et al.，2011），严重时会使植物出现各种中毒症状，包括生长发育迟缓、叶片枯黄甚至死亡（Kaur et al., 2014）。因此，硒对植物的有益和有害作用与硒的浓度和植物种类有关。

硒可以通过植物根系和叶片进入体内，其中根系吸收为主要途径。植物主要吸收无机态和有机态形式的硒，但不能吸收元素形态的单质硒或金属硒化合物（White and Broadley, 2009）。硒酸盐和亚硒酸盐是植物吸收无机态硒的两种主要形式，而吸收的有机态硒则主要为硒代氨基酸。一旦进入植物体内，无机硒将被还原同化成硒代氨基酸（如硒代半胱氨酸和硒代甲硫氨酸等），从而被植物进一步利用。本章分别介绍植物根系和叶片吸收硒的途径，总结目前已发现的、参与硒吸收和转运的相关蛋白质，并对植物硒同化代谢分子机制进行介绍，以期为作物硒生物强化和植物修复提供理论基础。

第一节 植物根系对硒的吸收

一、植物根系对硒酸盐的吸收

硒酸盐和亚硒酸盐是植物根系吸收无机硒的两种主要形式，其存在形态主要受土壤的氧化还原电位 pE 和 pH 的影响。在碱性和氧化条件良好的土壤中（pE+pH＞15），硒主要以硒酸盐的形式存在；而在酸性到中性的还原性土壤环境中（7.5＜pE+pH＜15），硒主要以亚硒酸盐形式存在（Elrashidi et al., 1987）。植物根系主要通过硫酸盐转运蛋白进行硒酸盐吸收，而亚硒酸盐则可能主要通过磷酸盐转运蛋白进行吸收。此外，植物还可以通过氨基酸转运蛋白对有机态硒进行吸收（图 4-1）。

植物根系吸收硒可以分为三种途径，包括：硫酸盐转运蛋白系统吸收硒酸盐（SeO_4^{2-}），磷酸盐转运蛋白系统、水通道蛋白和阴离子通道蛋白等吸收亚硒酸盐（SeO_3^{2-}、$HSeO_3^-$ 和 H_2SeO_3），氨基酸转运蛋白系统吸收硒代氨基酸等有机硒（SeCys 和 SeMet）。

硒和硫均位于元素周期表中的第ⅥA 族，具有类似的原子大小、键能、电离势与电子亲和性等。硒酸根 [SeO_4^{2-}] 与硫酸根 [SO_4^{2-}] 也具有相似的化学特性，两者的离子半径分别为 0.42 Å 和 0.30 Å。大多数植物无法区分硒酸盐与硫酸盐，因此，硒酸盐往往可通过硫酸盐转运系统为植物所吸收。硒酸盐与硫酸盐的吸收存在竞争关系，高浓度的硫酸盐会抑制硒酸盐的吸收，而植物缺硫则会促进硒酸盐的吸收。植物吸收硒酸盐是通过硫酸盐转运蛋白进行，这首先是在模式植物拟南芥（*Arabidopsis thaliana*）中发现的。通过筛选拟南芥硒酸盐耐受突变体，鉴定出了具有转运硒酸根与硫酸根活性的硫酸盐转运蛋白 SULTR1;2（Sulfate Transporter 1;2）（Shibagaki et al., 2002）。根据氨基酸序列的同源性，一般将拟南芥中硫酸盐转运蛋白分为四个家族（Gigolashvili and Kopriva, 2014; Buchner et al., 2004）。其中第Ⅰ家族为高亲和性转运蛋白，包括 SULTR1;1、SULTR1;2 和 SULTR1;3。硒酸盐主要通过高亲和性硫酸盐转运蛋白 SULTR1;1 和 SULTR1;2 进入植物根系，且 SULTR1;2 的贡献要大于 SULTR1;1（Barberon et al., 2008）。SULTR1;1 和 SULTR1;2 的蛋白序列同源性约为 70%，主要在根毛、根表皮和皮层中表达（Yoshimoto et al., 2002）。两者均定位在细胞质膜上，其基因表达均受植物体内硫稳态的调控（Schiavon et al., 2015）。在硫充足条件下，两者在根中的表达量均很低，虽然 SULTR1;2 的编码基因表达量高于 SULTR1;1，但在缺硫条件下 SULTR1;1 和 SULTR1;2 编码基因的

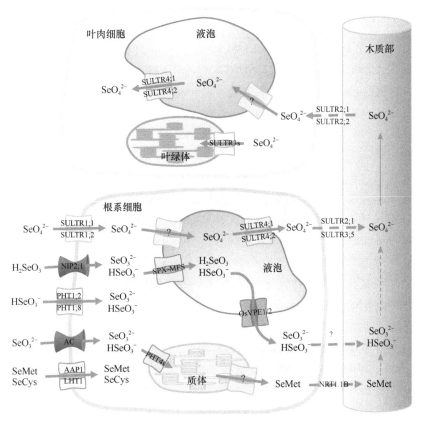

图 4-1　植物硒吸收和转运过程

SULTR，硫酸盐转运蛋白；NIP，水孔通道蛋白；PHT，磷酸盐转运蛋白；AC，阴离子通道蛋白；AAP，氨基酸通透酶；LHT，氨基酸通透酶同源蛋白；SPX-MFS，液泡磷转运蛋白；VPE，液泡磷外排蛋白；NRT1.1B，硝酸盐转运蛋白

表达均受强烈诱导，以保证植物在缺硫条件下吸收到足够的硫酸盐供植物生长发育所需（Rouached et al.，2008；Yoshimoto et al.，2002；Huang et al.，2016）。

虽然高亲和性硫酸盐转运蛋白 SULTR1;1 和 SULTR1;2 都可介导植物根部从外界环境吸收硒酸盐，但两者间存在不均衡的功能冗余性（Barberon et al.，2008）。通过比较拟南芥 *SULTR1;1* 和 *SULTR1;2* 基因功能丧失突变体 *sultr1;1*、*sultr1;2* 和两个基因功能同时丧失突变体 *sultr1;1 sultr1;2* 的硒酸盐耐性，发现 *sultr1;1 sultr1;2* 双突变体的硒酸盐耐性最高，其次是 *sultr1;2* 单突变体，而 *sultr1;1* 单突变体与野生型无显著差异（Barberon et al.，2008）。在不同浓度硒酸钠处理下测定根伸长的半抑制浓度，发现拟南芥野生型、*sultr1;1* 单突变体、*sultr1;2* 单突变体和 *sultr1;1 sultr1;2* 双突变体的根伸长半抑制浓度分别为 0.7 μmol/L、0.7 μmol/L、8 μmol/L 和 15 μmol/L。这些结果都表明硫酸盐转运蛋白 SULTR1;2 在拟南芥根系吸收硒酸盐中发挥主导作用。在 SULTR1;2 蛋白存在的情况下，SULTR1;1 基本不参与硒酸盐的吸收，而当 *SULTR1;2* 基因突变之后，SULTR1;1 也可以发挥硒酸盐的吸收功能。进一步测定不同突变体地上部硒积累量证实了上述结论。在培养基中以 12.5 μmol/L 甲硫氨酸为硫素来源，以保证不同硫酸盐转运蛋白突变体处于不缺硫状态，添加 30 μmol/L 硒酸钠处理 5 d 后测定不同基因型材料地上部总硒含量，结果显

示 *sultr1;2* 单突变体和 *sultr1;1 sultr1;2* 双突变体的总硒含量均显著低于野生型，仅分别约为野生型的 30% 和 17%，而 *sultr1;1* 单突变体与野生型无显著差异（Barberon et al.，2008）。另外，拟南芥硫酸盐转运蛋白基因家族其他成员（包括 SULTR1;3、SULTR2;1、SULTR2;2、SULTR3;1、SULTR3;2、SULTR3;3、SULTR3;4、SULTR3;5 和 SULTR4;1 等）突变之后，硒酸盐的耐受性均与野生型无显著差异（El Kassis et al.，2007），进一步说明拟南芥中主要由 SULTR1;2 负责根系吸收硒酸盐。

由于硒酸盐和硫酸盐均可以通过硫酸盐转运蛋白 SULTR1;1 和 SULTR1;2 进行吸收，因此可以通过提高 *SULTR1;1* 和 *SULTR1;2* 基因的表达，促进硒在植物体内的积累，尤其是作物可食部位硒的积累，实现硒的生物强化。拟南芥 *SULTR1;1* 和 *SULTR1;2* 基因的启动子区域均存在缺硫响应顺式作用元件 SURE（sulfur-responsive element），在缺硫条件下可以被快速诱导表达（Maruyama et al.，2005；Maruyama-Nakashita et al.，2005；Maruyama-Nakashita et al.，2007）。在缺硫条件下，拟南芥体内积累的硒显著升高（White et al.，2004），类似现象在小麦等植物中也被证实（Li et al.，2008）。拟南芥高硫突变体 *msa1-1* 中由于丝氨酸羟甲基转移酶基因 *AtSHM7/MSA1* 的突变造成甲基化主要供体 *S*-腺苷甲硫氨酸的含量降低，导致硫酸盐转运蛋白基因 *SULTR1;1* 和 *SULTR1;2* 的启动子区甲基化程度降低，使得两种蛋白在根系中的表达量升高，造成硫酸盐和硒酸盐的吸收增加，最终导致突变体地上部总硫含量和总硒含量均比野生型显著升高（Huang et al.，2016）。Chen 等（2020）利用正向遗传筛选方法在水稻中也鉴定到水稻富硒突变体 *cadt1*，基因定位克隆发现突变的 *OsCADT1* 基因为拟南芥 *AtSHM7* 基因的同源基因。该基因突变之后同样导致水稻体内产生强烈的缺硫响应，从而引起根部硫酸盐转运蛋白 *OsSULTR1;1* 基因的上调表达，最终使水稻突变体 *cadt1* 籽粒中总硫和总硒含量比野生型提高约 1 倍。由于 *cadt1* 突变体在大田种植条件下的主要农艺性状（包括株高、产量等）均与野生型无显著差异，有望应用于培育籽粒富硒水稻新品种。

植物对硫酸盐和硒酸盐的吸收存在明显的竞争作用。例如，随着环境介质中硫酸盐含量的降低，植物吸收硒酸盐的量增加（Li et al.，2008；White et al.，2004）。然而，外源施加硒酸盐并不能抑制硫酸盐的吸收，而是一定程度上促进硫酸盐的吸收，可能与硒酸盐处理上调 *SULTR1;1* 和 *SULTR1;2* 基因的表达有关（Takahashi et al.，2000；Zhang et al.，2006a）。硫酸盐转运蛋白 SULTR1;1 和 SULTR1;2 对硫酸盐和硒酸盐的选择性可能存在差异。这种差异性与硫的营养状态有关，同时不同物种也存在差别。研究表明，在外界环境中硫酸盐含量高的条件下，硫酸盐蛋白对硫酸盐的选择性较低；而在低硫条件下，硫酸盐蛋白对硫酸盐的选择性要显著高于硒酸盐（White et al.，2004）。另外，不同物种中硫酸盐转运蛋白对硫酸盐和硒酸盐的选择性也存在差异。在硒超富集植物沙漠王羽（*Stanleya pinnata*）中，硫酸盐转运蛋白基因的表达量通常远高于非硒超富集植物（Wang et al.，2018）。同时，硫酸盐浓度升高对硒超富集植物硒吸收的抑制作用要小于非硒超富集植物（Schiavon et al.，2015），说明硒超富集植物中的硫酸盐转运蛋白对硒酸盐的选择性要高于非硒超富集植物。然而，目前在植物中尚未发现特异性吸收硒酸盐的转运蛋白。

二、植物根系对亚硒酸盐的吸收

在酸性到中性的还原性土壤条件下（如淹水稻田土壤系统），亚硒酸盐是无机硒的主要存在形态。在水溶液中，低浓度的亚硒酸根存在三种不同的形态：SeO_3^{2-}、$HSeO_3^-$ 和 H_2SeO_3。这三者在不同 pH 条件下可相互转化，根据亚硒酸根的电离常数（$K_1=2.7\times10^{-3}$，$K_2=2.5\times10^{-7}$）可以计算出不同 pH 下三种不同亚硒酸根形态的比例（表4-1）。例如，在 pH 3 时，$HSeO_3^-$ 和 H_2SeO_3 分别占到约 73% 和 27%，几乎不存在 SeO_3^{2-}；当 pH 4～5 时，超过 96% 以上的形态为 $HSeO_3^-$；当 pH 6 和 7 时，$HSeO_3^-$ 和 SeO_3^{2-} 分别为主要形态，占比分别为 80% 和 70% 左右；而当 pH 8 时，超过 96% 以上的形态为 SeO_3^{2-}。因此，在不同 pH 条件下，植物根系可能通过不同的途径吸收亚硒酸根。在水稻、玉米等不同植物中，根系吸收亚硒酸根受 pH 的影响，当 pH 升高时，亚硒酸根的吸收显著下降（Zhang et al.，2010b；Zhang et al.，2006b）。

表4-1　三种不同形式亚硒酸根在不同 pH 下的占比

pH	SeO_3^{2-}/%	$HSeO_3^-$/%	H_2SeO_3/%
3	0.18	72.96	27.02
4	0.24	96.20	3.56
5	2.43	97.21	0.36
6	19.99	79.98	0.03
7	71.43	28.57	0.00
8	96.15	3.85	0.00

注：数据改自 Zhang 等（2006b）。

在 pH 为 5 的条件下，低温处理以及利用呼吸作用抑制剂处理均可显著抑制根系亚硒酸根的吸收，同时根系吸收亚硒酸根的速率符合米氏方程，表明 pH 5 时根系主要以 $HSeO_3^-$ 为主要形态，通过转运蛋白介导的消耗能量的主动运输吸收亚硒酸根（Zhang et al.，2010b；Zhang et al.，2006b）。进一步水培发现，磷酸盐可以显著抑制亚硒酸盐的吸收而缺磷则促进其吸收（Li et al.，2008；Hopper and Parker，1999），说明根系可能通过磷酸盐转运蛋白系统吸收亚硒酸盐。OsPHT1;2/OsPT2 是水稻磷酸盐转运蛋白，主要在根系皮层细胞和主根及侧根的中柱表达，负责根系无机磷酸盐的吸收（Ai et al.，2009）。过量表达或敲除 *OsPHT1;2* 基因均能显著升高或降低根系亚硒酸盐的吸收（Zhang et al.，2014）。在烟草中过量表达另外一个水稻高亲和磷酸盐转运蛋白基因 *OsPHT1;8/OsPT8* 同样可以促进硒在地上部的积累（Jia et al.，2011；Song et al.，2017）。因此，水稻根系可能主要通过磷酸盐转运蛋白 OsPHT1;2 和 OsPHT1;8 吸收亚硒酸盐。植物中的磷酸盐转运蛋白家族根据序列的同源性可以分为 PHT1、PHT2、PHT3 和 PHT4 等四个亚家族，其中 PHT1 亚家族成员大多在根系的表皮细胞和外皮层细胞表达（Raghothama and Karthikeyan，2005；Lopez-Arredondo et al.，2014）。然而，除了 OsPHT1;2 和 OsPHT1;8 之外，其他 PHT1 成员是否也参与根系亚硒酸盐吸收还有待进一步研究。

与 pH 5 的条件不同，pH 为 3 或者 8 时，低温处理和抑制根系呼吸作用对根系亚硒酸盐的吸收均没有显著影响，同时随着外界亚硒酸盐浓度的升高，根系吸收的亚硒酸根量呈线性增加，说明在这两种 pH 条件下，根系吸收亚硒酸盐可能是一个非耗能的被动吸收过程（Zhang et al.，2010b）。利用 $HgCl_2$ 和 $AgNO_3$ 特异性抑制根系水孔通道蛋白的活性，可以显著降低 pH 为 3 时根系对亚硒酸盐的吸收（Zhang et al.，2010b；Zhang et al.，2006b；Niemietz and Tyerman，2002）。另外，敲除水稻 NIP（nodulin 26-like intrinsic membrane protein）类型水孔通道蛋白基因 *OsNIP2;1/Lsi1* 的突变体，在 pH 3.5 和 4.5 的条件下，其根系亚硒酸根的吸收显著下降（Zhao et al.，2010）。*OsNIP2;1/Lsi1* 主要在水稻根系外皮层和内皮层细胞表达，其编码的蛋白质在细胞质膜的定位存在极性分布，主要定位在外皮层和内皮层细胞质膜的向外一侧，负责将硅酸和亚砷酸吸收进入细胞内（Ma et al.，2008；Ma et al.，2006）。因此，在低 pH 条件下，硒酸盐可能以 H_2SeO_3 的形式通过水通道蛋白被植物根系所吸收。除 OsNIP2;1/Lsi1 之外，NIP 水通道蛋白家族的其他成员对亚砷酸也具有不同程度的通透性，而对硅酸的通透性仅局限于少数几个成员（Bienert et al.，2008；Kamiya et al.，2009；Ma et al.，2008；Sun et al.，2018）。OsNIP2;2 已被证明不参与根系亚硒酸盐吸收（Zhao et al.，2010），其他 NIP 水通道蛋白是否参与植物根系亚硒酸盐吸收还有待深入研究。

在 pH 为 8 的条件下，溶液中的亚硒酸盐主要以 SeO_3^{2-} 的形态存在，但目前植物根系吸收 SeO_3^{2-} 的途径尚不清楚。早期研究表明，激活容积敏感型或者钙离子依赖型阴离子通道蛋白均可以促进亚硒酸盐的吸收（Arvy，1989），而利用化学抑制剂抑制不同阴离子通道蛋白（anion channel，AC）活性可以使根系亚硒酸盐的吸收降低 10%～25%。因此，根系阴离子通道蛋白可能负责部分 SeO_3^{2-} 的吸收，但具体的阴离子通道蛋白以及参与 SeO_3^{2-} 吸收的其他通道蛋白或者转运蛋白还需进一步鉴定。

植物对亚硒酸盐和硒酸盐的吸收速率相似（Hopper and Parker，1999），但植物吸收的硒酸盐更容易被转运至地上部，而亚硒酸盐更易在根部积累（Arvy，1993；De Souza et al.，1998；Li et al.，2008）。陈思杨等（2011）研究发现，在营养液培养条件下外源添加硒酸盐处理后，水稻和小麦吸收的硒分别约有 80% 和 62% 可从根部转移至地上部，而外源添加亚硒酸盐处理仅有约 10% 左右可转移至地上部。

三、植物根系对有机态硒的吸收

有机态硒主要以硒代氨基酸的形式存在，如硒代半胱氨酸（SeCys）和硒代甲硫氨酸（SeMet）等。虽然硒酸盐和亚硒酸盐是土壤中植物可利用的两种形态的硒，但某些土壤中也存在一定量的有机态硒，也可以被植物根系吸收利用（Abrams et al.，1990a；El Mehdawi et al.，2015）。植物根系吸收有机态硒是一个主动耗能过程，可以被代谢抑制剂所抑制（Abrams et al.，1990b）。在相同条件下，植物对 SeCys 和 SeMet 的吸收效率大于硒酸盐或亚硒酸盐。如在水培条件下供给相同浓度（5 μmol/L）的有机硒和无机硒时，硬粒小麦对 SeMet 的吸收速率分别是硒酸盐和亚硒酸盐的 100 倍和 25 倍，而对 SeCys 的吸收速率是硒酸盐的 20 倍、亚硒酸盐的 4 倍（Kikkert and Berkelaar，2013）。油菜虽然与硬粒小麦有所差异，但根系吸收 SeCys 和 SeMet 的效率也均高于硒酸盐和亚硒酸盐。

由于 SeCys 和 SeMet 与半胱氨酸和甲硫氨酸的化学结构非常相似，根系吸收 SeCys

和 SeMet 被认为是通过转运半胱氨酸和甲硫氨酸的氨基酸转运蛋白。拟南芥中氨基酸转运蛋白 AAP1 及其同源蛋白 LHT1 具有转运多种氨基酸转运的活性，可介导半胱氨酸和甲硫氨酸的吸收过程，可能参与了根系 SeCys 和 SeMet 的吸收过程（Hirner et al.，2006；Boorer et al.，1996）。

第二节　硒在植物中的迁移与转运

一、硒酸盐在植物中的迁移与转运

硒酸盐被吸收进入植物根系表皮细胞后，一部分会被转运至根部液泡进行储存，其余将通过共质体途径和质外体途径径向运输至中柱，进一步通过木质部向地上部转运。由于硒酸盐与硫酸盐的理化性质非常相似，硒酸盐不仅根系吸收由硫酸盐转运蛋白完成，从根部向地上部的转运以及在地上部的分配可能也是通过硫酸盐转运蛋白进行。以拟南芥为例，硫酸盐转运蛋白家族按照其氨基酸同源性可以分为 4 个亚家族（图 4-2），其中 SULTR1;1、SULTR1;2 和 SULTR1;3 属于第 I 亚家族，SULTR2;1 和 SULTR2;2 属于第 II 亚家族，SULTR3;1、SULTR3;2、SULTR3;3、SULTR3;4 和 SULTR3;5 属于第 III 亚家族，SULTR4;1 和 SULTR4;2 属于第 IV 亚家族（Gigolashvili and Kopriva，2014；Buchner et al.，2004；Kumar et al.，2011）。硫酸盐转运蛋白 II 家族成员 SULTR2;1 和 III 家族成员 SULTR3;5 负责根系硫酸盐的木质部装载，控制硫酸盐从根部向地上部的转运（Takahashi et al.，2000；Maruyama-Nakashita et al.，2015；Kataoka et al.，2004a）。SULTR2;1 是低亲和硫酸盐转运蛋白，主要在根系的中柱细胞和木质部薄壁细胞表达，负责将硫酸盐转运到木质部细胞，促进硫酸盐经木质部向地上部转运（Maruyama-Nakashita et al.，2015）。SULTR3;5 在酵母异源系统中并没有检测到硫酸盐转运活性，但其可以与 SULTR2;1 互作促进硫酸盐从根部向地上部的转运（Kataoka et al.，2004a）。硫酸盐转运蛋白 II 家族另一个低亲和硫酸盐转运蛋白 SULTR2;2 主要在叶片的维管束鞘细胞表达，负责将木质部导管中的硫酸盐进行卸载，从而转运到叶肉细胞中进行同化（Takahashi et al.，2000）。

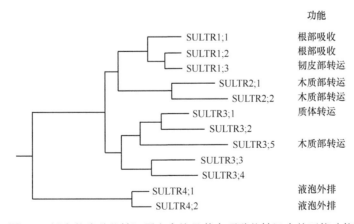

图 4-2　拟南芥硫酸盐转运蛋白家族及其在硒酸盐转运中的可能功能

利用拟南芥所有硫酸盐转运蛋白家族成员氨基酸序列进行进化树分析。右边显示各成员已知的转运功能

在拟南芥中，*SULTR2;1* 和 *SULTR2;2* 基因突变可显著影响根部硒往地上部的转运，

而 *SULTR3;5* 基因功能缺陷对该转运过程的影响明显弱于前者，该结果表明 *SULTR2;1* 和 *SULTR2;2* 基因在根部硒的木质部转运中起主导作用。同时，类似于 *SULTR1* 家族基因，当植物硫供给缺乏或外源添加硒酸盐处理时，*SULTR2;1* 和 *SULTR2;2* 基因的表达量上调但 *SULTR3;5* 基因表达不受缺硫诱导。这些结果暗示 SULTR2;1 和 SULTR2;2 可直接介导硒酸盐进入中柱薄壁细胞，SULTR3;5 可能调节 SULTR2;1 活性而非直接介导木质部转运。目前关于硒酸盐和亚硒酸盐通过韧皮部的再分配过程研究较少，早期报道定位于拟南芥根部和地上部韧皮部的硫酸盐转运蛋白 SULTR1;3 可参与硫酸盐再分配过程。外源硒酸盐处理可上调小麦中 *SULTR1;3* 同源基因的表达，但其具体机制仍需进一步深入解析（Boldrin et al., 2016）。

转运到叶肉细胞的硒酸盐一部分被转运到液泡中进行储存，另一部分则被转运到质体中进行同化。负责将硒酸盐或硫酸盐转运到液泡中的转运蛋白目前还不清楚。由于液泡相较于胞质带正电荷，硒酸盐和硫酸盐等阴离子可被动进入液泡。部分研究显示，硫酸盐和亚硫酸盐很可能通过阴离子通道进入到液泡内（Martinoia et al., 2000）。拟南芥中将硫酸盐或者硒酸盐从液泡中外排到胞质的过程则主要由定位于液泡膜上的硫酸盐转运蛋白Ⅳ家族成员 SULTR4;1 和 SULTR4;2 负责（Kataoka et al., 2004b）。转录组学分析结果显示，这两个转运蛋白基因表达量在高硒处理条件下上调（Zhang et al., 2006a）。该现象可能是由于硒与硫间存在竞争关系，在外源硒处理下，植物体内产生缺硫响应，从而诱导液泡膜上 *SULTR4;1* 和 *SULTR4;2* 基因表达以将液泡中储存的硫外排到胞质以便维持机体硫稳态。水稻硫酸盐转运蛋白Ⅳ家族仅有一个成员 OsSULTR4;1，其在硫酸盐和硒酸盐液泡外排中的功能还有待研究。植物液泡中的 pH 大多为 5～6（Ferjani et al., 2011; Krebs et al., 2010），硒存在的主要形态为 $HSeO_3^-$。因此，SULTR4 介导的硒酸盐从液泡转运到胞质中可能对硒的液泡外排贡献不大。

硫酸盐和硒酸盐的同化可以在根系细胞的质体中进行，但大部分被转运到地上部叶片的叶绿体中被同化。负责将硫酸盐转和硒酸盐运到质体和叶绿体中的转运蛋白主要是硫酸盐转运蛋白Ⅲ家族成员。拟南芥中的 5 个 SULTR3 蛋白均定位在叶绿体被膜上，单独敲除 *SULTR3;1*、*SULTR3;2*、*SULTR3;3* 和 *SULTR3;4* 基因导致叶绿体吸收硫酸盐分别下降 29%、26%、34% 和 31%，而敲除 *SULTR3;5* 则对硫酸盐转运到叶绿体中没有显著影响（Cao et al., 2013; Chen et al., 2019）。因此，叶绿体的硫酸盐和硒酸盐转运主要由 SULTR3;1～SULTR3;4 负责。然而，将这 5 个基因同时敲除仅导致叶绿体中硫酸盐的积累降低 50% 左右，说明硫酸盐和硒酸盐也可以通过被动吸收的方式或者其他途径进入叶绿体中。目前并未发现将叶绿体中的硫酸盐或硒酸盐外排的转运蛋白，这可能是因为胞质相较于叶绿体具有更低的总电荷，硫酸盐或硒酸盐可依赖于电化学梯度被动移动至胞质（Pottosin and Shabala, 2016）。同时，硫酸盐和硒酸盐主要在叶绿体中进行同化代谢，从无机态转化成有机形态，因此叶绿体的外排形式可能主要以有机态为主（Gigolashvili and Kopriva, 2014）。

二、亚硒酸盐在植物中的迁移与转运

与硒酸盐被植物根系吸收后大部分转运至地上部不同，亚硒酸盐吸收进入根部后很快被同化为有机硒化合物，因此硒通过木质部以硒酸盐和有机硒的形态从根部向地上部转运（Li et al.，2008；White et al.，2004）。植物根系细胞的细胞质 pH 为中性，通过水孔蛋白吸收进入根系细胞的 H_2SeO_3 大部分转化为 SeO_3^{2-} 和 $HSeO_3^-$。目前，有关 SeO_3^{2-} 转运至液泡和质体的研究很少，具体过程还不清楚。之前的研究结果表明，SeO_3^{2-} 可以以被动吸收的形式缓慢进入水稻或者玉米根系（Zhang et al.，2006b；Zhang et al.，2010b）。因此，SeO_3^{2-} 可能也以被动吸收的形式进入液泡和质体。

植物胞质中的 pH 通常接近中性，通过阴离子通道蛋白转运到胞质中的 SeO_3^{2-} 部分会转化为 $HSeO_3^-$，而通过水孔蛋白转运到胞质中的 H_2SeO_3 被转化为 SeO_3^{2-} 和 $HSeO_3^-$（表 4-1）。目前还不清楚 SeO_3^{2-} 如何被转运到液泡和质体，以及如何转载到木质部向地上部运输。$HSeO_3^-$ 通常认为是通过具有磷酸盐转运活性的 SPX-MFS [SYG1/PHO81/XPR1（SPX）domain and the Major Facilitator Superfamily（MFS）domain] 家族蛋白被转运到液泡中。拟南芥基因组中有三个 *SPX-MFS* 基因，其中 *AtSPX-MFS1/PHT5；1/VPT1* 编码一个定位在液泡膜上的磷酸根转运蛋白，负责将磷酸盐转入液泡中（Liu et al.，2016；Liu et al.，2015）。在分离纯化的烟草液泡中瞬时表达发现 AtSPX-MFS1 也具有转运 SO_4^{2-}、NO_3^- 和 Cl^- 等阴离子的作用（Liu et al.，2015）。因此，AtSPX-MFS1 很可能也具有转运 $HSeO_3^-$ 的活性。水稻 SPX-MFS 蛋白家族具有 4 个成员，其中 *OsSPX-MFS1* 和 *OsSPX-MFS3* 的表达受缺磷诱导，而 *OsSPX-MFS2* 在缺磷条件下则表达下调（Wang et al.，2012）。OsSPX-MFS1、OsSPX-MFS2 和 OsSPX-MFS3 均定位在液泡膜上，其中 OsSPX-MFS3 具有转运磷酸盐的活性（Wang et al.，2015）。Wang 等（2015）的研究表明 OsSPX-MFS3 可能具有磷酸根外排活性，但通过 *OsSPX-MFS3* 基因过量表达和敲除突变体的液泡磷酸盐含量测定，表明 OsSPX-MFS3 具有磷酸盐内流活性，负责将胞质中的磷酸根转入液泡中（Xu et al.，2019）。过量表达 *OsSPX-MFS3* 促进磷酸根在液泡中的积累，而敲除该基因则造成液泡中磷酸盐含量下降（Xu et al.，2019）。亚硒酸盐的液泡外排可能主要由液泡磷酸盐外排蛋白负责（图 4-1）。水稻 3-磷酸甘油转运蛋白家族成员 OsVPE1 和 OsVPE2 具有将磷酸根从液泡外排到胞质中的活性，过量表达 *OsVPE1* 或者 *OsVPE2* 基因均可显著降低液泡中磷酸盐的含量，而单独或同时敲除 *OsVPE1* 和 *OsVPE2* 均可促进液泡中磷酸盐的积累，导致在低磷条件下液泡中的磷无法外排，从而出现低磷敏感表型（Xu et al.，2019）。

亚硒酸盐转运到质体中进行同化可能由磷酸盐转运蛋白 4 亚家族成员 PHT4 负责（图 4-1）。拟南芥 PHT4 家族具有 6 个成员，其中 PHT4;2 到 PHT4;5 均定位于叶绿体被膜上，在酵母中表达时均具有磷酸盐的转运活性。其中，PHT4;3 和 PHT4;5 可能参与将亚硒酸盐转运到叶绿体中（Guo et al.，2008），而 PHT4;2 参与根部质体中磷酸根的外排，可能也负责将亚硒酸盐从质体中外排到胞质中（Mukherjee et al.，2015）。水稻 PHT4 家族具有 7 个成员，其中 OsPHT4;2、OsPHT4;3 和 OsPHT4;4 也定位在叶绿体被膜，具有磷酸盐转运活性，但是否参与质体亚硒酸盐转运还有待深入研究。

三、有机态硒在植物中的迁移与转运

有机态硒在植物中的迁移转运过程包括根系吸收和同化生成的有机态硒从根部向地上部的转运，以及地上部不同组织间有机态硒的转移与再分配。在水稻田中，亚硒酸盐是硒的主要存在形式。根系吸收的亚硒酸盐很快被同化为有机态硒。在水培条件下用亚硒酸盐处理3 d，水稻根系吸收的亚硒酸盐有46%被转化为SeMet，其余30%为亚硒酸盐、15%为SeCys。测定地上部叶片和叶鞘硒的形态发现，85%左右的硒均为SeMet，说明水稻中硒主要以SeMet的形式从根部向地上部转运（Zhang et al.，2019b）。

木质部装载是营养元素从根部向地上部转运的关键步骤。水稻根系中的SeMet主要由硝酸盐转运蛋白NRT1.1B装载到木质部中，从而被转运到地上部（图4-1）。水稻 *NRT1.1B* 基因主要在根系维管束及叶鞘、叶片和茎秆的维管束组织中表达，其编码的蛋白质具有硝酸盐转运活性，参与水稻硝酸盐从根部向地上部的转运过程（Hu et al.，2015）。在酵母和爪蟾蛙卵细胞中表达NRT1.1B均能检测到其具有转运SeMet的活性。敲除 *NRT1.1B* 基因的水稻突变体，其根部SeMet的吸收以及向地上部转运均显著下降，而过量表达 *NRT1.1B* 基因显著提高了地上部硒的含量，包括叶片、叶鞘和籽粒中硒的含量（Zhang et al.，2019b），表明NRT1.1B不仅参与SeMet从根部向地上部转运的过程，可能还参与水稻灌浆过程中SeMet从叶片向籽粒的转运。

转运到地上部的SeMet在蛋白质合成过程中可以替代甲硫氨酸（Met）从而作为含硒蛋白的形式存在。在水稻灌浆过程中，含硒蛋白降解从而释放SeMet并转运到籽粒中，这是提高水稻籽粒硒含量的关键。然而，人们对含硒蛋白降解调控过程的认识还非常有限，其中可能涉及多种蛋白酶，包括丝氨酸蛋白酶、苏氨酸蛋白酶、天冬氨酸蛋白酶、半胱氨酸蛋白酶等。衰老叶片中蛋白降解形成的SeMet可能通过多种氨基酸通透酶（amino acid permease，AAP）和多肽转运蛋白被转运到新生组织或者灌浆的籽粒。拟南芥氨基酸通透酶AtAAP1负责将氨基酸转运到胚中，对种子贮藏蛋白的合成具有重要作用（Sanders et al.，2009）。该家族的成员AtAAP2在叶片衰老过程中表达上调，参与衰老叶片中氮的再分配；而AtAAP8则在叶片的韧皮部细胞表达，参与氮的库源分配（Santiago and Tegeder，2016；Zhang et al.，2010a）。

水稻中多个氨基酸转运蛋白也参与氨基酸从叶片向籽粒的转运过程。OsAAP1定位于细胞质膜上，主要在叶片、穗部和根中表达，过量表达 *OsAAP1* 可显著提高中性氨基酸从秸秆向籽粒中的转运，而敲除 *OsAAP1* 基因则可显著降低籽粒中包括Met在内的绝大多数氨基酸的含量（Ji et al.，2020）。OsAAP6对籽粒Met含量的影响与OsAAP1相反。OsAAP6定位于内质网上，在根和节的维管束细胞、颖壳和灌浆过程的籽粒中均有表达，可调控植株籽粒Met含量升高，过量表达则使Met降低（Peng et al.，2014），表明两者调控籽粒中氨基酸积累的功能可能不同。*OsAAP3* 与 *OsAAP1* 的表达模式类似，其过量表达也提高了水稻籽粒多数氨基酸的含量，但Met的含量没有改变（Lu et al.，2018）。在爪蟾蛙卵细胞中表达OsAAP1、OsAAP3、OsAAP7和OsAAP16均能检测到对包括Met在内的多种氨基酸的转运活性，其中OsAAP1对Met的转运亲和力（$K_{0.5}$=0.57 mmol/L）要高于OsAAP3（$K_{0.5}$=2.24 mmol/L）（Taylor et al.，2015）。SeMet与Met的结构非常相似，具有转运Met的氨基酸通透酶，可能也具有转运SeMet的活性，但仍需进一步的研究进行验证。

第三节　植物叶片吸收转运硒的过程

受土壤中硒含量通常较低和植物根系吸收效率不高等因素的影响，在实际生产中，富硒农产品往往通过外源喷施含硒叶面肥来实现（王晓芳等，2016；徐聪等，2018；贺前锋等，2016）。叶面喷施硒肥是提高作物可食部位硒含量、实现硒生物强化的有效方法。一方面，与土壤施加硒肥相比，叶面喷施硒肥可以减少由于土壤吸附和化学或生物转化等引起的硒损失；另一方面，被叶片吸收的硒未经过根系吸收、根部向地上部转运等过程，减少了硒在根系和其他非可食部位的截留，可以更有效地将硒富集到可食部位。叶片喷施硒酸盐或者亚硒酸盐等无机态硒以及 SeCys 或者 SeMet 等有机态硒，均可以被叶片有效吸收，进一步被转运到蔬菜的叶片或者禾本科作物的籽粒中（Farooq et al.，2019；Germ et al.，2020；Prom et al.，2020；Zhang et al.，2019a）。

与植物根系直接吸收硒不同，叶片并非是植物吸收养分的直接器官，叶片表面存在的蜡质层和角质层等保护结构往往也会阻碍表皮细胞对硒等矿质元素的吸收。因此，叶片吸收硒可分为两个步骤。首先，硒需要突破植物叶片表面的特异性保护结构和特殊化合物进入到叶片内部；其次，进入表皮细胞的硒通过共质体途径进入到叶肉细胞内，进一步被转运到植株其他部位。

喷施到叶片表面的硒突破表皮细胞是叶片吸收硒的关键环节。植物叶片表面通常覆盖一层不亲水的角质层。叶片表面的角质层作为一种物理屏障，可以保护植物免受水分损失、辐射伤害、病原菌感染等，同时也在很大程度上影响了叶片表面的养分吸收能力。目前研究已发现角质层对有机态和无机态离子/分子具有一定的渗透能力，养分通过角质层的本质是一种扩散解吸过程。叶面可允许通过的分子普遍较小，有机硒与纳米硒离子等直径较大的分子很难透过角质层。为保证叶面硒肥可通过角质层渗透作用进入植物，其肥料需具备良好的水溶性和分散性。

叶片表面的气孔是植物光合作用过程中二氧化碳进入叶片和释放氧气的通道，同时在叶片养分吸收中也扮演极其重要的角色。溶解的养分离子或颗粒可通过气孔进入植物体内，且其进入叶片的速率也远高于角质层。纳米硒肥可借助表面活性剂作用，通过气孔通道直接进入植物体内。

透过角质层渗透到表皮细胞或者通过气孔进入叶片的硒需被转运到叶肉细胞，从而进一步被分配到其他组织或在叶肉细胞中进行同化。目前国内外关于植物叶肉细胞吸收、转运矿质营养元素的研究很少。一般认为硒或其他进入叶片表皮细胞的元素可以通过共质体途径被转运到叶肉细胞（Shahid et al.，2017）。进入叶肉细胞之后进一步通过韧皮部维管系统与光合产物一起向其他组织进行长距离运输，分配到库器官中。随同光合作用产物一起进行长距离运输的过程与库源强度有关，源和库的强度决定了韧皮部运输的方向和速率（De Schepper et al.，2013）。韧皮部装载是通过韧皮部进行长距离运输的关键环节，决定了进行长距离运输的养分的种类和数量。对于水稻、小麦等具有节的禾本科植物，硒从叶片向籽粒运输的过程中需在节将硒从韧皮部卸载，再转运到木质部薄壁细胞，再装载到木质部转运至穗部，最终转运到灌浆中的籽粒，完成硒从源到库的转运过程（Yamaji and Ma，2017）。

与从根系吸收硒并向籽粒转运的过程一样，叶片吸收的硒从叶片向籽粒转运过程中涉及的韧皮部装载、韧皮部卸载、节中的木质部-韧皮部转换、木质部卸载等过程，均需要许多转运蛋白或者通道蛋白参与。但是，目前人们对参与硒酸盐、亚硒酸盐及有机态硒长距离运输的蛋白质还了解得非常少。水稻和大麦中的 *SPDT* 基因编码一个定位于细胞质膜上有机磷-质子共转运蛋白，在节中高度表达，负责将节中有机磷分配到籽粒中（Yamaji et al., 2017；Gu et al., 2022）。SPDT 蛋白是水稻硫酸盐转运蛋白家族 3 成员，不具有硫酸盐的转运活性，但是否具有硒酸盐或者亚硒酸盐的转运活性以及是否参与硒从节向籽粒的分配还有待深入研究。

第四节 植物体内硒的同化与代谢

由于硒和硫之间具有相似的化学性质，在植物中，硒的同化代谢也被认为是通过硫的同化代谢通路转化成有机态硒，但目前相关的分子和遗传证据还相对缺乏。硒的同化可在根部进行，但是根系吸收的大部分硒酸盐被转运至地上部进入叶绿体中进行同化（图 4-3）。硒酸盐首先在 ATP 硫酸化酶（ATP sulfurylase，ATPS）作用下与 ATP 发生

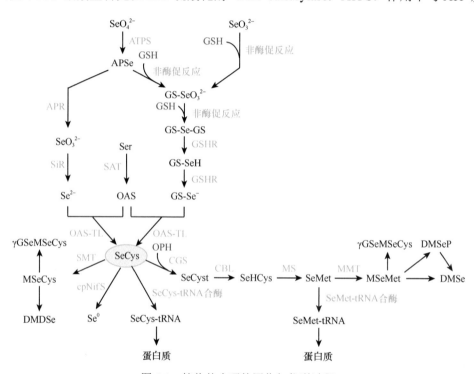

图 4-3 植物体内硒的同化与代谢过程

相关酶类全称：ATPS（adenosine triphosphate sulfurylase），腺苷三磷酸硫酸化酶；APR（adenosine 5′-phosphosulfate reductase），5′-磷硫酸腺苷还原酶；SiR（sulfite reductase），亚硫酸还原酶；SAT（serine acetyl transferase），丝氨酸乙酰转移酶；OAS-TL（O-acetylserine（thiol）lyase），O-乙酰丝氨酸（硫醇）裂合酶；GSHR（GSH reductase），GSH 还原酶；CGS（cystathionine γ-synthase），胱硫醚-γ-合酶；CBL（cystathionine β-lyase），胱硫醚-β-裂合酶；MS（methionine synthase），甲硫氨酸合酶；cpNifS（chloroplast SeCyslyase），叶绿体硒代半胱氨酸裂合酶；SMT（selenocysteine methyltransferase），硒代半胱氨酸甲基转移酶；MMT（S-adenosyl-methionine: methionine methyl transferase），硫-腺苷甲硫氨酸：甲硫氨酸甲基转移酶。相关化合物全称：APSe（adenosine 5′-phosphoselenate），5′-磷酸硒腺苷；Se^{2-}（selenide），硒化物；Ser（serine），丝氨酸；GSH（glutathione），还原型谷胱甘肽；SeCys（selenocysteine），硒代半胱氨酸；OPH（O-phosphohomoserine），O-磷酸高丝氨酸；SeCyst（secystathionine），硒代胱硫醚；SeHCys（selenohomocysteine），硒代同型半胱氨酸；SeMet（selenomethionine），硒代甲硫氨酸；Se0（elemental Se），元素硒；MeSeCys（Se-methylselenocysteine），甲基硒代半胱氨酸；γ-GSeMeSeCys（γ-glutamyl-SeMeSeCys），γ-谷氨酰-硒-甲基硒代半胱氨酸；MeSeMet（Se-methylselenomethionine），甲基硒代甲硫氨酸；γ-GSeMeSeMet（γ-glutamyl-Se-MeSeMet），γ-谷氨酰-硒-甲基硒代甲硫氨酸；DMeSeP（dimethylselenonium propionate），二甲基硒丙酸内盐；DMeSe（dimethylselenide），二甲基硒；DMeDSe（dimethyldiselenide），二甲基二硒

偶联，生成 5′-磷酸硒腺苷（APSe）（Terry et al.，2000）。该反应是无机态硫酸盐和硒酸盐同化的第一步，同时也是同化途径中的关键限速步骤（White，2018；Pilon-Smits and LeDuc，2009）。Reuveny（1977）在研究硫同化过程中首次发现 ATP 硫酸化酶可被硒酸盐去抑制化，证明了 ATP 硫酸化酶可介导硒酸盐的活化。拟南芥中存在 4 个 ATP 硫酸化酶（ATPS1-4），均定位在质体中（Rotte and Leustek，2000；Bohrer et al.，2015）。此外，通过采用可变的翻译起始位点，缺失叶绿体信号肽的 ATPS2 也可以定位在细胞质中（Bohrer et al.，2014）。由于 ATP 硫酸化酶催化反应是同化过程中的关键限速步骤，该基因已经成为基于植物基因工程技术开发富硒作物的目标基因。在芸薹属芥菜 *Brassica juncea* 中过表达拟南芥 *ATPS1* 基因，可显著促进硒酸盐的还原过程，其根部和地上部有机态硒化合物含量显著高于野生型（Pilon-Smits et al.，1999）。

硒酸盐通过 ATPS 活化生成 APSe 后，在质体中进一步在 5′-磷硫酸腺苷还原酶（APR）作用下被还原成亚硒酸盐，该反应也被认为是硒酸盐同化过程中另一个限速步骤。过表达 *APR* 基因也可上调植物体内有机态硒的水平（Sors et al.，2005a；Sors et al.，2005b；Setya et al.，1996）。由于 ATP 硫酸化酶的催化反应为可逆反应，且其催化逆反应的活性更高，因此 APSe 需被迅速还原成亚硒酸盐才能有效地进行后续的同化过程（Saito，2004）。在拟南芥中，*APR2* 基因突变导致突变体中硒酸盐含量增加而亚硒酸盐含量下降，表明 APR2 可催化 APSe 的还原过程（Grant et al.，2011）。同时，由于 *apr2* 突变体中含硫化合物如半胱氨酸和谷胱甘肽等含量受硫同化抑制而减少，其硒酸盐耐性显著降低（Grant et al.，2011）。

硒酸盐还原生成的亚硒酸盐以及根系直接吸收的亚硒酸盐在亚硫酸盐还原酶（SiR）的作用下进一步还原成硒化物（Se^{2-}）。SiR 可将亚硫酸盐转化为硫化物，但目前尚未有直接证据表明 SiR 可还原亚硒酸盐（Khan et al.，2010）。Se^{2-}进一步在半胱氨酸合酶复合体的作用下催化生成硒代半胱氨酸（SeCys）。半胱氨酸合酶复合体由丝氨酸乙酰转移酶（SAT，又简写为 SERAT）与 *O*-乙酰丝氨酸（硫醇）裂合酶（OAS-TL）组成，前者催化丝氨酸的乙酰化反应生成 *O*-乙酰丝氨酸（OAS），紧接着后者催化 OAS 与 Se^{2-}的接合反应，生成该合成通路中第一个有机态硒——SeCys（Terry et al.，2000）。该反应主要在植物的叶绿体内完成，但在胞质内也能发生（Ng and Anderson，1979；Heeg et al.，2008）。

APSe 除了通过硫同化通路被还原成 SeCys 之外，也可以通过还原型谷胱甘肽（GSH）介导的还原反应生成 SeCys（图 4-3），其中涉及两步非酶促反应和两步 GSH 还原酶催化的酶促反应。APSe 首先与一个分子的 GSH 螯合形成 $GS\text{-}SeO_3^{2-}$，该反应为非酶促反应。随后，$GS\text{-}SeO_3^{2-}$与另外一个 GSH 分子进一步螯合生成 GS-Se-GS，该反应同样为非酶促反应。接着，GS-Se-GS 在 GSH 还原酶（GSHR）催化下先后生成 GS-SeH 和 $GS\text{-}Se^-$。$GS\text{-}Se^-$进一步在 *O*-乙酰丝氨酸（硫醇）裂合酶的作用下与 OAS 裂合生成 SeCys，完成从 APSe 到 SeCys 的同化分支通路。根系吸收的 SeO_3^{2-}以及由 SeO_4^{2-}还原生成的 SeO_3^{2-}可以直接与 GSH 螯合形成 $GS\text{-}SeO_3^{2-}$，通过 GSH 介导的同化分支通路最终生成 SeCys（图 4-3）。

植物中虽然存在两条将 APSe 同化为 SeCys 的通路，即 APR 和 SiR 依赖的通路和 GSHR 依赖的通路，但哪条通路起主导作用还不明确（Terry et al.，2000；Trippe and Pilon-Smits，2021；White，2018）。拟南芥中 *SiR* 基因为单拷贝，其编码的蛋白定位于质体中（Bork et al., 1998）。抑制或沉默 *SiR* 基因表达并未影响亚硒酸盐还原成硒化物（Fisher et al., 2016; Ng and Anderson, 1979），表明在拟南芥中亚硒酸盐可能通过 GSHR 依赖的通路被还原成 SeCys（Terry et al., 2000; Hsieh and Ganther, 1975）。

生成的 SeCys 在植物体内可以被进一步代谢，其代谢途径至少存在 4 个分支（图 4-3）。首先，SeCys 中硒可以被释放出来形成零价硒（Se⁰）。该反应由定位在叶绿体中的硒代半胱氨酸裂合酶（cpNifS）所催化。拟南芥中 AtCpNifs 既可从半胱氨酸中释放出零价硫，也可从硒代半胱氨酸中释放出零价硒，并且催化后者释放零价硒的活性更强（Pilon-Smits et al., 2002）。通过 AtCpNifs 作用催化释放出来的零价硫可在叶绿体中形成铁硫簇（Fe-S），是植物光合作用电子传递过程中许多蛋白质的重要辅因子。被 AtCpNifs 催化释放出来的零价硒可能被错误地形成 Fe-Se，从而对光合系统产生损伤（Balk and Lobreaux, 2005）。另外，SeCys 在 SeCys-tRNA 合酶的作用下与 tRNA 结合，用于蛋白质的合成，生成硒蛋白（图 4-3）。

在第三个代谢分支中，SeCys 可以被硒代半胱氨酸甲基转移酶甲基化，生成甲基硒代半胱氨酸（MeSeCys）。MeSeCys 可以被进一步甲基化，形成可发挥的二甲基二硒（DMeDSe），或者被进一步谷氨酰化，生成 γ-谷氨酰-硒-甲基硒代甲硫氨酸（γ-GSeMeSeMet）。在最后一个也是最重要的代谢分支中，SeCys 通过一系列反应合成 SeMet，其中需要多种酶参与。SeCys 首先在胱硫醚-γ-合酶（CGS）的催化下与 O-磷酸高丝氨酸（OPH）反应生成硒代胱硫醚（SeCyst），随后被胱硫醚-β-裂合酶（CBL）转化成硒代同型半胱氨酸（SeHCys），最后在甲硫氨酸合酶（MS）的催化下将 SeHCys 转化成 SeMet（White, 2018; McCluskey et al., 1986; Dawson and Anderson, 1988; Zhou et al., 2000）。与 SeCys 一样，SeMet 也可以直接用于蛋白质合成，或者进行甲基化。SeMet 的甲基化反应是由硫-腺苷甲硫氨酸：甲硫氨酸甲基转移酶（MMT）催化进行的，生成甲基硒代甲硫氨酸（MeSeMet）。MeSeMet 可以直接被进一步甲基化生成具有挥发性的二甲基硒（DMeSe）。MeSeMet 也可以先被转化成二甲基硒丙酸内盐（DMeSeP），然后再转化成 DMeSe。利用 DMeSeP 和 SeMet 分别处理油菜时发现前者释放挥发性 DMeSe 的速率是后者的 6 倍，表明 DMeSeP 可能是 SeMet 转化为 DMeSe 的中间产物，可以很快被转化为 DMeSe（De Souza et al., 2000）。SeMet 的甲基化过程被认为与植物的硒敏感性或者硒解毒有关（Zhu et al., 2009）。与 MeSeCys 类似，MeSeMet 也可以被谷氨酰化，生成 γ-谷氨酰-硒-甲基硒代甲硫氨酸（γ-GSeMeSeCys）。

通过基因工程的方法改造植物硒同化和代谢途径中关键酶的表达或者活性，可以提高植物的富硒能力及硒的解毒能力。例如，在蔬菜中过量表达拟南芥的 *ATPS1* 基因可以显著促进硒吸收和无机硒向有机硒的转化（Pilon-Smits et al., 1999）。在拟南芥中过量表达内源的 *ATPS1* 基因或来源于铜绿假单胞菌的 *APR* 基因，均可以显著促进硒酸盐的还原，过量表达转基因植株还原的硒酸盐的量比对照植株高出 90%（Sors et al., 2005a）。在田间试验条件下，过量表达拟南芥 *ATPS1* 基因的芥菜转基因株系积累硒的量是野生型的 4 倍以上，显示出很好的富硒效果（Banuelos et al., 2005）。

半胱氨酸合酶复合体是硫/硒同化通路中的关键酶，敲除该复合体中的丝氨酸乙酰转移酶或者 O-乙酰丝氨酸（硫醇）裂合酶均可不同程度上影响硫的同化和代谢，以及对重金属胁迫的耐受性。拟南芥中有 5 个丝氨酸乙酰转移酶，分别定位在细胞质、叶绿体和线粒体中,其中定位在线粒体上的 SERAT2;2 对 OAS 的合成贡献较大,仅当 5 个基因中敲除其中 4 个基因之后的四突变体才显著影响 Cys 和 GSH 的合成,表明拟南芥中的丝氨酸乙酰转移酶基因家族存在明显的功能冗余性。拟南芥中有 9 个与 OAS-TL 具有序列同源的基因,但只有定位于细胞质的 OAS-TL A、定位于质体的 OAS-TL B 和定位于线粒体的 OAS-TL C 具有催化 OAS 与 S^{2-} 合成 Cys 的活性(Wirtz et al.,2004)。敲除拟南芥 OAS-TL 基因不同程度上影响了 Cys 和 GSH 的合成,造成突变体对镉胁迫更为敏感(Heeg et al.,2008;López-Martín et al.,2008)。通过基因编辑技术敲除水稻 OAS-TL 基因 OsOASTL-A1 同样显著降低了 Cys、GSH 和植物螯合素的合成,突变体表现出对砷胁迫的敏感性增加(Wang et al.,2020)。虽然上述基因的突变造成了硫同化和代谢的改变,但是否影响硒的同化代谢并不清楚。

Sun 等(2021)通过正向遗传学的方法筛选到了一株水稻耐砷突变体并克隆了突变的基因 OsASTOL1。该基因编码一个定位于叶绿体的 OAS-TL,具有 O-乙酰丝氨酸(硫醇)裂合酶活性。突变体中的 OsASTOL1 发生一个氨基酸的突变造成 OAS-TL 酶活降低,但由于有其他同工酶的存在,并不影响半胱氨酸合成。与野生型相比,突变的 OsASTOL1 蛋白与丝氨酸乙酰转移酶形成的半胱氨酸合酶复合体的稳定性增强,提高了水稻体内丝氨酸乙酰转移酶的活性,导致半胱氨酸合成的关键底物 OAS 积累增加,而 OAS 是调控硫吸收同化的信号物分子,正向调控硫和硒的吸收及同化,提高水稻体内硫和硒含量,同时增加 Cys、GSH 和植物螯合素等含硫代谢产物的合成,增强水稻对砷的解毒能力,并使更多的砷与植物螯合素进行螯合而被截留在根部,最终造成突变体表现出植株耐砷、稻米降砷富硒的表型,是培育水稻耐砷富硒品种的关键靶基因(Sun et al.,2021)。

表达参与催化 SeCys 进一步代谢相关酶类的关键基因同样影响硫和硒的代谢以及硒的解毒。Van Hoewyk 等(2005)研究发现在拟南芥中过表达叶绿体硒代半胱氨酸裂合酶基因 AtCpNifs 可提高其硒酸盐耐性。在芥菜中过量表达拟南芥胱硫醚-γ-合酶 CGS 基因的转基因植株显著提高了挥发性 DMeSe 的产生,同时对亚硒酸的耐性增强(Van Huysen et al.,2003)。在其他植物中过表达拟南芥 CGS 基因同样可以提高 Met 的合成,但表达胱硫醚-β-裂合酶 CBL 基因对 Met 的合成并没有显著影响,对硒的积累和解毒作用也还不清楚(Amir,2010;Maimann et al.,2001)。硒代半胱氨酸甲基转移酶和硫-腺苷甲硫氨酸:甲硫氨酸甲基转移酶对生成挥发性硒化物 DMeSe 和 DMeDSe 至关重要。在大肠杆菌中表达拟南芥硫-腺苷甲硫氨酸:甲硫氨酸甲基转移酶基因 MMT 可以有效地催化生成挥发性 DMeSe,产生量是未转化菌株的 10 倍。但是,在植物中表达编码这些酶的基因是否可以生成并从植株出挥发出硒化物 DMeSe 和 DMeDSe 还有待研究。

虽然关于植物中硒同化和代谢的通路及参与的酶类相对比较清楚,但编码相关酶类的基因功能及其表达调控机制亟需更深入的研究,以充分阐明植物硒吸收和同化的机制与途径,从而为开发作物的硒生物强化、植物修复和超积累植物提供基础依据。

参 考 文 献

陈思杨, 江荣风, 李花粉, 2011. 苗期小麦和水稻对硒酸盐/亚硒酸盐的吸收及转运机制. 环境科学, 32(1): 284-289.

贺前锋, 李鹏祥, 易凤姣, 等, 2016. 叶面喷施硒肥对水稻植株中镉、硒含量分布的影响. 湖南农业科学, (1): 37-42.

孙国新, 李媛, 李刚, 等, 2017. 我国土壤低硒带的气候成因研究. 生物技术进展, 7(5): 387-394.

王晓芳, 罗章, 万亚男, 等, 2016. 叶面喷施不同形态硒对草莓吸收和转运硒的影响. 农业资源与环境学报, 33(4): 334-339.

徐聪, 刘媛媛, 孟凡乔, 等, 2018. 农产品硒含量及与土壤硒的关系. 中国农学通报, 34(7): 96-103.

Abrams M M, Burau R G, Zasoski R J. 1990a. Organic selenium distribution in selected california soils. Soil Sci Soc Am J, 54(4): 979-982.

Abrams M M, Shennan C, Zasoski R J, et al. 1990b. Selenomethionine uptake by wheat seedlings. Agron J, 82(6): 1127-1130.

Aggarwal M, Sharma S, Kaur N, et al. 2011. Exogenous proline application reduces phytotoxic effects of selenium by minimising oxidative stress and improves growth in bean (*Phaseolus vulgaris* L.) seedlings. Biological Trace Element Research, 140(3): 354-367.

Ai P, Sun S, Zhao J, et al. 2009. Two rice phosphate transporters, OsPht1;2 and OsPht1;6, have different functions and kinetic properties in uptake and translocation. Plant J, 57(5): 798-809.

Amir R, 2010. Current understanding of the factors regulating methionine content in vegetative tissues of higher plants. Amino Acids, 39(4): 917-931.

Arvy M P, 1989. Some factors influencing the uptake and distribution of selenite in the bean plant(*Phaseolus vulgaris*). Plant Soil, 117(1): 129-133.

Arvy M P, 1993. Selenate and selenite uptake and translocation in bean plants(*Phaseolus Vulgaris*). J Exp Bot, 44(263): 1083-1087.

Balk J, Lobreaux S, 2005. Biogenesis of iron-sulfur proteins in plants. Trends in Plant Science, 10(7): 324-331.

Banuelos G, Terry N, Leduc D L, et al. 2005. Field trial of transgenic Indian mustard plants shows enhanced phytoremediation of selenium-contaminated sediment. Environ Sci Technol, 39(6): 1771-1777.

Barberon M, Berthomieu P, Clairotte M, et al. 2008. Unequal functional redundancy between the two *Arabidopsis thaliana* high-affinity sulphate transporters SULTR1;1 and SULTR1;2. New Phytol, 180(3): 608-619.

Bienert G P, Thorsen M, Schussler M D, et al. 2008. A subgroup of plant aquaporins facilitate the bi-directional diffusion of As(OH)$_3$ and Sb(OH)$_3$ across membranes. BMC Biol, 6: 26.

Bohrer A-S, Kopriva S, Takahashi H. 2015. Plastid-cytosol partitioning and integration of metabolic pathways for APS/PAPS biosynthesis in *Arabidopsis thaliana*. Front Plant Sci, 5: 751.

Bohrer A S, Yoshimoto N, Sekiguchi A, et al. 2014. Alternative translational initiation of ATP sulfurylase underlying dual localization of sulfate assimilation pathways in plastids and cytosol in *Arabidopsis thaliana*. Front Plant Sci, 5(750): 750.

Boldrin P F, De Figueiredo M A, Yang Y, et al. 2016. Selenium promotes sulfur accumulation and plant growth in wheat(*Triticum aestivum*). Physiol Plantarum, 158(1): 80-91.

Boorer K J, Frommer W B, Bush D R, et al. 1996. Kinetics and specificity of a H$^+$/amino acid transporter from *Arabidopsis thaliana*. J Biol Chem, 271(4): 2213-2220.

Bork C, Schwenn J D, Hell R. 1998. Isolation and characterization of a gene for assimilatory sulfite reductase from *Arabidopsis thaliana*. Gene, 212(1): 147-153.

Boyd R. 2011. Selenium stories. Nat Chem, 3(7): 570.

Buchner P, Takahashi H, Hawkesford M J. 2004. Plant sulphate transporters: co-ordination of uptake, intracellular and long-distance transport. J Exp Bot, 55(404): 1765-1773.

Cao M J, Wang Z, Wirtz M, et al. 2013. SULTR3;1 is a chloroplast-localized sulfate transporter in *Arabidopsis thaliana*. Plant J, 73(4): 607-616.

Chen J, Huang X, Salt D E, et al. 2020. Mutation in *OsCADT1* enhances cadmium tolerance and enriches selenium in rice grain. New Phytol, 226(3): 838-850.

Chen Z, Zhao P X, Miao Z Q, et al. 2019. SULTR3s function in chloroplast sulfate uptake and affect ABA biosynthesis and the stress response. Plant Physiol, 180(1): 593-604.

Dawson J C, Anderson J W. 1988. Incorporation of cysteine and selenocysteine into cystathionine and selenocystathionine by crude extracts of spinach. Phytochemistry, 27(11): 3453-3460.

De Schepper V, De Swaef T, Bauweraerts I, et al. 2013. Phloem transport: a review of mechanisms and controls. J Exp Bot, 64(16): 4839-4850.

De Souza M P, Lytle C M, Mulholland M M, et al. 2000. Selenium assimilation and volatilization from dimethylselenoniopropionate by Indian mustard. Plant Physiol, 122(4): 1281-1288.

De Souza M P, Pilon-Smits E A H, Lytle C M, et al. 1998. Rate-limiting steps in selenium assimilation and volatilization by Indian Mustard1. Plant Physiol, 117(4): 1487-1494.

El Kassis E, Cathala N, Rouached H, et al. 2007. Characterization of a selenate-resistant *Arabidopsis* mutant: Root growth as a potential target for selenate toxicity. Plant Physiol, 143(3): 1231-1241.

El Mehdawi A F, Lindblom S D, Cappa J J, et al. 2015. Do selenium hyperaccumulators affect selenium speciation in neighboring plants and soil? An X-ray microprobe analysis. Int J Phytoremediation, 17(8): 753-765.

Ellis D R, Salt D E. 2003. Plants, selenium and human health. Curr Opin Plant Biol, 6(3): 273-279.

Elrashidi M A, Adriano D C, Workman S M, et al. 1987. Chemical equilibria of selenium in soils: a theoretical development. Soil Science, 144(2): 274-280.

FAO, WHO. 2002. Human vitamin and mineral requirements: report of a joint FAO/WHO expert consultation, Bangkok, Thailand, Food and Nutrition Division, FAO Romo.

Farooq M U, Tang Z, Zeng R, et al. 2019. Accumulation, mobilization, and transformation of selenium in rice grain provided with foliar sodium selenite. J Sci Food Agric, 99(6): 2892-2900.

Ferjani A, Segami S, Horiguchi G, et al. 2011. Keep an Eye on PPi: The vacuolar-type H^+-pyrophosphatase regulates postgerminative development in *Arabidopsis*. Plant Cell, 23(8): 2895-2908.

Fisher B, Yarmolinsky D, Abdel-Ghany S, et al. 2016. Superoxide generated from the glutathione-mediated reduction of selenite damages the iron-sulfur cluster of chloroplastic ferredoxin. Plant Physiol Biochem, 106: 228-235.

Germ M, Kacjan-Marsic N, Kroflic A, et al. 2020. Significant accumulation of iodine and selenium in chicory(*Cichorium intybus* L. var. *foliosum* Hegi) leaves after foliar spraying. Plants, 9(12): 1766.

Gigolashvili T, Kopriva S. 2014. Transporters in plant sulfur metabolism. Front Plant Sci, 5: 442.

Grant K, Carey N M, Mendoza M, et al. 2011. Adenosine 5′-phosphosulfate reductase(APR2) mutation in *Arabidopsis* implicates glutathione deficiency in selenate toxicity. Biochemical Journal, 438(2): 325-335.

Gu M, Huang H, Hisano H, et al. 2022. A crucial role for a node-localized transporter, HvSPDT, in loading phosphorus into barley grains. New Phytol, 234(4): 1249-1261.

Guo B, Jin Y, Wussler C, et al. 2008. Functional analysis of the *Arabidopsis* PHT4 family of intracellular phosphate transporters. New Phytol, 177(4): 889-898.

Harthill M. 2011. Review: micronutrient selenium deficiency influences evolution of some viral infectious diseases. Biol Trace Elem Res, 143(3): 1325-1336.

Hartikainen H. 2005. Biogeochemistry of selenium and its impact on food chain quality and human health. J Trace Elem Med Biol 18(4): 309-318.

Heeg C, Kruse C, Jost R, et al. 2008. Analysis of the *Arabidopsis O*-acetylserine(thiol)lyase gene family demonstrates compartment-specific differences in the regulation of cysteine synthesis. Plant Cell, 20(1):

168-185.

Hirner A, Ladwig F, Stransky H, et al. 2006. *Arabidopsis* LHT1 is a high-affinity transporter for cellular amino acid uptake in both root epidermis and leaf mesophyll. Plant Cell, 18(8): 1931-1946.

Hopper J L, Parker D R. 1999. Plant availability of selenite and selenate as influenced by the competing ions phosphate and sulfate. Plant Soil, 210(2): 199-207.

Hsieh H S, Ganther H E. 1975. Acid-volatile selenium formation catalyzed by glutathione reductase. Biochemistry, 14(8): 1632-1636.

Hu B, Wang W, Ou S, et al. 2015. Variation in NRT1.1B contributes to nitrate-use divergence between rice subspecies. Nat Genet, 47(7): 834-838.

Huang X, Chao D, Koprivova A, et al. 2016. Nuclear localised more sulphur accmulation 1 epigenetically regulates sulphur homeostasis in *Arabidopsis thaliana*. PLoS Genetics, 12(9): e1006298.

Ji Y, Huang W, Wu B, et al. 2020. The amino acid transporter AAP1 mediates growth and grain yield by regulating neutral amino acid uptake and reallocation in *Oryza sativa*. J Exp Bot, 71(16): 4763-4777.

Jia H, Ren H, Gu M, et al. 2011. The phosphate transporter gene OsPht1;8 is involved in phosphate homeostasis in rice. Plant Physiol, 156(3): 1164-1175.

Kamiya T, Tanaka M, Mitani N, et al. 2009. NIP1;1, an aquaporin homolog, determines the arsenite sensitivity of *Arabidopsis thaliana*. J Biol Chem, 284(4): 2114-2120.

Kataoka T, Hayashi N, Yamaya T, et al. 2004a. Root-to-shoot transport of sulfate in *Arabidopsis*. Evidence for the role of SULTR3;5 as a component of low-affinity sulfate transport system in the root vasculature. Plant Physiol, 136(4): 4198-4204.

Kataoka T, Watanabe-Takahashi A, Hayashi N, et al. 2004b. Vacuolar sulfate transporters are essential determinants controlling internal distribution of sulfate in *Arabidopsis*. Plant Cell, 16(10): 2693-2704.

Kaur N, Sharma S, Nayyar H. 2014. Selenium in agriculture: A nutrient or toxin for crops? Archives of Agronomy and Soil Science, 60(12): 1593-1624.

Khan M S, Haas F H, Samami A A, et al. 2010. Sulfite reductase defines a newly discovered bottleneck for assimilatory sulfate reduction and is essential for growth and development in *Arabidopsis thaliana*. Plant Cell, 22(4): 1216-1231.

Kikkert J, Berkelaar E. 2013. Plant uptake and translocation of inorganic and organic forms of selenium. Arch Environ Contam Toxicol, 65(3): 458-465.

Kohler L N, Foote J, Kelley C P, et al. 2018. Selenium and type 2 diabetes: systematic review. Nutrients, 10(12): 1924.

Krebs M, Beyhl D, Görlich E, et al. 2010. *Arabidopsis* V-ATPase activity at the tonoplast is required for efficient nutrient storage but not for sodium accumulation. Proc Natl Acad Sci USA, 107(7): 3251-3256.

Kumar S, Asif M H, Chakrabarty D, et al. 2011. Differential expression and alternative splicing of rice sulphate transporter family members regulate sulphur status during plant growth, development and stress conditions. Funct Integr Genomics, 11(2): 259-273.

Lenz M, Lens P N L. 2009. The essential toxin: The changing perception of selenium in environmental sciences. Sci Total Environ, 407(12): 3620-3633.

Li H F, Mcgrath S P, Zhao F J. 2008. Selenium uptake, translocation and speciation in wheat supplied with selenate or selenite. New Phytol, 178(1): 92-102.

Liu J, Yang L, Luan M, et al. 2015. A vacuolar phosphate transporter essential for phosphate homeostasis in *Arabidopsis*. Proc Natl Acad Sci USA, 112(47): E6571-E6578.

Liu T Y, Huang T K, Yang S Y, et al. 2016. Identification of plant vacuolar transporters mediating phosphate storage. Nat Commun, 7(1): 11095.

Lopez-Arredondo D L, Leyva-Gonzalez M A, Gonzalez-Morales S I, et al. 2014. Phosphate nutrition:

improving low-phosphate tolerance in crops. Annu Rev Plant Biol, 65: 95-123.

López-Martín M C, Becana M, Romero L C, et al. 2008. Knocking out cytosolic cysteine synthesis compromises the antioxidant capacity of the cytosol to maintain discrete concentrations of hydrogen peroxide in Arabidopsis. Plant Physiol, 147(2): 562-572.

Lu K, Wu B, Wang J, et al. 2018. Blocking amino acid transporter OsAAP3 improves grain yield by promoting outgrowth buds and increasing tiller number in rice. Plant Biotechnol, 16(10): 1710-1722.

Ma J F, Tamai K, Yamaji N, et al. 2006. A silicon transporter in rice. Nature, 440(7084): 688-691.

Ma J F, Yamaji N, Mitani N, et al. 2008. Transporters of arsenite in rice and their role in arsenic accumulation in rice grain. Proc Natl Acad Sci USA, 105(29): 9931-9935.

Maimann S, Hoefgen R, Hesse H. 2001. Enhanced cystathionine beta-lyase activity in transgenic potato plants does not force metabolite flow towards methionine. Planta, 214(2): 163-170.

Martinoia E, Massonneau A, Frangne N. 2000. Transport processes of solutes across the vacuolar membrane of higher plants. Plant Cell Physiol, 41(11): 1175-1186.

Maruyama-Nakashita A, Nakamura Y, Saito K, et al. 2007. Identification of a novel cis-acting element in SULTR1;2 promoter conferring sulfur deficiency response in *Arabidopsis* roots. Plant Cell Physiol, 48: S34.

Maruyama-Nakashita A, Nakamura Y, Watanabe-Takahashi A, et al. 2005. Identification of a novel cis-acting element conferring sulfur deficiency response in *Arabidopsis* roots. Plant J, 42(3): 305-314.

Maruyama-Nakashita A, Watanabe-Takahashi A, Inoue E, et al. 2015. Sulfur-responsive elements in the 3′-nontranscribed intergenic region are essential for the induction of SULFATE TRANSPORTER 2;1 gene expression in *Arabidopsis* roots under sulfur deficiency. Plant Cell, 27(4): 1279-1296.

Maruyama A, Nakamura Y, Watanabe-Takahashi A, et al. 2005. Identification of a novel cis-acting element in SULTR1;1 promoter conferring sulfur deficiency response in *Arabidopsis* roots. Plant Cell Physiol, 46: S58.

McCluskey T J, Scarf A R, Anderson J W. 1986. Enzyme catalysed α, β-elimination of selenocystathionine and selenocystine and their sulphur isologues by plant extracts. Phytochemistry, 25(9): 2063-2068.

Mukherjee P, Banerjee S, Wheeler A, et al. 2015. Live imaging of inorganic phosphate in plants with cellular and subcellular resolution. Plant Physiol, 167(3): 628-638.

Ng B H, Anderson J W. 1979. Light-dependent incorporation of selenite and sulphite into selenocysteine and cysteine by isolated pea chloroplasts. Phytochemistry, 18(4): 573-580.

Niemietz C M, Tyerman S D. 2002. New potent inhibitors of aquaporins: silver and gold compounds inhibit aquaporins of plant and human origin. FEBS Lett, 531(3): 443-447.

Peng B, Kong H, Li Y, et al. 2014. OsAAP6 functions as an important regulator of grain protein content and nutritional quality in rice. Nat Commun, 5: 4847.

Pilon-Smits E A, Leduc D L. 2009. Phytoremediation of selenium using transgenic plants. Curr Opin Biotech, 20(2): 207-212.

Pilon-Smits E A, Quinn C F, Tapken W, et al. 2009. Physiological functions of beneficial elements. Curr Opin Plant Bio, 12(3): 267-274.

Pilon-Smits E A H, De Souza M P, Hong G, et al. 1999. Selenium volatilization and accumulation by twenty aquatic plant species. J Environ Qual, 28(3): 1011-1018.

Pilon-Smits E A H, Garifullina G F, Abdel-Ghany S, et al. 2002. Characterization of a NifS-like chloroplast protein from *Arabidopsis*. Implications for its role in sulfur and selenium metabolism. Plant Physiol, 130(3): 1309-1318.

Pottosin I, Shabala S. 2016. Transport across chloroplast membranes: optimizing photosynthesis for adverse environmental conditions. Molecular Plant, 9(3): 356-370.

Prom U T C, Rashid A, Ram H, et al. 2020. Simultaneous biofortification of rice with zinc, iodine, iron and selenium through foliar treatment of a micronutrient cocktail in five countries. Front Plant Sci, 11: 589835.

Pyrzynska K. 2002. Determination of selenium species in environmental samples. Microchimica Acta, 140(1-2): 55-62.

Raghothama K G, Karthikeyan A S. 2005. Phosphate acquisition. Plant Soil, 274(1): 37.

Rayman M P. 2012. Selenium and human health. Lancet, 379(9822): 1256-1268.

Rayman M P. 2020. Selenium intake, status, and health: a complex relationship. Hormones(Athens), 19(1): 9-14.

Reuveny Z. 1977. Derepression of ATP sulfurylase by the sulfate analogs molybdate and selenate in cultured tobacco cells. Proc Natl Acad Sci USA, 74(2): 619-622.

Rotte C, Leustek T. 2000. Differential subcellular localization and expression of atp sulfurylase and 5′-adenylylsulfate reductase during ontogenesis of *Arabidopsis* leaves indicates that cytosolic and plastid forms of atp sulfurylase may have specialized functions. Plant Physiol, 124(2): 715-724.

Rouached H, Wirtz M, Alary R, et al. 2008. Differential regulation of the expression of two high-affinity sulfate transporters, SULTR1.1 and SULTR1.2, in *Arabidopsis*. Plant Physiol, 147(2): 897-911.

Saito K. 2004. Sulfur assimilatory metabolism. The long and smelling road. Plant Physiol, 136(1): 2443-2450.

Sanders A, Collier R, Trethewy A, et al. 2009. AAP1 regulates import of amino acids into developing *Arabidopsis* embryos. Plant J, 59(4): 540-552.

Santiago J P, Tegeder M. 2016. Connecting source with sink: The role of *Arabidopsis* AAP8 in phloem loading of amino acids. Plant Physiol, 171(1): 508-521.

Schiavon M, Pilon M, Malagoli M, et al. 2015. Exploring the importance of sulfate transporters and ATP sulphurylases for selenium hyperaccumulation-a comparison of *Stanleya pinnata* and *Brassica juncea*(Brassicaceae). Front Plant Sci, 6: 2.

Setya A, Murillo M, Leustek T. 1996. Sulfate reduction in higher plants: molecular evidence for a novel 5′-adenylylsulfate reductase. Proc Natl Acad Sci USA, 93(23): 13383-13388.

Shahid M, Dumat C, Khalid S, et al. 2017. Foliar heavy metal uptake, toxicity and detoxification in plants: A comparison of foliar and root metal uptake. J Hazard Mater, 325: 36-58.

Shibagaki N, Rose A, Mcdermott J P, et al. 2002. Selenate-resistant mutants of *Arabidopsis thaliana* identify Sultr1;2, a sulfate transporter required for efficient transport of sulfate into roots. Plant J, 29(4): 475-486.

Song Z P, Shao H F, Huang H G, et al. 2017. Overexpression of the phosphate transporter gene OsPT8 improves the Pi and selenium contents in *Nicotiana tabacum*. Environ Exp Bot, 137: 158-165.

Sors T G, Ellis D R, Na G N, et al. 2005a. Analysis of sulfur and selenium assimilation in *Astragalus* plants with varying capacities to accumulate selenium. Plant J, 42(6): 785-797.

Sors T G, Ellis D R, Salt D E. 2005b. Selenium uptake, translocation, assimilation and metabolic fate in plants. Photosynth Res, 86(3): 373-389.

Sun G, Liu X, Williams P N, et al. 2010. Distribution and translocation of selenium from soil to grain and its speciation in paddy rice(*Oryza sativa* L.). Environ Sci Technol, 44(17): 6706-6711.

Sun S K, Chen Y, Che J, et al. 2018. Decreasing arsenic accumulation in rice by overexpressing OsNIP1;1 and OsNIP3;3 through disrupting arsenite radial transport in roots. New Phytol, 219(2): 641-653.

Sun S K, Xu X, Tang Z, et al. 2021. A molecular switch in sulfur metabolism to reduce arsenic and enrich selenium in rice grain. Nat Commun, 12(1): 1392.

Takahashi H, Watanabe-Takahashi A, Smith F W, et al. 2000. The roles of three functional sulphate transporters involved in uptake and translocation of sulphate in *Arabidopsis thaliana*. Plant J, 23(2): 171-182.

Tan H, Mo H, Lau A T Y, et al. 2018. Selenium species: current status and potentials in cancer prevention and therapy. Inter J Mol Sci, 20(1): 75.

Tan L C, Nancharaiah Y V, Van Hullebusch E D, et al. 2016. Selenium: environmental significance, pollution, and biological treatment technologies. Biotechnol Adv, 34(5): 886-907.

Taylor M R, Reinders A, Ward J M. 2015. Transport function of rice amino acid permeases (AAPs). Plant Cell Physiol, 56(7): 1355-1363.

Terry N, Zayed A M, De Souza M P, et al. 2000. Selenium in higher plants. Annu Rev Plant Physiol Plant Mol Biol, 51: 401-432.

Trippe R C, Pilon-Smits E A H. 2021. Selenium transport and metabolism in plants: Phytoremediation and biofortification implications. J Hazard Mater, 404: 124178.

Van Hoewyk D, Garifullina G F, Ackley A R, et al. 2005. Overexpression of AtCpNifS enhances selenium tolerance and accumulation in *Arabidopsis*. Plant Physiology, 139(3): 1518-1528.

Van Huysen T, Abdel-Ghany S, Hale K L, et al. 2003. Overexpression of cystathionine-gamma-synthase enhances selenium volatilization in *Brassica juncea*. Planta, 218(1): 71-78.

Wang C, Huang W, Ying Y, et al. 2012. Functional characterization of the rice SPX-MFS family reveals a key role of OsSPX-MFS1 in controlling phosphate homeostasis in leaves. New Phytol, 196(1): 139-148.

Wang C, Yue W, Ying Y, et al. 2015. Rice SPX-major facility superfamily 3, a vacuolar phosphate efflux transporter, is involved in maintaining phosphate homeostasis in rice. Plant Physiol, 169(4): 2822-2831.

Wang C, Zheng L, Tang Z, et al. 2020. OsOASTL-A1 functions as a cytosolic cysteine synthase and affects arsenic tolerance in rice. J Exp Bot, 71(12): 3678-3689.

Wang J, Cappa J J, Harris J P, et al. 2018. Transcriptome-wide comparison of selenium hyperaccumulator and nonaccumulator *Stanleya* species provides new insight into key processes mediating the hyperaccumulation syndrome. Plant Biotechnol J, 16(9): 1582-1594.

Wang J, Li H, Yang L, et al. 2017. Distribution and translocation of selenium from soil to highland barley in the Tibetan Plateau Kashin-Beck disease area. Environ Geochem Health, 39(1): 221-229.

White P J. 2018. Selenium metabolism in plants. BBA-Gen Subjects, 1862(11): 2333-2342.

White P J, Bowen H C, Parmaguru P, et al. 2004. Interactions between selenium and sulphur nutrition in *Arabidopsis thaliana*. J Exp Bot, 55(404): 1927-1937.

White P J, Broadley M R. 2009. Biofortification of crops with seven mineral elements often lacking in human diets - iron, zinc, copper, calcium, magnesium, selenium and iodine. New Phytol, 182(1): 49-84.

Williams P N, Lombi E, Sun G, et al. 2009. Selenium characterization in the global rice supply chain. Environ Sci Technol, 43(15): 6024-6030.

Wirtz M, Droux M, Hell R. 2004. O-acetylserine(thiol) lyase: an enigmatic enzyme of plant cysteine biosynthesis revisited in *Arabidopsis thaliana*. J Exp Bot, 55(404): 1785-1798.

Xu L, Zhao H, Wan R, et al. 2019. Identification of vacuolar phosphate efflux transporters in land plants. Nat Plants, 5(1): 84-94.

Yamaji N, Ma J F. 2017. Node-controlled allocation of mineral elements in Poaceae. Curr Opin Plant Biol, 39: 18-24.

Yamaji N, Takemoto Y, Miyaji T, et al. 2017. Reducing phosphorus accumulation in rice grains with an impaired transporter in the node. Nature, 541(7635): 92-95.

Yoshimoto N, Takahashi H, Smith F W, et al. 2002. Two distinct high-affinity sulfate transporters with different inducibilities mediate uptake of sulfate in Arabidopsis roots. Plant J, 29(4): 465-473.

Zhang H, Zhao Z, Zhang X, et al. 2019a. Effects of foliar application of selenate and selenite at different growth stages on Selenium accumulation and speciation in potato(*Solanum tuberosum* L.). Food Chem, 286: 550-556.

Zhang L, Hu B, Deng K, et al. 2019b. NRT1.1B improves selenium concentrations in rice grains by facilitating selenomethinone translocation. Plant Biotechnol J, 17(6): 1058-1068.

Zhang L, Hu B, Li W, et al. 2014. OsPT2, a phosphate transporter, is involved in the active uptake of selenite in rice. New Phytol, 201(4): 1183-1191.

Zhang L, Tan Q, Lee R, et al. 2010a. Altered xylem-phloem transfer of amino acids affects metabolism and leads to increased seed yield and oil content in *Arabidopsis*. Plant Cell, 22(11): 3603-3620.

Zhang L H, Abdel-Ghany S E, Freeman J L, et al. 2006a. Investigation of selenium tolerance mechanisms in *Arabidopsis thaliana*. Physiol Plantarum, 128(2): 212-223.

Zhang L H, Shi W M, Wang X C. 2006b. Difference in selenite absorption between high- and low-selenium rice cultivars and its mechanism. Plant Soil, 282(1-2): 183-193.

Zhang L H, Yu F Y, Shi W M, et al. 2010b. Physiological characteristics of selenite uptake by maize roots in response to different pH levels. J Plant Nutr Soil Sci, 173(3): 417-422.

Zhao F, McGrath S P. 2009. Biofortification and phytoremediation. Curr Opin Plant Biol, 12(3): 373-380.

Zhao X Q, Mitani N, Yamaji N, et al. 2010. Involvement of silicon influx transporter OsNIP2;1 in selenite uptake in rice. Plant Physiol, 153(4): 1871-1877.

Zhou Z S, Smith A E, Matthews R G. 2000. L-selenohomocysteine: one-step synthesis from L-selenomethionine and kinetic analysis as substrate for methionine synthases. Bioorganic Med Chem Lett, 10(21): 2471-2475.

Zhu Y, Pilon-Smits E A H, Zhao F, et al. 2009. Selenium in higher plants: understanding mechanisms for biofortification and phytoremediation. Trends Plant Sci, 14(8): 436-442.

第五章 硒生物营养强化技术与应用

生物营养强化被定义为提高粮食作物中矿物质和维生素含量以提高粮食作物营养价值的过程（Kiran，2020）。硒生物营养强化则是以矿物质硒作为强化因子的生物营养强化过程，通过选育作物品种、施用外源硒肥、调控农艺管理措施及采用现代生物工程等技术措施来增加植物中硒的含量及有效性（Gupta and Gupta，2017）。本章着重讨论硒生物营养强化的理论基础及实用技术，并通过案例分析，介绍硒生物营养强化技术的应用及效果，总结提出生物营养强化未来发展的方向，希望为硒生物营养强化研究提供有价值的参考。

第一节 硒生物营养强化的理论基础

硒是动物和人类必需的微量元素之一，目前已在人体和动物体中发现了 25 种硒蛋白，如硫氧还蛋白还原酶（TPx）、谷胱甘肽过氧化物酶（GSH-Px）等（Newman et al.，2019；Schiavon and Pilon-Smits，2017），在抗氧化及增强免疫力等方面发挥重要作用（Arakawa et al.，2013）。适量补充硒，能改善和预防慢性淋巴性甲状腺炎、自身免疫性甲状腺疾病、艾滋病、心血管疾病及癫痫等疾病（Rayman，2012）。研究发现，克山病及大骨节病与居民低硒摄入量密切相关，病区基本都处于低硒的地球化学环境中（袁丽君等，2016）。国家卫生健康委员会推荐成人日硒摄入量为 60 μg/d，而中国 72% 地区的居民仍处于硒摄入量不足的状态（中国营养学会，2014；Dinh et al.，2018）。为了改善缺硒地区居民硒摄入量不足带来的"隐性饥饿"问题，需要对农产品进行硒生物营养强化，提高农产品硒含量。

硒生物营养强化是指通过选育作物品种、施用外源硒肥、调控农艺管理措施及采用现代生物工程等技术措施来增加植物中硒含量及其有效性的方法（Gupta and Gupta，2017；AL-Ghumaiz et al.，2020）。硒元素围绕"岩石—土壤—肥料—作物/动物—食品—人体"链条（图 5-1）进行传递，硒生物营养强化的基础在于研究这一传输链条上硒元素的传输规律，特别是硒在土壤-作物系统中的吸收转化及代谢，根据农作物中硒含量的定量强化目标，逆向设计由土壤向作物中硒传输的调控技术。

一、土壤中的硒

土壤是作物中硒元素的主要来源，土壤的硒水平及其生物有效性直接决定了作物的硒水平。实施硒生物营养强化，首先必须了解土壤中硒的分布及其生物有效性，通过农艺技术提高土壤有效硒含量，可提高硒生物营养强化效率。

硒作为一种稀有分散元素，在地壳中的分布极不均匀。土壤中硒的含量在空间分布上变化范围很大，分布极不均匀。全球多数土壤的全硒含量范围为 0.1～2 mg/kg，平均

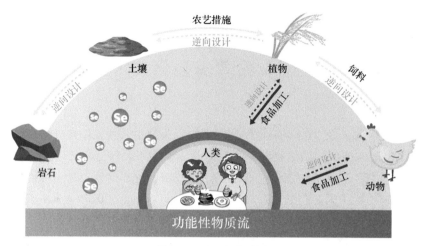

图 5-1 硒元素传输链

含量为 0.4 mg/kg（Plant et al.，2014）。世界卫生组织报道，中国是 40 个硒缺乏国家之一，中国缺硒区主要包括中国东北至西南的低硒地质带，土壤硒含量低于 0.2 mg/kg，横跨我国东北、太行山、秦岭、黄土高原直至青藏高原东部地区。而在紫阳和恩施等富硒地区，土壤硒含量最高可达 79.08 mg/kg 和 36.69 mg/kg，平均含量分别可达 27.81 mg/kg 和 17.29 mg/kg（Dinh et al.，2018）。导致土壤硒区域差异的可能原因是：①大气沉降导致表层土壤硒含量分布不均；②人类活动（如工业和电厂燃煤、施肥等）导致土壤硒含量分布差异；③基岩控制土壤硒含量的分布（Dinh et al.，2018）。

土壤中的硒按化学价态可分为硒化物（-2）、元素硒（0）、亚硒酸盐（+4）、硒酸盐（+6）。元素态硒和硒化物以矿物形态存在，难溶于水，难以被植物吸收。亚硒酸盐和硒酸盐易溶于水，分别以 SeO_3^{2-} 和 SeO_4^{2-} 形式存在于土壤溶液中（Saha et al.，2017）。根据硒的溶解性、移动性及有效性的大小，土壤硒又可以分为不同形态。目前有关硒的形态划分多为操作性定义，即以不同的提取剂，或者按与土壤组分的结合方式进行的划分（朱永官，2018）。较常见的形态分类为水溶态、可交换态、铁锰氧化物结合态、有机结合态和残渣态。水溶性硒和交换性硒在土壤溶液中游离存在，或吸附在土壤颗粒表面，吸附力较弱，因此被认为是可被植物和土壤微生物利用的生物有效态硒。有机结合硒吸附于土壤有机质中，生物利用度普遍较低，但这部分硒在土壤溶液中可以缓慢释放转化成生物有效硒。铁锰氧化物结合硒和残渣态硒，被铁铝氧化物等土壤成分强力固定，难以被植物吸收利用（Li et al.，2017）。

硒的生物有效性是一个没有通用定义的概念，为了准确评估土壤中硒的生物有效性，研究人员致力于建立量化土壤硒生物有效性的方法，包括化学浸提法和 DGT 测定法（周菲等，2022）。化学浸提法是测定土壤有效态硒的传统方法，分为单一提取法和连续提取法。单一提取法利用水、KH_2PO_4、$CaCl_2$ 及 EDTA 等作为浸提剂，萃取得到的上清液硒浓度定义为土壤生物有效硒（Zhao et al.，2005）。连续提取法则使用多种选择性浸提剂将土壤中不同结合形态的硒依次提取出来再进行分析评价。化学浸提法受到浸提剂选择、浸提时间等因素的影响，存在一定的局限性。Peng 等（2017）将植物吸收生物有效态硒定义为在一定时间内通过植物细胞膜从土壤中自由获取硒的过程，并提出利用梯度薄膜扩散技术（diffusive gradients in thin-films，DGT）测定土壤硒的生物有效性。DGT（图 5-2）是一种模拟植物细胞膜吸收过程的方法，它主要由滤膜、扩散膜和吸附膜组成。相比化学浸提法，DGT 测定法可以模拟植物的吸收过程，更准确地反映植物对金属的吸收，是目前较为先进的土壤硒生物有效性评价方法（周菲等，2022）。

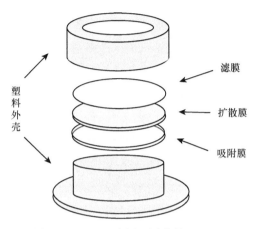

图 5-2　DGT 示意图（周菲等，2022）

硒的生物有效性受到多种因素影响，土壤性质是影响土壤硒元素生物有效性的关键因素。土壤 pH 和氧化还原电位等可以影响土壤中硒的化学形态，从而影响土壤硒的溶解度，例如，在酸性土壤中，硒与含铁或铝的矿物质或有机质结合形成稳定化合物，从而降低了硒的溶解度，进而降低硒的有效性，提高土壤 pH 可以有效地提高土壤硒的生物有效性。还原条件有利于元素硒和硒化物的形成，氧化环境下土壤中的硒以硒酸盐、亚硒酸盐为主，因此 Eh 的降低会导致土壤中硒的有效性降低（Natasha et al.，2018）。土壤黏土和有机质含量也是重要的影响因素，硒对黏土矿物具有较高的亲和性，因此在黏土矿土中其生物有效性较低。有机质对硒生物有效性的影响较为复杂，一方面，低分子质量有机酸能溶解和释放固定在土壤固相上的硒，提高硒的生物有效性，土壤有机质还可以通过产生表面负电荷来提高硒的有效性；另一方面，土壤中的高分子质量有机酸可以与水溶态硒复合或螯合，降低硒的有效性（Dinh et al.，2018）。土壤中聚居的大量微生物，包括细菌、放线菌、真菌等，参与土壤中硒元素的迁移转化，如微生物可将有机结合态硒通过矿化过程转化为无机硒（Wang et al.，2012），改变土壤硒的生物有效性。除了土壤性质，植物、气候条件和不同农艺措施等因素也在不同程度上影响着硒的生物有效性（Dinh et al.，2018）。硒生物营养强化过程中，通过调节土壤理化性质，如施加硒肥的同时施加少量石灰（提高酸性土壤 pH）、提高土壤硒的生物有效性，可进一步提高硒生物营养强化效率。

土壤中人为添加的硒和土壤固有的硒在有效性方面有着明显的差别（朱永官，2018），利用硒的固/液分配系数 K_d（K_d=土壤残留 Se/水提取态 Se，单位 L/kg）研究了在性质方面有较大差异的 26 种土壤样品，结果表明，背景硒的 K_d 值大于添加硒（Tolu et al.，2014），即添加硒的有效性较高。这种差异与原始地质阶段硒的含量无关，而与硒在土壤中的缓慢持留作用有关。环境条件的变化调控有机-无机复合体的聚合程度，从而改变着硒的移动性和有效性。老化亦改变着土壤中硒形态的分配和有效性（Tullo et al., 2016; Li et al., 2016）。

二、植物对硒的吸收、转运与代谢

硒生物营养强化的对象是植物，了解植物对硒的吸收、转运与代谢过程，通过对其中关键酶及转运蛋白的调节，可实现植物对硒的高吸收及高积累，从而达到硒生物营养强化的目的。

（一）常见作物硒含量及形态

硒虽然不是作物必需元素，但对作物的生理特征存在显著影响，主要体现在调节作物光合作用、提高作物抗逆性及作物生理调控三个方面（Elkelish et al., 2019; Semenova et al., 2017）。以小麦为例，Rady 等（2020）发现施用外源硒显著提高了小麦叶片组织的肉质性、叶绿素含量、光合效率等。在干旱胁迫、高温、低温等逆境条件下，作物体内会产生大量的游离自由基。兰敏等（2018）研究表明硒能够增加抗氧化酶 SOD、POD 活性，降低体内活性氧含量和膜脂的过氧化损伤，缓解干旱胁迫对小麦幼苗的不良影响。Sattar 等（2017）的研究也表明外源施硒在干旱条件下能够改善小麦水分利用效率、提高叶绿素含量、清除叶片 ROS。

不同植物对土壤中硒的吸收存在一定的差异（朱永官，2018），农作物硒含量一般在 0.05~1.50 mg/kg，大部分低于 30 mg/kg。在北欧的一些地区，由于硒易于从空气沉降而被植物吸收，苔藓中硒含量变化范围为 470~590 μg/kg（干重）。无根蘑菇美味牛肝菌 *Boletus edulis* 也可从空气沉降中高度富集硒，吸收量可达 17 mg/kg（干重）。饲料植物的研究表明，多数国家草本植物中硒含量范围为 2~174 μg/kg（干重，平均值 33 μg/kg），三叶草和紫苜蓿为 5~880 μg/kg（干重，平均值 99 μg/kg），其他饲料植物为 4~870 μg/kg（干重，平均值 67 μg/kg）。农产品中平均硒含量不超过 100 μg/kg（干重）。对全球市场上包括中国、美国、印度和埃及等国家的 1092 个稻米（主要为精米）样品硒含量的分析发现，精米中硒的含量范围为 2~1370 ng/g（干重）（表 5-1）。

表 5-1 全球市场上不同国家产精米中硒含量的描述性统计（Williams et al., 2009）

国家	精米中硒含量/（ng/g）				占全球份额	
	中位数	平均值	最小值-最大值	样本数	精米总产量/%	精米出口总量/%
中国	65	88	2~1370	523	28.53	3.1
埃及	6	9	6~87	102	1.03	2.9
法国	70	94	53~241	13	0.01	—
加纳	66	92	21~254	45	0.04	—
印度	141	157	35~371	50	21.71	16.5
意大利	59	66	32~158	25	0.23	2.3
日本	50	55	26~109	26	1.69	—
菲律宾	86	101	56~241	18	2.46	—
西班牙	47	48	6~104	57	0.11	1.2

续表

国家	精米中硒含量/（ng/g）				占全球份额	
	中位数	平均值	最小值-最大值	样本数	精米总产量/%	精米出口总量/%
泰国	58	96	6～487	72	4.28	34.5
美国	176	180	6～406	161	1.38	10.6
总计	67	95	2～1370	1092	61.47	71.10

注："—"表示未收集到相关数据。

在水稻体内，硒的主要存在形态以硒代甲硫氨酸（SeMet）和硒代胱氨酸（SeCysCysSe）为主，其中硒代甲硫氨酸占总硒含量的76.2%～94.3%；小麦中硒主要以硒代甲硫氨酸（SeMet）、硒代胱氨酸（SeCysCysSe）及六价硒形态存在，其中硒代甲硫氨酸占总硒含量的68.8%～92.7%，六价硒形态占总硒含量的0～16.2%；马铃薯中硒的主要存在形态为硒代胱氨酸（SeCysCysSe）、硒代甲硫氨酸（SeMet）、六价硒及甲基硒代半胱氨酸（MeSeCys），硒代甲硫氨酸占总硒含量的39.0%～85.2%，六价硒形态占比较高，为0～47.5%。而在大部分作物中，硒常以硒代甲硫氨酸（SeMet）形态存在，硒富集能力较强的作物种类会将硒代甲硫氨酸（SeMet）转变为甲基硒代半胱氨酸（MeSeCys）（张泽洲，2019）。一般来说，无机硒对人类健康的毒性大，而SeMet和SeCys更安全（Hatfield et al.，2014）。

（二）硒超积累植物

植物在自然环境中积累硒的能力不同，可分为三大类：①普通植物（非积累植物），即累积硒浓度小于100 μg/kg（干重）的植物，包括花生、高粱等在内的农作物都属于非积累植物；②硒积累植物，如芥菜（*Brassica juncea*）和甘蓝型油菜（*Brassica napus*），硒含量可达100～1000 μg/kg（干重），在非硒和含硒土壤上均能生长；③硒超积累植物，某些植物（如十字花科植物）和豆科植物黄芪属（*Astragalus*）在富硒土壤上生长时，各器官中积累的硒含量超过1000 mg/kg，硒含量可达植物干重的0.1%～1.5%，这些植物被称为硒超积累植物（Prins et al.，2019；Lima et al.，2018）。目前国内外科学家已发现约50种硒超积累植物，分布在7科，但大部分分布在豆科、菊科和十字花科（Reynolds and Pilon-Smits，2018）。主要的硒超积累植物及其特征见表5-2。

表5-2 常见硒超积累植物（White，2016）

中文名	拉丁名	科	分布地	硒含量/（mg/kg）
双钩黄芪	*Astragalus bisulcatus*	豆科	美国柯林斯堡，美国新墨西哥州，古巴	13 685
迦南金合欢	*Acacia cana*	豆科	澳大利亚昆士兰	1 121
	Oonopsis wardii	菊科	美国纽约州奥尔巴尼	9 120
	Dieteriaca nescens	菊科	美国中西部	1 600
沙漠王羽	*Stanleya pinnata*	十字花科	美国柯林斯堡，美国犹他州，美国拉勒米市	2 490
壶瓶碎米荠	*Cardamine hupingshanensis*	十字花科	中国湖北恩施	1 965
密叶滨藜	*Atriplex confertifolia*	苋科	美国犹他州汤普森	1 734

硒超积累植物可作为原材料生产富硒食品，改善人体硒摄入量不足带来的"隐性饥饿"问题。Yuan 等（2013）于湖北恩施硒矿区发现一种硒超积累植物——壶瓶碎米荠，经检测发现其根部硒含量高达 8000 mg/kg，叶片中硒含量高达 3000 mg/kg，且壶瓶碎米荠中的硒主要以硒代胱氨酸形态存在。Cui 等（2018）研究了壶瓶碎米荠中硒的含量及形态特征，发现该种植物富集大量有机硒，可作为原材料加工食品，改善癌症和其他疾病患者健康状况。此外，硒超积累植物能够超量吸收土壤中的硒并将其转运到地上部分累积起来，可用于植物修复，降低土壤硒浓度，改善硒污染地区污染状况（Zhu et al., 2009）。

（三）植物对硒的吸收转运及代谢

土壤中的硒酸盐、亚硒酸盐或有机硒化合物［硒代半胱氨酸（SeCys）和硒代甲硫氨酸（SeMet）］等有机硒，可以被植物根部吸收。硒元素与硫元素化学性质相似，植物可以通过硫酸盐转运蛋白吸收硒酸盐并转运到不同组织。参与根细胞从根际吸收硒酸盐这一过程的关键转运蛋白是高亲和性硫酸盐转运体 AtSULTR1;1 和 AtSULTR1;2（White, 2016）。硒酸盐和硫酸盐通过相似的途径运输，硫酸盐可以通过竞争性离子吸附显著抑制硒酸盐的吸收，Deng 等（2021）研究表明，向土壤中施硫后，大豆植株对硒酸盐的吸收明显受到抑制。植物通过磷酸盐转运蛋白吸收根际土壤中的亚硒酸盐。在 OsPT2 过量表达和敲除该基因的水稻植株中，亚硒酸盐吸收率分别显著升高和降低，说明 OsPT2 在亚硒酸盐吸收过程中起着至关重要的作用（Zhang et al., 2014）。有机硒在土壤中含量较少，主要以有机小分子的形式存在，目前针对有机硒在植物体内吸收机制的研究较少，植物体内半胱氨酸和甲硫氨酸转运体可能参与了有机硒的吸收转运过程。

向土壤中施硒，当外源硒以硒酸盐为主时，硒极易被吸收并转运至地上部分；若外源硒以亚硒酸盐为主，则无机硒在植物体内主要被根部同化，形成有机硒形式，如硒代氨基酸等，而不再向上运输。因此，在植物叶片中含有大量的硒酸盐，而亚硒酸盐含量较少。SeMet 和 SeMeSeCys 仅通过韧皮部运输，而无机硒则通过韧皮部和木质部运输，SeMet 和 SeMeSeCys 比无机硒更快速地到达韧皮部并运输到籽粒中。有机硒更易在籽粒中积累，大部分无机硒则会滞留在维管束末端（Carey et al., 2012）。在硒积累植物中，在植物生长的营养阶段，硒主要存在于幼叶中；在生殖阶段，硒主要存在于种子中。在普通植物中，作物的籽粒和根中的硒含量大致相同，茎和叶中硒的含量较少（Terry et al., 2000）。对富硒稻米样品进行的定位研究发现，硒的含量在稻米的不同部位有明显不同（Williams et al., 2009）。稻米横截面外层富集了较多的硒，特别是种皮和糊粉层部分。含硒量最高的部分为合点区，并且在合点区可以很明显地看出由内向外硒含量分布逐渐增加。进一步利用 X 射线吸收近边结构（μ-XANES）技术对富硒稻米样品进行形态表征发现，稻米不同部分的硒形态不同：在稻壳和米糠中，无机硒的相对比例高于胚乳中的 10 倍。在稻壳中检测到 SeMet、SeCys 和 MeSeCys，但以 SeMet 为主。而在米糠中，有机硒只以 MeSeCys 形态存在，占其总硒量的 47%。胚乳中无机硒几乎很少，有机硒占其总硒量的 94.5%，这些有机硒中 55% 是 MeSeCys（表 5-3，图 5-3）。

表 5-3　稻米中硒形态的 μ-XANES 分析（Williams et al.，2009）

样品	硒酸盐	亚硒酸盐	SeMet	SeCys	MeSeCys	无机硒/%	拟合度（χ^2）
处理前各部分							
谷壳	14.7（1.3）	30.2（1.7）	32.7（8.1）	8.3（5.7）	14.1（5.8）	45	0.16
米糠	25.0（2.3）	28.2（2.5）			46.8（12.2）	53	0.12
胚乳		5.5（1.3）	39.2（1.2）		55.3（1.4）	5	0.07
微观分析							
横向点 1		7.8（0.9）			92.2（1.5）	8	1.11
横向点 2		8.1（0.9）			91.9（1.3）	8	0.23
横向点 3		7.5（1.4）			92.5（2.1）	8	0.27
横向点 4					100	0	0.24
纵向点 1		13.3（1.2）		6.7（0.8）	80.7（1.4）	13	0.31
纵向点 2		3.7（0.6）			96.3（1.5）	4	0.33
纵向点 3		7.2（1.1）			92.8（1.3）	7	0.33
纵向点 4		14.4（1.7）			85.6（2.3）	14	0.41

注：括号中为正负误差，99 个数据点。

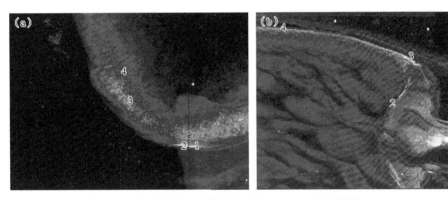

图 5-3　中国水稻硒元素定位的 X 射线荧光光谱（μ-XRF）元素图（Williams et al.，2009）
(a) 横断面；(b) 纵断面标签 1~4 表示由 μ-XANES 表征的硒物种形成的区域

植物中已知的主要硒代谢过程是硒酸盐的还原、硒代氨基酸的合成以及挥发性硒的形成（陈松灿等，2014）。硒酸盐通过 ATP 硫酰化酶（APS）被活化为腺苷磷酸硒酸盐（APSe），腺苷磷酸硒盐在 5′-磷硫酸腺苷还原酶（APR）的催化作用下，还原产生亚硒酸盐。研究证实将拟南芥编码 APR 的基因敲除后，突变体植株内硒酸盐含量明显增加，而亚硒酸盐则减少（Grant et al.，2011）。当硒酸盐或亚硒酸盐在植物体内积累时，可引起氧化应激，植物快速地将亚硒酸盐还原为硒化物，硒化物可以与 O-乙酰丝氨酸结合形成硒代半胱氨酸（SeCys）。当硒代半胱氨酸非特异性地结合到蛋白质中时，会破坏蛋白质的功能。植物通过酶促反应将硒代半胱氨酸转化为硒代甲硫氨酸（SeMet），避免其累积产生毒害作用，SeMet 可进一步转化为挥发性二甲基硒化物（DMeSe）。一些植物还通过 SeCys 甲基转移酶（SMT）将 SeCys 甲基化成挥发性二甲基二硒（DMeDSe）（Reynolds and Pilon-Smits，2018）。二甲基硒化物和二甲基二硒最终会挥发到大气中。此外，植物还可以将硒代胱氨酸分解成元素硒和丙氨酸，减轻硒的毒性（Gupta and Gupta，2017）。通过对植物硒的吸收、转运与代谢过程的研究，研究人员发现参与这些过程的重要转运体（AtSULTR1;1、AtSULTR1;2 及 OsPT2 等）和酶（APS 和 SMT），可利用基因工程技术调节相关蛋白表达，从而实现作物硒生物营养强化。

第二节 硒生物营养强化的实用技术

基于对硒元素传输链条及传输规律的研究，全球科学家们进行了大量硒生物营养强化技术研究与实践，以应对硒营养缺乏带来的"隐性饥饿"问题。本节内容将讨论以下五种硒生物营养强化的技术，包括土壤强化、叶面强化、传统育种、基因工程及天然富硒区强化（图 5-4）。

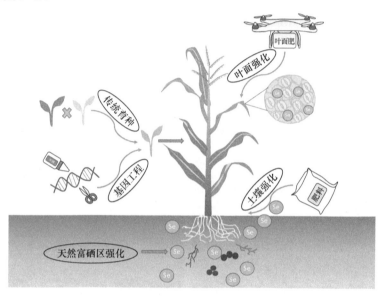

图 5-4 硒生物营养强化技术

一、土壤强化

作物吸收的硒元素主要来自于土壤，作物硒含量与土壤总硒及有效硒水平呈显著正相关。土壤硒生物营养强化，就是指通过向土壤中添加土壤改良剂，以提高土壤总硒或有效硒水平，改良土壤作物根际环境，提高农产品中硒含量水平。

在土壤总硒含量低的地区，往无机肥料中添加硒，最常见的是硒酸盐或亚硒酸盐，然后将含硒肥料直接施用于土壤，可显著提高作物硒含量。由于操作简便，土壤施加无机硒肥料是使用最为广泛的硒生物营养强化手段，在美国、英国、澳大利亚、新西兰及芬兰等地均进行了成功试验。自 2012 年以来，芬兰政府率先应用含无机硒肥料作为生物营养强化项目，硒被广泛施用于土壤中，硒用量为 10 mg/hm^2 时可显著提高农产品硒含量，现在芬兰人的每日硒摄入量达到 70～80 μg（Ebrahimi et al.，2019）；在非洲马拉维地区，政府围绕玉米作物开展了硒生物营养强化项目，将硒和 N、P、K 共同施用于玉米地，与对照组相比，每公顷施加 1 g 硒，玉米籽粒硒含量平均提高 21 μg/kg，取得良好的强化效果（Chilimba et al.，2012）（图 5-5）。同样，土壤硒生物营养强化在小麦、水稻、蘑菇及豌豆上也取得了成功（Mora et al.，2015）。王晓丽等（2020）在南宁对淮山进行硒生物营养强化试验，在块茎初期施加 250 g Se/hm² 时能显著提高淮山硒含量，达到 767 μg/kg，为对照组的 31.5 倍。

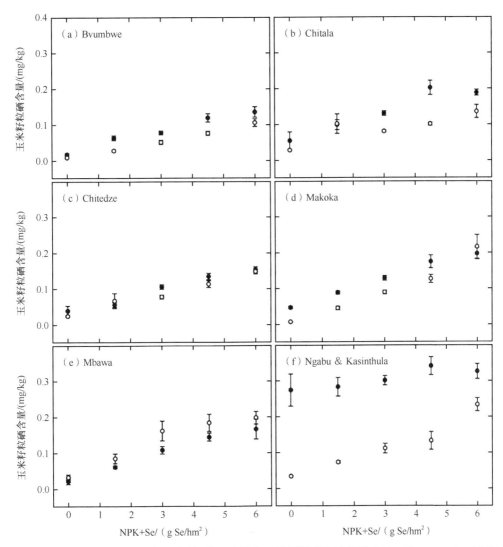

图 5-5 连续两年 NPK+Se 施肥对马拉维 6 个地点玉米籽粒硒含量的影响（Chilimba et al.，2012）
实心圆为 2008 年 9 月采样；空心圆为 2009 年 10 月采样

植物吸收无机硒后，将其中的大部分转化为 SeCys 和 SeMet（Pedrero et al.，2007；Cubadda et al.，2010）。对在 0.5 mg Se/kg 土壤上种植的水稻进行 HPLC-ICP-MS 和原位 X 射线吸收近边结构（XANES）分析，结果表明硒代甲硫氨酸是大米硒的主要存在形态，占总硒含量的 65%，其次是硒代半胱氨酸，无机硒比例较低（Li et al.，2010）。土壤施硒后，除了能显著提高作物硒含量之外，也能提高作物产量与品质。水稻硒生物营养强化试验结果显示，施加适量硒，水稻植株高度、干物质量、千粒重及单株产量均显著提高，表明低硒对植物生长有促进作用（Wang et al.，2012）。适当施用硒可以促进植物的光合作用，促进植物氨基酸和蛋白质的合成，缓解叶片衰老的不利影响（Yin et al.，2019）。

影响土壤硒生物营养强化的因素很多，主要包括施硒量、施肥时间及土壤性质。一般来说，施硒量越多，作物硒含量越高，作物硒含量和施硒量呈正比。土壤施硒量从 0.5 mg/kg 增加到 20 mg/kg，水稻地上部硒含量相应增加，施硒量最高时，糙米硒含量能达到 2.73 mg/kg（Dai et al.，2019）。但是作物硒含量并不是越高越好，该研究也指出，施加高剂量的硒会降低水稻的抗氧化能力，抑制水稻植株生长，还会导致大米中无机硒比例上升、有机硒比例下降（图5-6）。土壤性质，如 pH、有机质含量等能通过影响硒的有效性进而影响作物硒含量。张艳玲等（2002）的研究结果表明土壤中水溶态硒的含量与土壤 pH 呈极显著正相关，土壤 pH 升高，土壤中水溶态硒的含量显著增加。土壤有机质通过吸附/解吸作用影响土壤有效硒含量，研究表明，土壤有机质对硒的吸附/解吸在更大程度上与有机质的组分含量相关，富里酸含量高时，土壤硒元素的有效性高；胡敏酸含量高时，土壤有效硒含量低；土壤有效硒含量越高，作物硒含量越高。此外，施肥时期也很重要，为了让作物高效吸收施加的外源硒，需要根据作物生长发育规律，选择合适的生育期施加硒肥。宋佳平等（2020）利用硒肥对'桂淮7号'和'紫玉淮山'两个品种进行硒生物营养强化研究发现，在块茎初期添加硒肥（100 g Se/hm^2），效果优于其他施肥时期，淮山硒含量可达（193.4±21.3）μg/kg，为对照组硒含量的 4.5 倍。

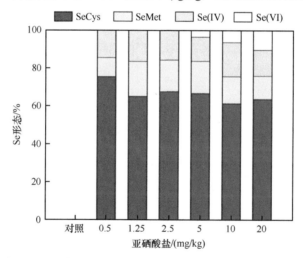

图 5-6　土壤施硒对糙米硒形态的影响（Dai et al.，2019）

除了土壤施硒外，土壤微生物也能显著促进作物对硒的吸收。土壤中的微生物，如丛枝菌根真菌（AMF）、假单胞菌、芽孢杆菌、根瘤菌等，对于提高植物对硒的生物营养强化利用率具有潜在的作用，其作用机制可归纳为以下四个方面：①微生物改变土壤性质，参与硒的氧化还原反应，提高其在土壤中的生物有效性；②有益微生物通过释放某些分泌物促进植物硒吸收，调节根系形态，促进植物发育；③微生物上调了植物硒代谢相关基因和蛋白质的表达；④接种微生物可使植物产生某些有助于硒吸收的代谢物（Yang et al.，2021）。在小麦根际土壤接种智利南部筛选的耐硒窄养单胞菌、芽孢杆菌、肠杆菌和假单胞菌，与未接种小麦相比，接种小麦的根、叶中硒含量均显著提高。这些微生物能将无机硒还原为纳米硒和其他有机硒，并携带转移到小麦植株的叶片中（Acua et al.，2013）。同样，Durán 等（2013）的研究评估了硒细菌和 AMF 在小麦植株硒吸收、转运过程中的协同作用，研究结果表明，硒细菌与AMF混合接种可显著提高籽粒硒含量，AMF有利于真菌菌丝附近细菌菌群的生长，加速其代谢活性和养分循环，并影响植物对磷、类金属和重金属等元素的获取。

二、叶面强化

叶面强化是指通过叶面肥料，将外源硒稀释后喷洒在作物叶面，使作物硒含量提高的硒生物营养强化方式。与土壤施硒类似，叶面施硒是另一重要农业硒生物营养强化技术，被广泛应用于硒生物强化相关研究及富硒农产品开发。在水稻、马铃薯、大豆、胡萝卜和小麦（Poblaciones et al.，2014）等多种作物中，已成功报道了这种硒生物营养强化方法的有效性。例如，杨舒添等（2020）研究发现60 g/hm² 的叶面喷硒处理能分别使4个品种玉米籽粒硒含量提高146.09 µg/kg、171.05 µg/kg、231.56 µg/kg和69.77 µg/kg（表5-4）。

表5-4 品种、浓度不同处理组合的籽粒硒含量

处理		硒含量/(µg/kg)			硒平均含量/(µg/kg)	差异显著性	
		I	II	III		0.05	0.01
JN18	CK	52.5436	57.8449	56.6381	55.6755	m	N
	Se15	81.2652	88.2677	78.1913	82.5747	kl	KL
	Se30	115.9222	122.4530	128.4236	122.2663	j	J
	Se60	206.8584	195.2252	203.2031	201.7622	g	G
	Se120	397.9171	408.2791	412.9172	406.3711	c	C
JN20	CK	52.1781	63.0341	62.7255	59.3126	m	N
	Se15	88.0098	90.9425	94.9402	91.2975	k	K
	Se30	142.6160	134.7950	144.8740	140.7617	i	I
	Se60	227.1170	223.2380	240.7243	230.4598	f	F
	Se120	471.7990	466.7050	475.7584	471.4208	b	B
CTN1965	CK	74.8007	69.8241	76.8990	73.8413	l	LM
	Se15	116.6499	117.1902	123.5731	119.1377	j	J
	Se30	162.1538	168.5851	173.0592	167.9237	h	H
	Se60	301.4582	303.6181	311.1160	305.3974	e	E
	Se120	508.7759	520.9003	523.3904	517.6889	a	A
JDN41	CK	61.8650	66.5550	56.3001	61.5734	m	MN
	Se15	80.7225	84.1943	80.5510	81.8226	Kl	KL
	Se30	108.1684	115.6622	118.9321	114.2542	j	J
	Se60	153.7345	166.6039	163.6930	161.3438	h	H
	Se120	318.3698	325.6980	330.0211	324.6963	d	D

注：不同大、小写字母分别表示差异极显著（$P<0.01$）和显著（$P<0.05$）。

叶面施肥是利用叶片角质层的渗透性、角质层裂缝和气孔的吸收功能将吸收进来的硒经过各种生物化学变化，合成为复杂的有机硒，再输送到植株各器官。因此，通过叶

面施硒与土壤施硒类似，获得的富硒作物中的硒以有机硒为主，无机硒比例较低。例如，在水稻、大豆和甘薯叶面喷施 40 mg/L 亚硒酸钠溶液后，SeMet 是富硒粮食作物中主要的硒形态，占总硒含量的 77%～90%（Wei et al.，2021）。

影响叶面施硒生物营养强化效果的因素包括施硒量、作物类型及气候条件等。和土壤施硒类似，叶面施硒量越高，硒生物营养强化效果越好。因为叶面施肥主要作用于叶面，所以作物叶表面积越大，叶面硒生物营养强化效果越好，因此，水稻、玉米及各种果树常施用叶面硒肥。此外，气候条件对叶面肥肥效有较大的影响，雨水及风会加速叶面硒肥的流失、降低肥效，不应在下雨或刮风的天气下喷洒含硒叶面肥。

直接施用于土壤的硒约有 12% 能积累于小麦籽粒，当叶面施硒时，这个比例更高，可达到 17.3%（Broadley et al.，2010）。紫粒小麦品种（202w17）和普通小麦品种（山农129）的田间研究表明（Xia et al.，2020），土壤和叶面施硒均提高了两个小麦品种根系、地上部和籽粒中的硒含量，叶面施硒时，两个品种籽粒中硒含量较高（图 5-7）。这主要是由于根际施加的硒进入土壤后，容易被土壤有机质固定，从有效态硒转变为不能被植物根系吸收的硒结合形态。叶面施肥能避免养分在土壤中固定或转化，直接供给作物吸收，提高了硒的利用率。

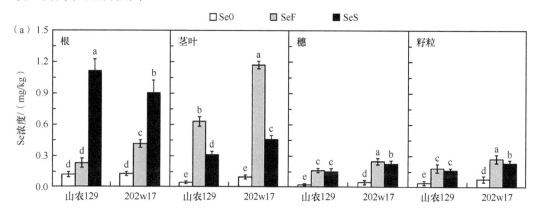

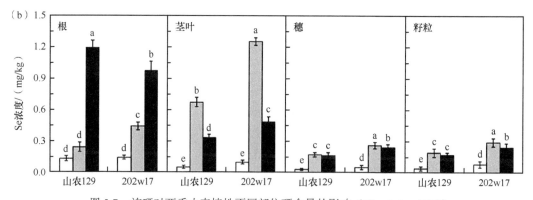

图 5-7　施硒对两季小麦植株不同部位硒含量的影响（Xia et al.，2020）

(a) 2017～2018 年；(b) 2018～2019 年。Se0，对照组；SeF，叶面施硒；SeS，土壤施硒

三、天然富硒区自然富硒

土壤中硒的分布极不均匀。例如，中国存在横跨东北到西南的马鞍形土壤缺硒带，

有72%国土为缺硒地区；但同时也存在安徽石台、陕西安康等多个富硒区，以及湖北恩施等高硒区。安徽石台表层土壤平均硒含量为0.56 mg/kg，硒含量最高达54.20 mg/kg，富硒区域（硒含量高于0.4 mg/kg）占全县面积的1/3（夏琼等，2017）。陕西安康被称为"中国硒谷"，土壤平均硒含量为0.57 mg/kg，含硒量0.2～3.0 mg/kg的土壤面积约占总面积的51.90%（王浩东等，2013）。湖北恩施作为"世界硒都"，不仅拥有范围广的富硒土壤资源（富硒地区面积达到5990 km^2），而且存在异常高硒土壤，如渔塘坝地区土壤硒含量高达2018 mg/kg（朱建明等，2008）。

Long等（2020）在恩施鱼塘坝村的研究显示，该地耕地土壤样品总硒平均含量为1.75 mg/kg，对应生产的大米和玉米平均硒含量分别达到62.90 μg/kg和320.30 μg/kg，均达到富硒农产品硒含量标准。廖彪和金华峰（2020）在陕西安康紫阳县汉王镇对可耕作土壤及所生长的农作物进行硒含量检测分析，发现该地区55.25%的土壤样品达到富硒标准（大于0.4 mg/kg），农作物平均硒含量为0.28 mg/kg，34.64%的样品达到陕西省富硒标准规定。廖彪和张振（2021）对安康市农产品硒含量做了专项调查，发现安康市农产品中达到富硒标准（DB61/T556—2018）的有572个，占检测样本的62.93%，含硒的农产品共计266个，占总量的29.26%，其中达到富硒标准的以粮食类及蔬菜水果为主，粮食类占富硒总样品的59.62%，蔬菜水果类占17.31%。在恩施种植的茶叶平均硒含量为1.07mg/kg（刘芳，2016），在紫阳种植的富硒茶平均硒含量为0.65 mg/kg，达到了富硒茶硒含量的标准要求。农业开发中利用这些含硒土壤及水资源可增加植物组织中硒的积累，因此，利用天然富含硒的土壤种植食物可能是一种可行的、低成本的自然生物强化农艺策略（Winkel et al.，2015）

天然硒生物营养强化具有成本低（不需外源硒添加）、环境友好等特点，但是实施起来也面临很多问题。首先，中国是一个缺硒大国，已查明的富硒土壤资源较少，因此有必要进行大面积的地质普查，以发掘更多富硒土壤资源。其次，由于作物硒吸收能力及土壤硒浓度存在差异，天然富硒区土壤种植的作物硒含量不稳定，对不能达到富硒农产品标准的作物，需要结合土壤或叶面施肥强化，使其硒含量达标；对于硒含量超标的农作物可进行深加工，制成富硒补充剂或混合饲料，避免直接食用。此外，天然富硒区常发现重金属（镉、汞等）伴生现象，为此，需要进一步研究富硒降镉和富硒降汞等技术，以降低农作物重金属超标的风险（臧华伟等，2021）。

四、传统生物育种

农作物生物育种是以转基因技术为核心，融合了分子标记、杂交选育等常规手段的先进技术（黄大昉，2013）。传统的生物育种方法有引种与选择育种、杂交育种、回交育种、诱变育种、远缘杂交育种、倍性育种、杂种优势利用和群体改良等8种方法；现代育种技术主要有植物细胞工程、转基因育种和分子标记辅助选择育种等3种（席章营等，2016）。生物育种技术已被应用于硒生物强化领域（Ramalho et al.，2020），通过改变作物基因及其组成以获得具有优良性状的作物品种，提高作物可食部分硒含量。

传统生物育种方法，在水稻、小麦、玉米、胡萝卜等作物的有益元素（Fe、I、Zn）强化上发挥重要作用（Cakmak，2008；Garcia-Casal et al.，2017）。在富硒作物培育方面，中国河北省农林科学院通过杂交选育的方式，选育出黑小麦'冀资麦 16 号'，其硒含量为 0.0492 mg/kg，相较亲本'冀紫 439'硒含量增幅约为 208%，与普通白粒小麦品种'冀麦 5265'相比，硒含量增幅为 556%（表 5-5）；在 2016 年实际生产过程中，产量较对照'冀紫 439'增产 7.5%，稳产性较好（吕亮杰等，2020）。中国农业科学院油料作物研究所利用聚合杂交、小孢子培育、分子标记辅助选择等育种手段选育了全国首个高效蔬菜——杂交油菜新品种'硒滋圆 1 号'，该品种油菜可以在非富硒土壤中生产出硒含量高于 0.01 mg/kg（鲜重）的富硒油菜薹（刘晟等，2020）。传统生物育种强化得到的富硒作物，适宜在世界范围内进行推广应用，以克服发展中国家生产技术及设备落后等问题。然而，大部分作物富硒性状不显著，导致传统生物育种方法缺少优质种质资源。通过在全球种质库搜索作物硒浓度遗传变异植株，可为富硒作物育种提供优质的种质资源。田间试验发现野生小麦（*Triticum dicoccum* 和 *Triticum spelta*）及其亲族小麦（*Aegilops tauschii*）的硒含量明显高于栽培小麦，可以纳入育种计划中（White and Broadley，2009）。俄罗斯的另一项研究也表明，商品小麦不同品种积累硒的能力存在差异（Seregina et al.，2001）。通过确定小麦籽粒硒含量性状是否存在显著的遗传变异，可对这一性状进行杂交育种，培育富硒小麦品种。杜前进等（2009）通过水稻品种的筛选，从 14 个优良水稻品种中筛选出 3 个硒生物富集系数大于 2 的水稻品种，可作为优质种质资源，结合各自品种的农艺性状，列入今后育种计划中，通过后续筛选能够适应环境变化的育种材料，为今后水稻硒生物强化中杂交育种等常规育种方法提供优质的种质资源。

表 5-5 '冀资麦 16 号'的 5 种矿质元素含量（吕亮杰等，2020）

矿质元素	含量/（mg/kg）			较对照增幅/%	
	冀资麦 16 号	冀紫 439	冀麦 5265	冀紫 439	冀麦 5265
Ca	854	247	340	247.75**	151.18**
Mg	1180	1020	1302	84.31**	44.39**
Fe	40.4	37.2	30.1	8.60*	34.22**
Se	0.0492	0.016	0.0075	207.50**	556.00**
Zn	30.4	27.4	23.3	10.95*	30.47**

注：'冀紫 439'、'冀麦 5265'为对照品种；* 和 ** 分别表示在 0.05、0.01 水平上差异显著

传统生物育种技术的应用，能够有效提高作物硒积累量。但生物育种技术不仅仅受到遗传效应的影响，还会受到种植方式、密度、外界因素等其他因素影响，导致作物中硒含量的变化。首先，作物富硒这一农艺性状的转移很容易受到种间生殖隔离的限制，筛选出的不同作物种间的优良基因难以相互利用。其次，利用有性杂交转移基因的成功与否，一般需要依据硒含量测定来判断，检出效率易受环境因素的影响。此外，用传统育种所获得的富硒特性，易受作物品种和地域环境的影响。例如，在北方培育的富硒作物品种，种植到南方，可能富硒特性就表现不明显。因此，在今后的生物育种硒营养强化研究中，需加强外界环境因素如温度、降水量、土壤 pH 等对作物富硒特性的长期影响研究，以筛选出普适性富硒作物品种。同时，可通过田间试验优化富硒作物栽培技术，形成特定富硒作物品种的标准化生产技术规程。

五、基因工程

基因工程（genetic engineering）又称基因拼接技术或 DNA 重组技术，是指利用分子生物学和微生物学等现代方法，将不同来源的基因按照预先设计的蓝图，在体外构建杂种 DNA 分子，然后将构建的杂种 DNA 分子导入体细胞中，改变生物原有的遗传特性、获得新品种的遗传技术，是提高作物产量和质量的有效途径（Baltes et al., 2017）。

基因工程已被证明可以提高植物对硒的积累能力，目前的研究主要集中在调节与硒相关的酶表达上。ATP 硫酰化酶（APS）是硒酸盐向亚硒酸盐转变的关键酶，导致硒酸盐还原增强。将编码 APS 的基因 *APS1* 导入印度芥菜（*Brassica juncea*）中过量表达，与野生型对照相比，转基因印度芥菜植物地上部硒积累量增加 2～3 倍，根部硒积累量增加 1.5 倍，且转基因植物中硒形态以有机硒为主，而野生型对照积累了较高比例的无机硒（Pilon-Smits et al., 1999）。在拟南芥（*Arabidopsis thaliana*）中导入双钩黄芪（*Astragalus bisulcatus*）硒代半胱氨酸甲基转移酶（SMT）基因，增加了植物茎秆中硒-甲基硒代半胱氨酸和 γ-谷氨酰基硒半胱氨酸的表达，导致硒积累增加（Pilon-Smits and LeDuc, 2009）。LeDuc 等（2006）通过单转基因植株杂交，开发了同时过量表达 SMT 和 APS 基因的双转基因印度芥菜，试验结果表明，相比野生对照，双转基因植株中硒积累量显著提高 9 倍。中国水稻研究所种质创新课题组构建了对籽粒高锌、高硒、低镉积累有显著贡献的主要数量性状基因座（qtl）$CSSL^{GZC6}$、$CSSL^{GSC5}$ 和 $CSSL^{GCC7}$，以低镉替换系 $CSSL^{GCC7}$ 为核心材料，与高锌稻米替换系 $CSSL^{GZC6}$ 和高硒稻米替换系 $CSSL^{GSC5}$ 分别进行杂交，获得了高锌低镉水稻品种 $CSSL^{GCC7+GZC6}$ 及高硒低镉水稻品种 $CSSL^{GCC7+GSC5}$（图 5-8），在实际生产过程中具有广泛的应用前景（Liu et al., 2020）。

除植物体内的酶以外，硫酸盐转运体可能是基因工程的潜在靶点，来自硒超积累植物的硒酸盐转运体将是潜在的研究重点。基因工程还可以通过提高 β-胡萝卜素、维生素 C 和富含半胱氨酸的蛋白质等物质的表达，促进植物对微量营养素的吸收，最终提高植物中矿物质元素的含量（Usmani et al., 2019）

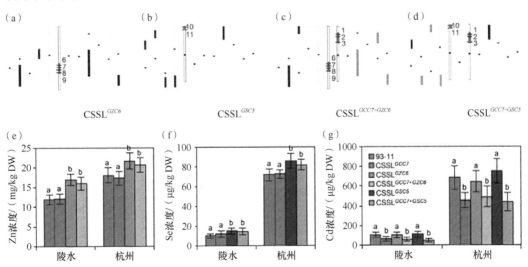

图 5-8 高锌低镉 CSSL$^{GCC7+GZC6}$ 及高硒低镉 CSSL$^{GCC7+GSC5}$ 水稻品种开发

(a)~(d) CSSL^{GZC6}、CSSL^{GSC5}、CSSL$^{GCC7+GZC6}$ 及 CSSL$^{GCC7+GSC5}$ 遗传背景;(e)~(g) 亲本及杂交水稻籽粒种 Zn、Se 和 Cd 的浓度,这些植物分别在陵水(2018 年)和杭州(2019 年)的稻田中种植,数据为均值 SD (n=6),不同字母表示 Tukey 检验的显著性水平为 5%;(h) 亲本及杂交水稻的成熟期植株形态比较

传统的育种方法存在盲目性较大、效率低等缺陷,随着分子生物学的发展,通过基因工程能够很有效地解决上述缺陷。基因工程不仅能够避免表现型的盲目性、提高效率,还能够缩短试验年限,克服远缘杂交不亲和性,从而更好地实现富硒农作物的培育与生产,促进富硒功能农业在基因层面的发展。然而,实施基因生物强化项目所需要的技术水平是许多国家所缺乏的。仅基于基因工程生物强化的策略在很大程度上依赖于每个地区农业土壤中的硒库,以及土壤中硒的生物有效性。如果土壤硒水平极低,种植的转基因作物也难以富集硒元素。此外,转基因食品的市场接受度较低,转基因作物的环境安全性和食用安全性有待验证。

第三节 硒生物营养强化案例研究

一、芬兰的硒生物营养强化

芬兰位于欧洲北部,土壤 pH 低、黏土组分含量高、秋冬季 Eh 低,基岩和土壤硒严重缺乏(Koljonen,1975)。Yläranta(1983)发现,芬兰的黏土平均总硒含量为 290 μg/kg,粗矿物土壤为 172 μg/kg,有机土壤为 464 μg/kg,土壤硒的生物有效性低。在该土壤环境中,生产的农副产品硒含量低于其他国家报告值,其中谷物硒含量≤0.01 mg/kg(干重),青饲料平均硒浓度为 0.02 mg/kg,牛肉和猪肉的平均硒浓度分别为 0.05 mg/kg 和 0.2 mg/kg,牛奶硒浓度为 0.02~0.03 mg/L(Oksanen and Sandholm,1970;Varo et al.,1980;Alfthan et al.,2015)。而居民日常硒摄入量(0.02~0.03 mg/d)(Mutanen,1984)远低于北欧(男性 0.06 mg/d,女性 0.05 mg/d)和欧盟(0.055 mg/d)的推荐摄入量,健康成年人的血浆硒浓度仅为 0.63~0.76 μmol/L(Alfthan,1988),远低于 1.0 μmol/L(满足硒蛋白 P 及谷胱甘肽过氧化物酶合成需求最低浓度)。为此,自 1984 年秋开始,芬兰官方农业部门决定在农业多营养肥料中添加硒酸钠,以提高农作物产品硒含量,其中用于粮食生产的肥料中硒含量为 16 mg/kg,用于青饲料生产的肥料中硒含量为 6 mg/kg(Yläranta,1985)。

为监测硒添加对土壤、植物、饲料和动植物源食品的影响,以及人体对硒的吸收情况，芬兰农业和林业部设立了硒工作组，对动植物和人体硒含量情况进行了长期的研究和监测，评估外源添加硒对动植物及人体的影响，并在必要时提出修改添加硒量的建议。为更好地规范生产，芬兰农业和林业部门于1993年修订了《芬兰肥料法》，1994年又颁布了相关规定，将肥料中硒的最大允许量调整为6 mg/kg，后又经过1998年及2006年两次修改，将肥料中硒的最大允许含量提高为15 mg/kg。

Thompson和Amoroso（2011）调查结果显示：经过土壤施加硒肥进行生物营养强化，2006年谷物（小麦、燕麦和大麦）的硒含量增加到1984年的20～30倍，牛肉和猪肉中硒的平均含量分别增加到原值的6倍和2倍，牛奶中硒的平均含量增加到原值的3倍，平均膳食摄入量增加到0.07 mg/d，人血浆硒平均浓度提高60%，达到最佳状态。Alfthan等（2015）发表的最新数据指出，谷物（小麦、燕麦和大麦）的硒含量在1984～2010年增加了20～30倍，牛肉和猪肉的硒浓度分别约为施硒肥前的6倍和3倍，青饲料硒的平均浓度增加为原值的10倍（达到0.2 mg/kg）。2010年芬兰居民血浆硒平均水平已达到1.4 μg/L（图5-9）。到2013年，全脂牛奶硒含量增加到0.22 mg/L，芬兰居民日平均硒摄入量增加到0.08 mg/d，芬兰农业土壤总硒含量在20年间平均增加了20%。趾甲硒含量能够反映人体6～12个月期间硒的综合摄入量（Longnecker et al.，1993），为了更好地探究外源补充硒对人体硒摄入的影响，Ovaskainen等（1993）、Virtanen等（1996）通过对成人脚趾甲硒平均含量进行检测，发现芬兰人脚趾甲硒平均含量从硒强化前的0.45 mg/kg增加到1995年的0.72 mg/kg。肝硒反映了几周内的膳食硒摄入量，对芬兰交通事故受害者肝脏中的硒浓度进行分析，结果显示1983～1985年平均肝硒含量为0.95 mg/kg，3～4年后增加到1.58 mg/kg（Alfthan et al.，2015）。

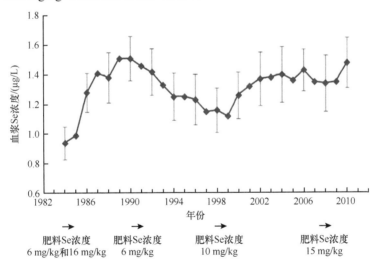

图5-9　健康芬兰人（n=60）在施硒前和施硒期间的血浆平均硒浓度（Alfthan et al.，2015）

芬兰作为第一个通过强制性在肥料中添加硒酸盐进行硒生物营养强化的国家，这一举措不仅提高了农作物和动物源食品的硒含量，而且使居民日硒摄入量也提高至推荐摄入标准，很好地解决了芬兰居民因硒日均摄入不足导致的人体健康问题。芬兰的硒生物营养强化举措是一个成功的案例，有许多亮点值得借鉴。

在技术方面，芬兰选择土壤强化，而非叶面强化，前文中提到，叶面施硒植物吸收的效率比土壤施硒效率更高，但这种方法存在一定的风险，如叶面施用硒肥时，作物对硒的吸收取决于喷施条件、植株生长阶段以及喷施期间和喷施后的气候等多个不可控因素喷施过程需要额外的操作成本，且不能保证强化后的作物硒含量均匀。不同于叶面强化，土壤强化是一种相对安全和经济的强化技术。基于芬兰常见的酸性土壤条件，为了更好地促进植物对硒的吸收，芬兰选用硒酸盐作为外源硒而不是亚硒酸盐或有机硒，因为亚硒酸盐容易在根中被同化为有机硒不再向上运输，而有机硒价格昂贵，不适宜作为外源硒添加至肥料中。根据动、植物硒含量的定量强化目标及硒在植物各器官中的分配规律，芬兰规定了肥料中硒的上限使用量，在作物和青饲料所使用的肥料中添加不同含量的硒，并成立专门的工作组，进行长期的监控，适时调整肥料中硒的上限使用量。因此，在制定硒强化方案时，因地制宜、因时制宜就显得尤为重要。依据硒元素的转运吸收规律和农作物中硒含量的定量强化需求，进行逆向调控设计，能够达到良好的硒的生物营养强化效果。

在管理方面，政府层面颁布强制性规定；科研机构对农作物、动物及人体组织硒含量进行定期检测，建立长期且全面的监测系统，评价硒肥施加的潜在影响；企业层面依据规定生产符合标准的产品，三方联动，相互配合，稳步推进，是一个典型的"政产学研"结合的案例。

二、中国宁夏硒砂瓜的硒生物营养强化

中国科学技术大学尹雪斌博士研究组一直进行以硒生物营养强化为支撑的功能农业技术研究，在江苏、陕西、湖北、河北、安徽等15个省份开展了30余个大宗品种（水稻、小麦、茶、果蔬、食用菌、水产、禽蛋等）的定量硒营养强化技术实践，示范推广面积超过200万亩[①]。

研究组对宁夏中卫市沙坡头区、中宁县及海原县三个地区的耕作土壤进行硒素资源调查，并绘制了宁夏中卫市土壤硒元素分布图，其中，富硒土壤（0.222 mg/kg以上）占比仅为16.1%，足硒土壤（0.175～0.222 mg/kg）占比25.9%，低硒土壤（0.116～0.175 mg/kg）占比37.9%，缺硒土壤（0.116 mg/kg以下）占比20.1%（参考宁夏富硒土壤标准DB64/T 1220—2016）。研究区域土壤硒含量较低，83.9%的土壤为非富硒土壤。中卫硒砂瓜产业是宁夏地区极具区域性优势的特色产业，也是中部干旱地区产业扶贫的成功范例，但是由于土壤硒含量低，种植的硒砂瓜硒含量不稳定，难以达到富硒农产品标准。为进一步巩固提升硒砂瓜品质，研究组以中卫市万亩产区建设为载体，开展硒砂瓜硒生物营养强化研究与示范。

采取土壤施加硒肥方式进行硒砂瓜的硒生物营养强化。硒肥采用苏州硒谷科技有限公司提供的富硒硒砂瓜专用有机硒肥（固体粉末态），施肥用量以每亩硒砂瓜植株可吸收硒12.04～16.05 mg计。其中，富硒区用量12.04 mg/亩，足硒区用量13.27 mg/亩，低硒区用量14.85 mg/亩，贫硒区用量16.05 mg/亩。

2018年7月，于中卫市众赢果蔬专业合作社的富硒硒砂瓜种植基地（10 000亩）采集成熟硒砂瓜样品39个（包含1个未施肥样品和38个施肥样品），检测总硒和硒

---

① 1亩≈666.67m²。

形态（表5-6），结果显示：未施用硒肥强化的对照硒砂瓜总硒含量为 3.5 μg/kg，有机硒占比 97.10%；经硒肥强化的成熟硒砂瓜总硒含量范围为 4.0~321.5 μg/kg，中位值为 53.25 μg/kg，平均硒含量为 84.10μg/kg。对成熟硒砂瓜中的硒存在形态分析结果显示：成熟硒砂瓜中未检出无机硒（包括+4硒、+6硒），有机硒形态以硒代甲硫氨酸为主，占比达到 69.10%~100%，中位值为 82.60%，平均有机硒占比为 84.25%。部分硒以未知形态存在。和不施肥的硒砂瓜相比，施肥后的硒砂瓜硒含量大幅提高。参考宁夏中卫富硒硒砂瓜合格标准为：硒砂瓜含硒量≥10 μg/kg，仅有两个样品硒含量不达标，经施肥后的样品富硒达标率为 94.7%。

表 5-6　中卫市富硒硒砂瓜总硒及硒形态

处理	总硒含量/（μg/kg，鲜重）			有机硒比例/%		
	范围	平均值	中位值	范围	平均值	中位值
不施硒肥		3.5			97.10	
施加硒肥	4.0~321.5	84.10	53.25	69.10~100	84.25	82.60

推广硒生物营养强化能满足人们"吃出健康"的新需求，有利于农产业的转型升级，使农业由吃饭产业转为健康产业，具备"农业增效、农民增收、人民增寿"的三增效应。中卫硒砂瓜案例就是一个硒生物营养强化助力乡村振兴的成功案例。

三、天然硒生物营养强化国际合作计划

为更好地开发富硒地区硒资源，中国科学技术大学尹雪斌博士研究组联合国际硒研究学会于 2014 年提出了天然硒生物营养强化国际合作计划（Natural Selenium Biofortification Program，NBP），该计划重点关注天然富硒区硒资源分布及硒在食物链中的传递规律，研究膳食硒摄入对人体健康水平产生的影响，寻找硒与人体健康之间的实证关系，为天然硒生物强化技术提供基础数据（龙泽东，2020）。

湖北恩施、安徽石台、江西宜春等地土壤和作物中硒含量显著高于其他区域，当地政府对硒资源调查及硒营养健康产业发展高度重视，大力支持开展相关科研项目及产业合作，具备开展 NBP 计划的基础条件。NBP 项目研究组以湖北恩施、安徽石台、江西宜春三个天然富硒区作为首批研究地，通过采集当地土壤、水和主要作物样品，分析土壤和作物的硒含量及形态组成特征，对当地居民的基本信息和膳食结构进行问卷调查，计算居民每日膳食硒摄入量，采集居民的头发和静脉血液样品，检测头发、血液硒含量及相关健康指标。

Long 等（2018）在安徽石台选择地理环境相似但土壤硒含量差距显著的大山村、仙寓村、永福村和源头村四个村落开展 NBP 研究。研究发现这四个村落土壤平均硒含量从高到低分别为 1607 μg/kg、1149 μg/kg、521 μg/kg 和 363 μg/kg，当地种植的谷物、豆类及蔬菜硒含量也呈相同下降趋势，根据膳食问卷调查计算所得的居民日硒摄入量分别为 298.4 μg、47.6 μg、40.0 μg 及 46.1 μg。发硒含量检测结果显示，居民日硒摄入量最高的大山村，发硒平均含量也达到最高水平（667.6 μg/kg）。龙泽东（2020）在湖北恩施开展 NBP 研究，恩施鱼塘坝耕地土壤样品总硒平均含量为 1.75 mg/kg，对应生产的大米和

玉米平均硒含量分别达到 62.9 μg/kg 和 320.3 μg/kg，居民每日膳食硒摄入量为 81.61 μg，达到中国营养学会推荐摄入量（60 μg/d）水平，血浆硒平均含量为 98.4 μg/kg。王晓丽等（2021）在宜春潭下村的研究显示，宜春潭下村土壤总硒含量平均值为 309.1 μg/kg，竹笋、韭菜和蒜总硒含量平均值分别为 258.4 μg/kg、101.1 μg/kg、102.6 μg/kg，居民膳食硒摄入量为 40.4 μg/d，血浆硒平均含量达到 100.8 μg/kg。以上研究结果说明，相比同区域的非富硒区，天然富硒区生产的作物及蔬菜含量较高，基本达到富硒农产品硒含量标准，居民每日膳食硒摄入量达到中国营养学会提出的成人推荐硒摄入量（60 μg/kg），血硒、发硒等人体组织硒水平较高，富硒区居民硒缺乏风险较低。

国际 NBP 计划正处于探索阶段，还有很多问题尚未阐明，研究涉及区域较少，且均为横断面调查，未来的研究中还需引入干预试验，加深相关结论说服力。目前的研究结果表明，天然富硒区土壤硒资源能够通过食物链传输到达人体，增加人体组织硒水平。天然富硒区强化技术具有操作简单、成本低廉、效果显著的特点，适宜在富硒地区进行推广。

第四节 硒生物营养强化发展展望

硒生物强化技术主要包括传统生物育种、基因工程、叶面强化、土壤强化及天然富硒区强化，在不同的地区选择适当的硒生物营养强化技术，均能取得较好的强化效果。作物硒生物营养强化是改善隐性饥饿的重要举措，相关研究在国内外得到广泛关注，富硒产品也逐渐被大家接受和认可。

关于硒生物营养强化的研究，未来应该重点关注以下几个方面。

（1）标准化：加强标准体系建设，促进富硒产业发展。

我国是富硒产品标准最多、最全面的国家，但目前来看，大部分地方或团体标准缺乏有机硒含量的规定，仅全国供销总社发布的行业标准《富硒农产品》（GH/T 1135—2017），对农产品中的硒含量及硒代氨基酸的含量占比进行了要求。因此，加强富硒农产品标准体系的建设至关重要，一方面，能够促进富硒农产品产业的有序发展；另一方面，能够划定富硒农产品硒含量及形态合理范围，保障人体生命安全。

（2）精准化：精准富硒生物强化，推进富硒产品发展。

据 2013 年中国居民膳食营养素参考摄入量报告，不同年龄段的推荐硒摄入量及可耐受最高硒摄入量均不相同，变化区间为 25～400 μg/d（程义勇，2014）。而我国富硒地区成人日硒摄入量可达到 4990 μg/d，远远超过可耐受最高摄入量。针对这一问题，在今后的发展过程中，硒生物强化应加强精准控制，对富硒产品的硒含量进行精确调控，从而满足不同年龄段人体需求，扩大富硒产品市场，推动富硒产品的发展。

（3）多元化：推进多元作物发展，扩大富硒产品种类。

目前，我国富硒产品以农作物为主，如小麦、水稻、杂粮等谷物，富硒产品市场种类单一，无法满足人们日常硒摄入量的需求。因此，未来硒生物营养强化技术的发展应面向更多的产品，而不仅仅局限于谷类作物。对于蔬菜、畜禽肉类及鱼类等产品，可以有针对性地研发硒生物强化新技术，或通过基因工程、生物育种等技术，培育或选育一批硒含量较高的蔬菜或畜禽，以满足不同膳食结构人群的需要。

（4）智慧化：加强智慧产品建设，提高智慧化生产。

随着科技的进步发展，生产加工方式的智慧化越来越普遍。未来硒产业的发展将突出智慧信息技术的应用，着力构建"从农田到餐桌"的智慧保障体系。以富硒产品信息、数据、物流、配送、仓储、冷链、质量安全溯源等核心，构建起基于"富硒农产品进销存+供应链管理+第三方电子支付+市场管理+物流配送+仓储服务"的整合应用平台，实现集标准化、信息化、数据化、智能化为一体的智慧销地新市场体系，提高我国硒产业智慧化生产。

（5）适度化：增强居民日硒摄入量监控，避免无差别补硒。

任何必需营养素与生物之间的剂量反应模型不是呈线性的，而是呈U形，包括缺乏、低优、最优、高优和毒性五个状态。硒缺乏和硒毒害作用之间的范围很狭窄，当硒摄入量充足的人群继续摄入过量硒时，也会发生硒中毒现象。因此，实施硒生物营养强化时，必须密切关注居民每日的硒摄入量情况，避免盲目补硒对人体健康产生不利影响。

（6）临床化：推进临床化研究，验证硒的人体效应。

研究者大量使用动物及细胞模型等验证了硒的有益影响，然而硒的人体研究数据较少，作为硒营养强化的出发点和目标，硒对人体健康的有益作用尚不清晰。因此，在未来的研究中，应加强科研单位、医疗机构和疾控中心的合作，进行硒营养强化富硒产品的临床研究，获取人体研究数据，以验证硒对人体健康产生的有益影响。

（7）高效化：开展影响因子研究，提高硒强化效率。

包括土壤叶面施肥等硒生物营养强化技术的发展已经较为成熟，实施起来也较为简单，然而却始终存在硒利用率不高的问题。因此，在未来的研究中，应重点探究相关因子（如土壤理化性质、农艺操作等）对硒生物营养强化的影响，以优化硒生物营养强化效果，提高硒资源的利用率。

参 考 文 献

陈松灿, 孙国新, 陈正, 等. 2014. 植物硒生理及与重金属交互的研究进展. 植物生理学报, 50 (5): 612-624.
程义勇. 2014.《中国居民膳食营养素参考摄入量》2013 修订版简介. 营养学报, 36(4): 313-317.
杜前进, 张永发, 曾宾, 等. 2009. 海南富硒地区水稻富硒品种的筛选. 中国土壤与肥料, (1): 46-49.
黄大昉. 2013. 我国农作物生物育种发展战略思考. 中国科学院院刊, 28(3): 315-321.
兰敏, 尹美强, 芦文杰, 等. 2018. 干旱胁迫下外源硒对小麦幼苗抗旱性的影响. 土壤, 50(6): 1182-1189.
廖彪, 金华峰. 2020. 紫阳县汉王镇土壤, 农作物硒含量测定及规律性研究. 湖北农业科学, 651(6): 159-161.
廖彪, 张振. 2021. 安康市农产品硒含量调查与研究. 陕西农业科学, 67(5): 82-85.
刘晟, 顿小玲, 金莉, 等. 2020. 硒高效菜用油菜杂交品种硒滋圆 1 号的适应性研究. 长江蔬菜, (18): 34-36.
刘芳. 2016. 硒都恩施与富硒茶. 农业与技术, 36(14): 154.
龙泽东. 2020. 硒在天然富硒区恩施与石台土壤-作物-人体系统中的分布特征和健康效应研究. 合肥: 中国科学技术大学博士学位论文.
吕亮杰, 孟雅宁, 张业伦, 等. 2020. 富硒富锌黑小麦冀资麦 16 号的选育及其营养成分分析. 河北农业科学, 24(2): 71-75.
宋佳平, 张泽洲, 曹升, 等. 2020. 土壤和叶面施硒对淮山硒累积的对比研究. 农业科学进展, 2(4): 134-148.
王浩东, 张建东, 谢娟平, 等. 2013. 安康地区土壤硒资源分布规律研究. 安康学院学报, 25(6): 8-11.
王晓丽, 张泽洲, 曹升, 等. 2020. 土壤施硒对淮山硒含量的影响. 宜春学院学报, 298(9): 90-94.
王晓丽, 张泽洲, 王张民, 等. 2021. 江西宜春市明月山地区土壤和多种作物中硒的含量及形态分布特征.

科学通报, 67(6): 11-29.

席章营, 陈景堂, 李卫华. 2016. 作物育种学. 北京: 科学出版社.

夏琼, 高东升, 高雅, 等. 2017. 安徽省石台县富硒土壤资源调查与评价. 安徽地质, 27(4): 314-316.

杨舒添, 杜天庆, 翟红梅, 等. 2020. 叶面喷硒对糯玉米生理特性及子粒硒含量的影响. 玉米科学, (1): 117-123.

袁丽君, 袁林喜, 尹雪斌, 等. 2016. 硒的生理功能、摄入现状与对策研究进展. 生物技术进展, 6(6): 396-405.

臧华伟, 张泽洲, 龙泽东, 等. 2021. 中国硒资源利用中若干关键科学问题的探讨. 生物技术进展, 11(4): 542-549.

张艳玲, 潘根兴, 李正文, 等. 2002. 土壤-植物系统中硒的迁移转化及低硒地区食物链中硒的调节. 生态环境学报, (4): 388-391.

张泽洲. 2019. 典型农作物中硒形态分析及其硒-镉相互作用研究. 武汉: 中国地质大学博士学位论文.

中国营养学会. 2014. 中国膳食参考摄入量 (2013 版). 北京: 科学出版社: 250-257.

周菲, 彭琴, 王敏, 等. 2022. 土壤-植物体系中硒生物有效性评价研究进展. 科学通报, 67(6): 461-472.

朱建明, 左维, 秦海波, 等. 2008. 恩施硒中毒区土壤高硒的成因: 自然硒的证据. 矿物学报, 28(4): 397-400.

朱永官. 2018. 土壤-植物系统中的硒、氟和碘及其环境行为. 见: 环境土壤学 (第 3 版), 北京: 科学出版社: 130-166.

Acua J, Jorquera M, Barra P, et al. 2013. Selenobacteria selected from the rhizosphere as a potential tool for Se biofortification of wheat crops. Biology and Fertility of Soils, 49(2): 175-185.

Alfthan G. 1988. Longitudinal study on the selenium status of healthy adults in Finland during 1975-1984. Nutrition Research, 8(5): 467-476.

Alfthan G, Eurola M, Ekholm P, et al. 2015. Effects of nationwide addition of selenium to fertilizers on foods, and animal and human health in Finland: From deficiency to optimal selenium status of the population. Journal of Trace Elements in Medicine and Biology, (31): 142-147.

AL-Ghumaiz N S, Motawei M I, Abd-Elmoniem E M, et al. 2020. Selenium and zinc concentrations in spring wheat (*Triticum aestivum*) genotypes under organic and inorganic fertilization. Journal of Plant Nutrition, 1-8.

Arakawa T, Deguchi T, Sakazaki F, et al. 2013. Supplementary seleno-L-methionine suppresses active cutaneous anaphylaxis reaction. Biological and Pharmaceutical Bulletin, 36(12): 1969-1974.

Baltes N J, Gil-Humanes J, Voytas D F. 2017. Genome engineering and agriculture: opportunities and challenges. Progress in Molecular Biology and Translational Science, 149: 1-26.

Broadley M R, Alcock J, Alford J, et al. 2010. Selenium biofortification of high-yielding winter wheat (*Triticum aestivum* L.) by liquid or granular Se fertilisation. Plant and Soil, 332(s1-2): 5-18.

Cakmak I. 2008. Enrichment of cereal grains with zinc: agronomic or genetic biofortification? Plant and Soil, 302(1): 1-17.

Carey A M, Scheckel K G, Lombi E, et al. 2012. Grain accumulation of selenium species in rice (*Oryza sativa* L.). Environmental Science and Technology, 46(10): 5557-5564.

Chilimba A, Young S D, Black C R, et al. 2012. Agronomic biofortification of maize with selenium (Se) in Malawi. Field Crops Research, 125: 118-128.

Cubadda F, Aureli F, Ciardullo S, et al. 2010. Changes in selenium speciation associated with increasing tissue concentrations of selenium in wheat grain. Journal of Agriculture and Food Chemistry, 58(4): 2295-2301.

Cui L W, Zhao J T, Chen J Y, et al. 2018. Translocation and transformation of selenium in hyperaccumulator plant *Cardamine enshiensis* from Enshi, Hubei, China. Plant and Soil, 425(1-2): 577-588.

Dai Z H, Imtiaz M, Rizwan M, et al. 2019. Dynamics of selenium uptake, speciation, and antioxidant response in rice at different panicle initiation stages. Science of the Total Environment, 691: 827-834.

Deng X, Zhao Z, Lv C, et al. 2021. Effects of sulfur application on selenium uptake and seed selenium speciation in soybean (*Glycine max* L.) grown in different soil types. Ecotoxicology and Environmental Safety, 209: 111790.

Dinh Q T, Gui Z W, Huang J, et al. 2018. Selenium distribution in the Chinese environment and its relationship with human health: a review. Environment International, 112: 294-309.

Dinh Q T, Wang M, Tran T A T, et al. 2019. Bioavailability of selenium in soil-plant system and a regulatory approach. Critical Reviews in Environmental Science and Technology, 49: 443-517.

Durán P, Acuña J, Jorquera M A, et al. 2013. Enhanced selenium content in wheat grain by co-inoculation of selenobacteria and arbuscular mycorrhizal fungi: A preliminary study as a potential Se biofortification strategy. Journal of Cereal Science, 57(3): 275-280.

Ebrahimi N, Stoddard F L, Hartikainen H, et al. 2019. Plant species and growing season weather influence the efficiency of selenium biofortification. Nutrient Cycling in Agroecosystems, 114: 11-124.

Elkelish A A, Soliman M H, Alhaithloul H A, et al. 2019. Selenium protects wheat seedlings against salt stress-mediated oxidative damage by up-regulating antioxidants and osmolytes metabolism. Plant Physiology and Biochemistry, 137: 144-153.

Garcia-Casal M N, Pena-Rosas J P, Giyose B, et al. 2017. Staple crops biofortified with increased vitamins and minerals: considerations for a public health strategy. Annals of the New York Academy of Sciences, 1390(1): 3-13.

Grant K, Carey N M, Mendoza M, et al. 2011. Adenosine 5′-phosphosulfate reductase (APR2) mutation in *Arabidopsis* implicates glutathione deficiency in selenate toxicity. Biochemical Journal, 438(2): 325-335.

Gupta M, Gupta S. 2017. An overview of selenium uptake, metabolism, and toxicity in plants. Frontiers in Plant Science, 7: 2074.

Hatfield D L, Tsuji P A, Carlson B A, et al. 2014. Selenium and selenocysteine: roles in cancer, health, and development. Trends in Biochemical Sciences, 39 (3): 112-120.

Kiran K. 2020. Advanced Approaches for Biofortification. In: Sharma T R, Deshmukh R, Sonah H. Advances in Agri-Food Biotechnology. Singapore: Springer: 29-55.

Koljonen T. 1975. The behavior of selenium in Finnish soils. Annales Academiae Scientiarum Fennicae, 14: 240-247.

LeDuc D L, AbdelSamie M, Móntes-Bayon M, et al. 2006. Overexpressing both ATP sulfurylase and selenocysteine methyltransferase enhances selenium phytoremediation traits in Indian mustard. Environmental Pollution, 144(1): 70-76.

Li H F, Lombi E, Stroud J L, et al. 2010. Selenium speciation in soil and rice: influence of water management and Se fertilization. Journal of Agricultural and Food Chemistry, 58(22): 11837-11843.

Li J, Peng Q, Liang D L, et al. 2016. Effects of aging on the fraction distribution and bioavailability of selenium in three different soils. Chemosphere, 144: 2351-2359.

Li Z, Liang D, Peng Q, et al. 2017. Interaction between selenium and soil organic matter and its impact on soil selenium bioavailability: A review. Geoderma, 295: 69-79.

Lima L W, Pilon-Smits E A H, Schiavon M. 2018. Mechanisms of selenium hyperaccumulation in plants: A survey of molecular, biochemical and ecological cues. Biochimica et Biophysica Acta (BBA)-General Subjects, 1862(11): 2343-2353.

Liu C, Ding S, Zhang A, et al. 2020. Development of nutritious rice with high zinc/selenium and low cadmium in grains through QTL pyramiding. Journal of Integrative Plant Biology, 62(3): 349-359.

Long Z, Xiang J, Song J, et al. 2020. Soil selenium concentration and residents daily dietary intake in a selenosis area: A preliminary study in Yutangba village, Enshi City, China. Bulletin of Environmental

Contamination and Toxicology, 105: 798-805.

Long Z, Yuan L, Hou Y, et al. 2018. Spatial variations in soil selenium and residential dietary selenium intake in a selenium-rich county, Shitai, Anhui, China. Journal of Trace Elements in Medicine and Biology, 50: 111-116.

Longnecker M P, Stampfer M J, Morris J S, et al. 1993. A 1-y trial of the effect of high-selenium bread on selenium concentrations in blood and toenails. The American Journal of Clinical Nutrition, 57: 408-413.

Mora M L, Durán P, Acuña J, et al. 2015. Improving selenium status in plant nutrition and quality. Journal of Soil Science and Plant Nutrition, 15(2): 486-503.

Mutanen M. 1984. Dietary intake and sources of selenium in young Finnish woman. Human Nutrition Applied Nutrition, 38(4): 265.

Natasha, Shahid M, Niazi N K, et al. 2018. A critical review of selenium biogeochemical behavior in soil-plant system with an inference to human health. Environmental Pollution, 234: 915-934.

Newman R, Waterland N, Moon Y, et al. 2019. Selenium biofortification of agricultural crops and effects on plant nutrients and bioactive compounds important for human health and disease prevention—a review. Plant Foods for Human Nutrition, 74(4): 449-460.

Oksanen H E, Sandholm M. 1970. The selenium content of Finnish forage crops. Agricultural and Food Science, 42(4): 250-253.

Ovaskainen M L, Virtamo J, Alfthan G, et al. 1993. Toenail selenium as an indicator of selenium intake among middle-aged men in a low-soil selenium area. American Journal of Clinical Nutrition, 57: 662-665.

Pedrero Z, Elvira D, Cámara C, et al. 2007. Selenium transformation studies during broccoli (*Brassica oleracea*) growing process by liquid chromatography-inductively coupled plasma mass spectrometry (LC-ICP-MS). Analytica Chimica Acta, 596(2): 251-256.

Peng Q, Wang M, Cui Z, et al. 2017. Assessment of bioavailability of selenium in different plant-soil systems by diffusive gradients in thin-films (DGT). Environmental Pollution, 225: 637-643.

Pilon-Smits E A H, Hwang S, Mel Lytle C, et al. 1999. Overexpression of ATP sulfurylase in Indian mustard leads to increased selenate uptake, reduction, and tolerance. Plant Physiology, 119(1): 123-132.

Pilon-Smits E A, LeDuc D L. 2009. Phytoremediation of selenium using transgenic plants. Current Opinion in Biotechnology, 20(2): 207-212.

Plant J, Bone J, Voulvoulis N, et al. 2014. Arsenic and Selenium. In: Holland H D, Turekian K K. Treatise on Geochemistry (2nd Ed). Oxford: Elsevier: 13-57.

Poblaciones M J, Rodrigo S, Santamaria O, et al. 2014. Agronomic selenium biofortification in *Triticum durum* under Mediterranean conditions: From grain to cooked pasta. Food Chemistry, 146: 378-384.

Prins C N, Hantzis L J, Valdez-Barillas J R, et al. 2019. Getting to the root of selenium hyperaccumulation-localization and speciation of root selenium and its effects on nematodes. Soil Systems, 3(3): 47.

Rady M O A, Semida W M, Abd El-Mageed T A, et al. 2020. Foliage applied selenium improves photosynthetic efficiency, antioxidant potential and wheat productivity under drought stress. International Journal of Agriculture and Biology, 24(5): 1293-1300.

Ramalho J C, Roda F A, Pessoa M F G, et al. 2020. Selenium agronomic biofortification in rice: improving crop quality against malnutrition. The future of rice demand: quality beyond productivity. Cham: Springer: 179-203.

Rayman M P. 2012. Selenium and human health. The Lancet, 379(9822): 1256-1268.

Reynolds R J B, Pilon-Smits E A H. 2018. Plant selenium hyperaccumulation-Ecological effects and potential implications for selenium cycling and community structure. Biochimica et Biophysica Acta (BBA)-General Subjects, 1862(11): 2372-2382.

Saha U, Fayiga A, Sonon L. 2017. Selenium in the soil-plant environment: a review. International Journal of

Applied Agricultural Sciences, 3: 1-18.

Sattar A, Cheema M A, Abbas T, et al. 2017. Separate and combined effects of silicon and selenium on salt tolerance of wheat plants. Russian Journal of Plant Physiology, 64(3): 341-348.

Schiavon M, Pilon-Smits E A H. 2017. The fascinating facets of plant selenium accumulation-biochemistry, physiology, evolution and ecology. New Phytologist, 213(4): 1582-1596.

Semenova G A, Fomina I R, Kosobryukhov A A, et al. 2017. Mesophyll cell ultrastructure of wheat leaves etiolated by lead and selenium. Journal of Plant Physiology, 219: 37-44.

Seregina I I, Nilovskaya N T, Ostapenko N O. 2001. The role of selenium in the formation of the grain yield in spring wheat. Agrokhimiya, 1: 44-50.

Shahid M, Niazi N K, Khalid S, et al. 2018. A critical review of selenium biogeochemical behavior in soil-plant system with an inference to human health. Environmental Pollution, 234: 915-934.

Terry N, Zayed A M, De Souza M P, et al. 2000. Selenium in higher plants. Annual Review of Plant Biology, 51(1): 401-432.

Thompson B, Amoroso L. 2011. Combating micronutrient deficiencies: food-based approaches. UK by the MPG Books Group.

Tolu J, Thiry Y, Bueno M, et al. 2014. Distribution and speciation of ambient selenium in contrasted soils, from mineral to organic rich. Science of the Total Environment, 479-480: 93-101.

Tullo P D, Pannier F, Thiry Y, et al. 2016. Field study of time-dependent selenium partitioning in soils using isotopically enriched stable selenite tracer. Science of the Total Environment, 562: 280-288.

Usmani Z, Kumar A, Ahirwal J, et al. 2019. Scope for Applying Transgenic Plant Technology for Remediation and Fortification of Selenium. In: Transgenic Plant Technology for Remediation of Toxic Metals and Metalloids. Chennai, India: Academic Press: 429-461.

Varo P, Nuurtamo M, Saari E, et al. 1980. Mineral element composition of Finnish foods. III. Annual variations in the mineral element composition of cereal grains. Acta Agriculturae Scandinavica Supplementum, 22: 27-35.

Virtanen S M, Veer P, Kok F, et al. 1996. Predictors of adipose tissue tocopherol and toenail selenium levels in nine countries: the EURAMIC study. European Journal of Clinical Nutrition, 50: 599-606.

Wang S S, Liang D L, Wang D, et al. 2012. Selenium fractionation and speciation in agriculture soils and accumulation in corn (Zea mays L.) under field conditions in Shaanxi Province, China. Science of the Total Environment, 427-428: 159-164.

Wang Y D, Wang X, Wong Y S. 2013. Generation of Se-enriched rice with enhanced grain yield, Se concentration and bioavailability through fertilisation with selenite. Food Chemistry, 141: 2385-2393.

Wei Y, Zhang J, Qiu S, et al. 2021. Selenium species determination in Se-enriched grain crops with foliar spray of sodium selenite by IP-RP-HPLC-UV-HG-AFS. Food Analytical Methods, 14: 1345-1358.

White P J, Broadley M R. 2009. Biofortification of crops with seven mineral elements often lacking in human diets—iron, zinc, copper, calcium, magnesium, selenium and iodine. New Phytologist, 182(1): 49-84.

White P J. 2016. Selenium accumulation by plants. Annals of Botany, 117(2): 217-235.

Williams P N, Lombi E, Sun G X, et al. 2009. Selenium characterisation in the global rice supply chain. Environmental Science and Technology, 43(15): 6024-6030.

Winkel L H E, Vriens B, Jones G D, et al. 2015 Selenium cycling across soil-plant-atmosphere interfaces: a critical review. Nutrients, 7(6): 4199-4239.

Xia Q, Yang Z P, Shui Y, et al. 2020. Methods of selenium application differentially modulate plant growth, selenium accumulation and speciation, protein, anthocyanins and concentrations of mineral elements in purple-grained wheat. Frontiers in Plant Science, 21(11): 1114.

Yang D, Hu C, Wang X, et al. 2021. Microbes: a potential tool for selenium biofortification. Metallomics,

13(10): mfab054.

Yin H Q, Qi Z Y, Li M Q, et al. 2019. Selenium forms and methods of application differentially modulate plant growth, photosynthesis, stress tolerance, selenium content and speciation in *Oryza sativa* L. Ecotoxicology and Environmental Safety, 169: 911-917.

Ying X, Liu Y, Liang P X, et al. 2019. Form and interconversion factor of selenium in soil. Agricultural Biotechnology, 8(1): 116-120.

Yläranta T. 1983. Selenium in Finnish agricultural soils. Annales Agriculturae Fenniae, 22: 122-136.

Yläranta T. 1985. Increasing the selenium content of cereal and grass crops in Finland. Helsinki: Yliopistopaino, 1985.

Yuan L X, Zhu Y Y, Lin Z Q, et al. 2013. A novel selenocystine-accumulating plant in selenium-mine drainage area in Enshi, China. PLoS One, 8(6): e65615.

Zhang L, Hu B, Li W, et al. 2014. OsPT2, a phosphate transporter, is involved in the active uptake of selenite in rice. New Phytologist, 201: 1183-119.

Zhao C Y, Ren J H, Xue C Z, et al. 2005. Study on the relationship between soil selenium and plant selenium uptake. Plant Soil, 277: 197-206.

Zhu Y G, Pilon-Smits E A H, Zhao F J, et al. 2009. Selenium in higher plants: understanding mechanisms for biofortification and phytoremediation. Trends in Plant Science, 14(8): 436-442.

第六章 植物硒的超积累及其机制

硒（selenium，Se）是人体和动物的必需营养元素，是植物的有益元素，但过量的硒对人体健康和植物都有害。近年来，人们发现少数植物能够超积累硒。本文从硒超积累植物的发现、种类和分类出发，论述了硒在超积累植物中的含量、形态、分布及其影响机制，评述了植物硒超积累伴生的生态效应和硒超积累植物的利用途径，提出了硒超积累植物研究和利用存在的问题，并对今后的研究进行了展望。

第一节 概 述

1818 年，Jöns Jakob Berzelius 首次从硫酸厂铅室的红色粉末中发现并制备出硒。此后，人们发现硒是一种天然存在于环境中的微量元素，其土壤含量差别很大。一般土壤含硒量为 0.1～2.0 mg/kg，而富硒地区土壤含硒量可达到 30～324 mg/kg 或更高（Girling，1984）。土壤是食物链中人和动植物硒的基本来源，其总硒含量主要受成土母质等土壤理化性质的影响（Xiao et al.，2020）。研究表明，土壤总硒含量不一定能反映该地区生长的植物是否受毒害或缺硒（Lakin，1972），其结果很大程度上还要取决于植物种类和土壤硒的赋存状态及生物有效性（Hasanuzzaman et al.，2020）。植物吸收硒的主要形态是硒酸盐（SeO_4^{2-}）和亚硒酸盐（SeO_3^{2-}）。硒酸盐多存在于氧化性的土壤中，亚硒酸盐多存在于偏酸性的土壤中，而硒化物则主要分布在高还原性土壤环境中（Bitterli et al.，2010）。

作为一种人和动物必需的微量营养元素，硒在环境中的丰缺状况与人和动物的健康状况密切相关。已有研究表明，硒低剂量有益、高剂量有毒，造成毒害的硒摄入量通常比引起硒营养缺乏的摄入量高一个数量级（Girling，1984）。动物硒中毒的一般症状与砷中毒症状相似（Tunnicliffe，1901）。取食高硒含量的植物引发的碱毒病（alkali disease）和蹒跚病（blind stagger）是早期发现的硒中毒症。美国加州圣乔奎谷凯斯特森（Kesterson）水库曾出现大批水禽的畸变、死亡，也是因为高硒含量的毒害作用（Presser，1994）。人类摄入超量硒同样会受到毒害，可能会诱发心脏病和甲状腺功能衰退等疾病（Rayman，2012；Mehdi et al.，2013；Hatfield et al.，2014）。对于合适的剂量范围，欧洲食品安全局（EFSA）推荐每天摄入硒 50～200 μg（EFSA，2014）。

硒营养摄入不足也会致病，发生在中国的克山病（Keshan disease）是首次报道的一种缺硒病，呈现出多灶性心肌坏死病理特征，主要发病于幼儿和育龄妇女（Chen et al.，1980）。缺硒还可导致白肌病、大骨节病等疾病，也会增加患心血管病、糖尿病、癌症、关节炎等多种疾病的风险（Rayman，2000；Rayman，2012）。硒是 25 种功能蛋白的组成元素（Dai et al.，2019）。膳食硒主要通过硒蛋白起作用，主要是参与多种细胞功能的氧化还原酶（Hatfield et al.，2014）。含硒的抗氧化酶或蛋白质可消除活性氧自由基对机体造成的氧化损伤（Tinggi，2003）。适量的硒还能提高人和动物体的免疫力，有效防止坏死性肝脏退化（Schwarz and Foltz，2002）。含硒的代谢产物（如 SeCys、MeSeCys 等）具有抗癌作用（Combs，2001）。由于硒能与多种金属发生拮抗作用，因此，硒还可以预防和减轻汞、砷、镉、铬等有毒重金属元素对机体的伤害。

科学补硒是缺硒地区居民营养健康的重要保障之一。为解决隐性饥饿问题，合理利用富硒地区资源优势或在贫硒地区大力发展富硒农业具有重要意义。膳食硒是人体主要的硒补充途径。植物性产品有富硒茶、富硒水稻；动物性产品有通过富硒饲料养殖生产的富硒猪肉和富硒鸡蛋等。

在受金属污染或高背景值土壤中生长的植物中发现了一种稀有特征——金属超积累（hyperaccumulation），这个积累水平远远超过大多数物种（Shier，1994）。超积累植物可以积累和耐受超高浓度的重金属而不会受到毒害。生长在富硒土壤上的大多数植物积累硒可以达到 100 mg/kg 干重，但不能耐受较高的硒浓度。然而，一些进化出硒耐受性的植物通常可以积累硒含量＞100 mg/kg 干重，这些植物被认为是富硒植物；另外一些物种甚至能积累 1000～15 000 mg/kg 干重的硒，被称为超积累植物（Rascio and Navari-Izzo，2011；White，2016）。有关超积累植物的基本信息，沈振国曾有总结（沈振国，2002），表 6-1 为常见重金属在土壤和普通植物中的平均浓度以及超量积累植物的临界标准。

表 6-1　重金属在土壤和普通植物中平均浓度以及超量积累植物的临界标准（沈振国，2002）

元素	土壤/（mg/kg，干重）	植物/（mg/kg，干重）	临界标准/（mg/kg，干重）
Cd		0.1	100
Co	10	1	1 000
Cr	60	1	1 000
Cu	20	10	1 000
Mn	850	80	10 000
Ni	40	2	1 000
Pb	10	5	
Se		0.1	1 000
Zn	50	70	10 000

硒超积累植物不仅可以作为一种有效的硒植物修复品种，还可以作为有机富硒原料生产富硒农产品。对硒超积累植物的研究有助于培育具有硒特异性吸收以及将无机硒高效转化为低毒有机硒化合物的作物。本章论述了硒超积累植物的发现、种类、硒超积累机制、硒抗性和解毒特征，以及硒超积累植物的生态特征与人类健康的关系等有关内容，以探讨利用硒超积累植物进行植物修复和发展富硒功能农业的可能性。

第二节　硒超积累植物的发现和分类

一、硒超积累植物的发现和特点

硒超积累植物在美国和澳大利亚非常普遍，一般认为硒含量 1000 mg/kg 干重是硒超积累植物的标准（Mehdi et al.，2013；Krzciuk and Galuszka，2014；White，2016；Pilon-Smits，2019）。然而，由于植物中的正常硒水平低于 2 mg/kg 干重，我们也有理由认为超过 100 mg/kg 干重的植物是硒超积累植物（van der Ent et al.，2013）。

事实上，除考虑元素的积累量外，还有另外两个更严格的标准来筛选超积累植物：①生物富集系数（bioconcentration factor）（即地上部与土壤元素浓度的比率）>1；②转运系数（translocation factor）（即地上部与根部元素浓度的比率）>1（McGrath and Zhao，2003）。由于不同种类的硒超积累植物具有广泛的硒浓度范围，而且在同一物种内，种群间和种群内的硒含量水平也有很大差异（Reynolds et al.，2020），因此，我们参考 Li 等（2018）的建议，遵循 Baker 和 Whiting（2002）以及 van der Ent 等（2013）提出的微量元素超积累含量标准，即生长在自然生境的植物可以顺利完成其生活史，且至少一个标本的叶片元素含量（干重）达到标准：砷（As）、铬（Cr）、钴（Co）、铜（Cu）、镍（Ni）和稀土元素（REE）超过 1000 mg/kg，锌（Zn）为 3000 mg/kg，锰（Mn）为 10 000 mg/kg，而镉（Cd）、硒（Se）和铊（Tl）超过 100 mg/kg（Baker et al.，1992；Baker and Whiting，2002；van der Ent et al.，2013）。曾有研究表明，至少有 20 多种植物能够超量积累硒（>1000 mg/kg）；如将硒超积累植物的临界标准定为>100 mg/kg，则硒超积累植物的数量将会大大增加，包括黄芪属（*Astragalus*）、紫菀属（*Aster*）、滨藜属（*Atriplex*）、胶草属（*Grindelia*）、蛇黄花属（*Gutierrezia*）中的一些种及牧豆树属（*Prosopis*）等。由于硒超积累植物分布在不同的植物科、属中，推测不同的硒超积累植物可能是独立进化而来（沈振国，2002）。

二、植物超积累硒的一般假说

重金属超积累是在植物中发现的一种稀有特征。植物对硒的过度积累可能是对相关生态胁迫，如对病原体和草食动物的防护，或者是作为元素化感作用的组成部分进化而来的（Lima et al.，2018）。目前有两个假说解释重金属超积累特性的现象（Cappa and Pilon-Smits，2014）：一是元素防御假说（elemental defense hypothesis），二是元素化感作用假说（elemental allelopathy hypothesis）。元素防御假说认为重金属超积累起到防御草食动物的作用（Boyd，2007），而元素化感作用假说则认为植物超积累重金属的目的是抑制邻近生物的生长（Mohiley et al.，2020）。植物硒的超积累发生在相对较少的属和种（约 40 个科）中，这很可能是一种衍生特征：超积累是由非超积累演变而来的。另外，也有人提出硒的富集是一种古老的特征，在硒质土壤广泛分布的时期更为普遍，但随着土壤硒浓度的降低而逐渐丧失（El Mehdawi and Pilon-Smits，2012）。

有大量证据支持元素防御假说和元素化感作用假说。例如，植物对硒的积累可能对群落中其他物种产生正面或负面影响（Valdez Barillas et al.，2012），硒在迁移过程中被生物吸收，低浓度时使它们受益，但在高浓度时会产生毒性。因此，硒的超积累可以保护植物免受对硒敏感的草食动物和病原体的侵害（元素防御），并通过高硒凋落物的沉积（元素化感）来减少周围的植被覆盖（Pilon-Smits，2019）。超积累植物分泌物淋溶或凋落物分解时，将硒释放到植物冠层下的表层土壤中，此过程可能会产生元素化感作用的现象，在这种情况下，部分定植物种的生长可能受到抑制或无法生长，从而使得超积累植物取得一定的生态竞争优势（Boyd，2013）。El Mehdawi 和 Pilon-Smits（2012）的研究证实，超积累植物周围土壤硒含量增加（生物富集）可影响附近硒敏感植物的生长。

硒向最有价值的器官——幼叶和生殖器官的转移富集符合超积累元素的防御假说，

该假说认为超积累硒可以起到保护植物免受草食动物侵害的作用（Freeman et al.，2006a；Freeman et al.，2006b；Quinn et al.，2011；Cappa et al.，2014；Statwick et al.，2016）。硒过量的毒害作用主要在牲畜摄入植物幼叶、花和种子时产生，这种威胁在春季和初夏最大（Galeas et al.，2007）。Vesk 和 Reichman（2009）进行了 31 项取食选择试验，通过在贝叶斯统计框架下进行元分析（Bayesian meta-analysis），推测植物超积累特性的进化过程中可能会有不同的选择压力，这取决于植物自然接触的一系列草食动物种类。Freeman 等（2007）发现硒超积累植物沙漠王羽（*Stanleya pinnata*）和硒富集植物印度芥菜（*Brassica juncea*）可积累硒产生毒性和威慑力以防御直翅目昆虫（蟋蟀和蝗虫）的取食（Freeman et al.，2007），另外还可以防御二斑叶螨（*Tetranychus urticae*）和西花蓟马（*Frankliniella occidentalis*）等细胞破坏性昆虫（Quinn et al.，2010b），以及取食印度芥菜韧皮部的青桃蚜虫（*Myzus persicae*）（Hanson et al.，2004）。硒可以保护印度芥菜免受真菌感染和毛虫的食草性侵袭（Hanson et al.，2003），还有研究表明硒有助于植物防御草原黑尾土拨鼠（*Cynomys ludovicianus*）的食草作用（Freeman et al.，2009），驱动了沙漠王羽和双钩黄芪（*Astragalus bisulcatus*）硒超积累能力的进化（Quinn et al.，2008）。

当然，这种超积累能力的进化也会付出代价。Quinn 等（2011）首次研究发现，印度芥菜硒积累的进化代价是减少了花粉的萌发率。同样的，当硒浓度大于 0.05%~0.1% 时，印度芥菜生物量、花粉萌发率、单粒种子重和种子总重、种子产量和种子发芽率均下降（Prins et al.，2011）。另有研究发现，花数与对植食者的防御能力呈负相关（Steven and Culver，2019），这表明繁殖能力和防御成本之间可能存在生态权衡。虽然积累硒可以起到抵御草食动物和病原体的作用，但也可能导致植物的氧化应激和生长减慢。Steven 和 Culver（2019）调查了硒积累成本和草食防御效益，测定了在亚硒酸钠处理和空白处理条件下 3 个富硒芥菜品种的硒水平、植株大小和花量，还用专性草食者菜粉蝶对两种处理的叶片进行了非选择性的食草性试验。结果表明，硒处理叶片数比对照略有增加，但部分品种的组织硒浓度与花量呈负相关。随着叶片中硒浓度的增加，开花量减少（图 6-1；相关系数=−0.0063，$F_{1,22}$=5.59，P=0.0273）。虽然硒

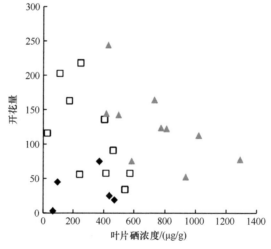

图 6-1　亚硒酸钠（0.1 mmol/L）处理下三个芥菜品种叶片硒浓度与开花量之间的关系

（Steven and Culver，2019）

图中三种形状（菱形、方形、三角形）表示三个芥菜品种

酸盐处理不会对整体生长产生负面影响，但叶片中硒浓度的增加与开花量的减少有关。在取食试验中，经硒酸盐处理的叶片被取食较少。在硒积累量最高的品种中，表现出开花量减少的硒酸盐处理植株的叶片也被少量取食。更多硒积累的保护优势可能会被对生殖的负面影响所抵消，与更极端的超积累植物相比，富硒芥菜的硒积累水平相对较低，这可能反映出强化草食性防御所需的最低硒积累水平（Steven and Culver，2019）。

三、硒超积累植物的种类

现今已有很多对硒超积累植物的调查报告或相关研究，在多数情况下可以直接查到硒含量情况。但某种植物在野外不同地点和不同生长阶段的实际含硒量结果可能仅具有定性意义，因为取样时间可能不是处于含硒量最高的阶段。例如，菊科（Asteraceae）典型含硒植物聚果卵莞（*Oonopsis condensata*），其硒含量要到 8 月盛花期才能达到最高值 4800 mg/kg，而在 12 月苗期为 1585 mg/kg（Beath et al.，1937）。已知的硒超积累植物大约有 50 种，横跨 7 个科，但大多数属于 3 个科：豆科（Fabaceae）、菊科（Asteraceae）和十字花科（Brassicaceae）（Reynolds and Pilon-Smits，2018）。我们参考现有文献报道，收录了部分自然生长在野外生境中的富硒植物品种及其硒含量，见表 6-2。

第三节　植物硒的超积累及其影响机制

一、土壤中硒的含量与形态对植物硒积累的影响

天然土壤硒的浓度差异悬殊，范围为 0~212 mg/kg（Statwick et al.，2016）。土壤中硒的生物可利用形态主要是硒酸盐（SeO_4^{2-}）（Reynolds et al.，2019）。缺硒土壤地区通常与富硒土壤邻近（Barceló and Poschenrieder，2010）。土壤中硒的缺乏和过量不仅与土壤硒的浓度有关，而且受土壤硒的化学形态影响（Lima et al.，2018）。硒对植物的生物有效性受土壤系统中硒的形态控制，这些形态与 pH、氧化还原电位、矿物学、有机质和竞争阴离子的存在有关（Favorito et al.，2020）。一项研究分析了科罗拉多州西部 32 个富硒和非富硒生境的土壤（Statwick and Sher，2017），发现在正常的低硒土壤中，总硒和生物有效态硒之间大致呈线性关系，但是在富硒土壤中（总硒＞2 mg/kg），总硒和生物有效态硒之间没有明显的相关性。由于硒会强烈吸附到黏土矿物中，并可能被水生生物富集，因此通常在富含黏土的沉积岩（包括泥岩和页岩，尤其是火山活动时期沉积的沉积岩）中发现高含量硒（Statwick and Sher，2017）。但是，矿物学特性并不是影响环境中硒含量的唯一因素。硒的离子形式特别易溶，可以很容易地被水浸出或沉积在土壤中（Statwick and Sher，2017）。土壤中硒的溶解度在一定程度上受土壤固相吸附（或滞留）的调节。在高渗透、碱性土壤条件下，亚硒酸盐可以迅速氧化成硒酸盐（Favorito et al.，2020）。碱性土壤对硒的吸附能力较低，这很可能是由于碱性条件提高了硒溶解度的缘故。一般认为，在大多数土壤条件下，硒酸盐比亚硒酸盐更易溶，因此更容易被生物利用（Spallholz，1994）。例如，在亚硒酸盐和硒酸盐处理的土壤中，小白菜对硒的最

表 6-2 硒超积累植物发现地点和硒含量

科名	属名	拉丁学名	地点	硒含量（mg/kg）	参考文献
菊科 Asteraceae	灰菀属 Dieteria	Dieteria canescens	美国中西部	1 600	Beath et al., 1939a
	胶菀属 Grindelia	Grindelia squarrosa	美国南达科他州 Lower Brule Reservation	930	Lakin and Byers, 1941
	蛇黄花属 Gutierrezia	Gutierrezia microcephala	美国犹他州 Thompson	1 287	Beath, 1943
	卵菀属 Oonopsis	Oonopsis condensata	美国中西部	3 250	Beath et al., 1939a
		Oonopsis foliosa	美国怀俄明州	888	Beath et al., 1937
	联毛紫菀属 Symphyotrichum	Symphyotrichum ascendens	美国科罗拉多州 Lascar	3 630	Beath et al., 1939b
		Symphyotrichum ericoides	美国爱达荷州 Soda Springs	4 455	Pfister et al., 2013
	木根菊属 Xylorhiza	Xylorhiza Parryi	美国科罗拉多州 Fort Collins, Pine Ridge	1 378	El Mehdawi et al., 2015b
		Xylorhiza venrtusta	美国怀俄明州	2 300	Beath et al., 1937
			美国犹他州 Thompson	648	Beath, 1943
十字花科 Brassicaceae	碎米荠属 Cardamine	Cardamine enshiensis	中国恩施渔塘坝	3 329	Cui et al., 2018
		Cardamine hupingshanensis	中国恩施渔塘坝	1 965	Yuan et al., 2013
		Cardamine sp.	中国恩施渔塘坝	635	Li et al., 2021
		Cardamine violifolia	中国恩施渔塘坝	1 427	Shao et al., 2014
			中国恩施渔塘坝	3 900	Both et al., 2018
			中国恩施渔塘坝	261	Both et al., 2020
	金尾芥属 Stanleya	Stanleya bipinnata	美国怀俄明州 Laramie	2 490	Beath et al., 1941
		Stanleya bipinnata	美国怀俄明州	1 320	Beath et al., 1937
		Stanleya integrifolia	美国爱达荷州	364	Beath et al., 1939b
			美国犹他州 Vernal	977	Beath et al., 1941
		Stanleya pinnata	美国中西部	1 252	Beath et al., 1939a
			美国犹他州 Moab	1 456	Beath et al., 1941
			美国科罗拉多州 Fort Collins, Pine Ridge	>4 200	Galeas et al., 2007

续表

科名	属名	拉丁学名	地点	硒含量 (mg/kg)	参考文献
十字花科 Brassicaceae			美国科罗拉多州 Fort Collins, Pine Ridge	3 713	Sura-de Jong et al., 2015
			美国新墨西哥州	1 110	Beath et al., 1939b
			美国犹他州	787	Beath et al., 1939b
藜科 Chenopodiaceae	滨藜属 Atriplex	Atriplex confertifolia	美国犹他州 Thompson	1 734	Beath, 1943
		Atriplex nuttallii	美国中西部	300	Beath et al., 1939a
			美国怀俄明州	930	Beath et al., 1937
			美国犹他州 Thompson	611	Beath, 1943
豆科 Fabaceae	黄芪属 Astragalus	Astragalus beathii	美国亚利桑那州 Cameron	1 034	Beath et al., 1941
		Astragalus beckwithii var. purpureus	美国内布拉斯加州 Clark County	970	Lakin and Byers, 1941
		Astragalus bisulcatus	美国中西部	5 330	Beath et al., 1939a
			美国怀俄明州 Laramie	1 326	Beath et al., 1941
			美国蒙大拿州 Culbertson	1 468	Beath et al., 1941
			美国蒙大拿州 Conrad	1 432	Beath et al., 1941
			美国怀俄明州	5 965	Beath et al., 1937
			美国科罗拉多州 Fort Collins, Pine Ridge	12 700	Galeas et al., 2007
			美国科罗拉多州 Fort Collins, Pine Ridge	13 685	Sura-de Jong et al., 2015
			美国新墨西哥州	1 321	Beath et al., 1939b
			美国蒙大拿州 Lascar	1 550	Beath et al., 1939b
			美国科罗拉多州	1 620	Beath et al., 1939b
			美国爱达荷州	755	Beath et al., 1939b
			美国怀俄明州 Lost Creek, Sweetwater Co.	962	Beath et al., 1941
			美国怀俄明州 Como Ridge, Carbon Co.	4 040	Beath et al., 1941
		Astragalus confertiflorus	美国新墨西哥州 Aztec	1 361	Beath et al., 1941

续表

科名	属名	拉丁学名	地点	硒含量（mg/kg）	参考文献
豆科 Fabaceae			美国科罗拉多州 Lascar	888	Beath et al., 1939b
			美国犹他州 Thompson	1 322	Beath, 1943
			美国内华达州	631	Beath et al., 1939b
			美国科罗拉多州 Lascar	2 148	Beath et al., 1939b
			美国新墨西哥州 Cuba	2 377	Beath et al., 1941
		Astragalus limatus	美国加利福尼亚州 Truckhaven	2 175	Beath et al., 1941
		Astragalus osterhoutii	美国科罗拉多州 Kremmling	2 678	Beath et al., 1940
			美国科罗拉多州 Kremmling	1 356	Beath et al., 1941
		Astragalus pattersonii	美国新墨西哥州 Gallup	1 008	Beath et al., 1941
			美国犹他州	2 154	Beath et al., 1939b
			美国新墨西哥州	676	Beath et al., 1939b
			美国科罗拉多州 Lascar	5 042	Beath et al., 1939b
			美国犹他州 Thompson	8 512	Beath, 1943
		Astragalus pattersonii var. praelongus	美国犹他州 Salina Expt. Sta. Road	3 090	Beath et al., 1941
			美国亚利桑那州 Leupp	4 835	Beath et al., 1941
			美国新墨西哥州 Cuba	2 370	Beath et al., 1941
		Astragalus pectinatus	美国中西部	1 642	Beath et al., 1939a
			美国怀俄明州 Lusk	3 190	Beath et al., 1941
			美国怀俄明州 Laramie	2 650	Beath et al., 1941
			美国蒙大拿州 Wolf Point	2 140	Beath et al., 1941
			美国怀俄明州	1 062	Beath et al., 1937
			美国蒙大拿州	1 815	Beath et al., 1939b

续表

科名	属名	拉丁学名	地点	硒含量 (mg/kg)	参考文献
豆科 Fabaceae		Astragalus praelongus	美国科罗拉多州 Lascar	2 114	Beath et al., 1939b
			美国怀俄明州 Lusk	2 840	Beath et al., 1941
			美国怀俄明州 Lusk	1 984	Beath et al., 1941
			美国蒙大拿州 Culbertson	1 375	Beath et al., 1941
			美国犹他州	1 284	Beath et al., 1939b
			美国新墨西哥州	4 474	Beath et al., 1939b
			美国亚利桑那州	4 500	Beath et al., 1939b
			美国犹他州 Moab	1 438	Beath et al., 1941
		Astragalus preussii	美国内布拉斯加州 Franklin	1 727	Beath et al., 1941
		Astragalus racemosus	美国怀俄明州	13 900	Beath et al., 1937
			美国南达科他州	3 920	Beath et al., 1939b
			美国怀俄明州 Lusk	6 801	Beath et al., 1941
			美国南达科他州 Bonesteel	2 600	Beath et al., 1941
		Astragalus sabulosus	美国犹他州 Thompsons	2 210	Beath et al., 1941
			美国亚利桑那州	1 734	Beath et al., 1939b
			美国犹他州 Thompsons	2 025	Beath et al., 1941
		Astragalus scobinatulus	美国怀俄明州 Baggs	1 282	Beath et al., 1941
	假含羞草属 Neptunia	Neptunia amplexicaulis	澳大利亚昆士兰州 Richmond	4 334	Knott and McCray, 1959
茜草科 Rubiaceae	穴果木属 Coelospermum	Coelospermum decipiens	澳大利亚昆士兰州 Cape York Peninsula	1 141	Knott and McCray, 1959
玄参科 Scrophulariaceae	火焰草属 Castilleja	Castilleja angustifolia	美国犹他州 Thompson	428	Beath, 1943

大提取效率分别为 4.91% 和 31.90%（Li et al., 2015）。同样的，Bañuelos 和 Meek（1989）通过土培甘蓝（*Brassica oleracea* var. *capata*）、花椰菜（*Brassica oleracea* var. *botrytis*）、羽衣甘蓝（*Brassica oleracea* var. *cephada*）和瑞士甜菜（*Beta vergaris* var. *Cicla*）四种植物发现，硒酸盐处理的植物硒吸收量比亚硒酸盐处理的高约 10 倍，且四种蔬菜的干重产量未受任何硒处理的显著影响。

有研究表明，植物特别是超积累植物对土壤硒的吸收与土壤硒浓度密切相关（刘亚峰等，2018）。在鄂西南渔塘坝硒矿区，自然生境中的碎米荠（*Cardamine violifolia*）茎叶的生物富集系数（植物硒/土壤硒）范围为 2.8～43.8，而且大部分都大于 10（表 6-3），说明碎米荠地上部组织对土壤硒的吸收富集能力比较强。同时，生物转移系数（植物地上部硒/根硒）在 0.46～1.88，大部分＞1，说明碎米荠具有从根部向地上部转移硒的生物性能。樊俊等（2014）采用 Na_2SeO_3 和硒矿粉进行盆栽试验的结果表明，施用硒对堇叶碎米荠硒含量和硒积累量的增加效果显著，植株各部位硒积累规律为叶＞茎＞根。当土壤添加 Na_2SeO_3 浓度为 50 mg/kg 时，叶片硒含量最高为 128.82 mg/kg。

表 6-3 碎米荠对土壤硒的吸收和转移（刘亚峰等，2018）

采样点	土壤 Se (mg/kg)	植物 Se/ (mg/kg)			生物富集系数		生物转移系数	
		叶	茎	根	叶	茎	叶	茎
农田 1	5.2	29	30	16	5.6	5.8	1.82	1.88
农田 2	6.6	105	85	83	15.8	12.9	1.26	1.02
农田 3	7.9	184	89	192	23.2	11.3	0.96	0.46
小溪流域 1	12.2	290	320	360	23.8	26	0.81	0.89
小溪流域 2	39.3	360	600	570	9.2	15.3	0.63	1.05
小溪流域 3	28	1226	512	841	43.8	18.4	1.46	0.7
矿渣堆 1	66.6	280	380	370	4.2	5.7	0.76	1.03
矿渣堆 2	140	390	500	330	2.8	3.6	1.2	1.52
矿渣堆 3	185	1365	700	1168	7.4	3.8	1.2	0.6

二、硒超积累植物的吸收、转化与分配机制

了解超积累植物对硒的吸收和代谢机制，对于环境、农业以及人和动物营养均具有广泛的意义，有助于培育具有硒特异性吸收能力，以及将无机硒酸盐高效转化为毒性较低并具有抗癌能力的有机硒的作物（Lima et al., 2018）。

（一）硒超积累植物中的硒含量与分布

十字花科 *Stanleya* 属是美国西部特有的一个小属，由 7 个种组成。在一项研究中，Cappa 等（2014）收集了 7 个种中的 6 个种群和所有 4 个变种的多个种群，测试了叶片、果实和土壤中的原位硒及硫浓度，调查了在富硒土壤上田间和温室环境中 *Stanleya* 属植物对硒的积累程度。研究确定沙漠王羽是其中唯一的硒超累积植物。而在天然富硒草

地上调查发现，两种多年生硒超累积植物沙漠王羽和双钩黄芪根的硒含量为100～1500 mg/kg，硒主要集中分布在根的皮层和表皮，而中柱中含量较低（Prins et al., 2019）。

Cui 等（2018）测定了恩施碎米荠（*Cardamine enshiensis*）的根、茎、叶干重平均硒含量分别为2985 mg/kg、3329 mg/kg 和2491 mg/kg。同步辐射微束X射线荧光（μ-SRXRF）分析发现，根中硒主要分布在皮层、内胚层和维管柱，而在地上部主要分布于茎的表皮、皮层和维管束，以及叶片的叶脉和叶缘（Cui et al., 2018）。Yuan 等（2013）报道，壶瓶碎米荠（*Cardamine hupingshanesis*）（图6-2）对硒的积累能力为：根（239±201）mg/kg，茎（316±184）mg/kg，叶（380±323）mg/kg，为次级硒超积累植物。而在硒尾矿附近，其叶、茎、根的干重硒含量分别达到（1965±271）mg/kg、（1787±167）mg/kg 和（4414±3446）mg/kg（Yuan et al., 2013）（图6-3）。

图6-2 壶瓶碎米荠（*Cardamine hupingshanensis*）（Xiang et al., 2019）
（a）花期全株；（b）根系

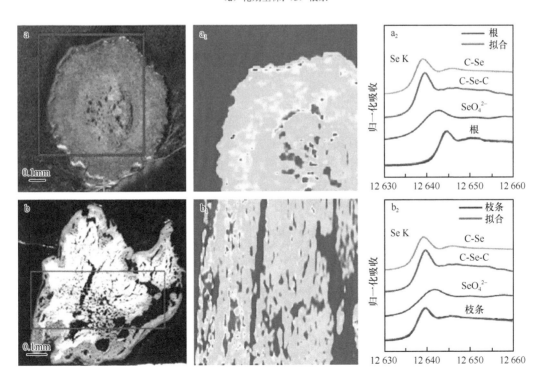

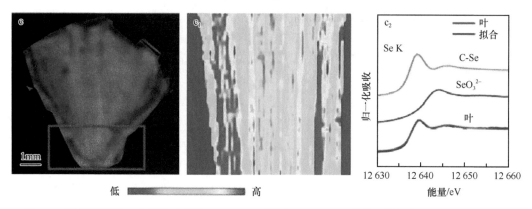

图 6-3　恩施碎米荠三个部位中硒的 μ-SRXRF 成像和 μ-XANES 谱的光学图像（Cui et al.，2018）

根横截面（a，a_1，a_2）、枝条（横截面 b，b_1，b_2）和叶（c，c_1，c_2）。(a)~(c) 中的红色方框区域是（a_1）~（c_1）中的 μ-SRXRF 映射的区域。在（a_2）~（c_2）中，用 SeO_3^{2-}（亚硒酸盐）、SeO_4^{2-}（硒酸盐）、$C-Se^-$（以硒代半胱氨酸为模型化合物）和 C-Se-C（以甲基硒半胱氨酸为模型化合物）进行最小二乘拟合

Jaiswal 等（2012）通过比较印度富硒地区超富硒印度芥菜种子、油和油饼中的硒含量发现，与榨油前种子（110±3.04 mg/kg）和油（3.50±0.66 mg/L）相比，硒的储存和富集主要集中在油饼中（143±5.18 mg/kg）。

Galeas 等（2007）以硒非超积累植物绢毛黄芪（*Astragalus sericoleucus*）、丝毛棘豆（*Oxytropis sericea*）和高山蒌蒿（*Thlaspi montanum*）为对照，在两个生长季节中，对硒超积累植物双钩黄芪和沙漠王羽中硒和硫浓度的季节变化进行监测，发现硒和硫含量在硒超积累植物中呈负相关，在硒非超积累植物中呈正相关。硒超积累植物中的硒在整个生长季节中呈现出一种特殊的转运过程：春季从根到幼叶，然后在夏季从衰老的叶片再活化转移到生殖组织，最后在秋季回到根。

El Mehdawi 等（2014）对柳叶白菀（*Symphyotrichum ericoides*）、硒超积累植物双钩黄芪、硒富集植物芥菜和艾菊叶翠菊（*Machaeranthera tanacetifolia*）的超积累硒、富硒和耐硒性进行了平行分析。柳叶白菀和艾菊叶翠菊对硒的累积量分别为 3000 mg/kg 和 1500 mg/kg。柳叶白菀的叶片 Se/S 和地上部/地下部硒含量比都很高，与双钩黄芪相似，均高于艾菊叶翠菊和芥菜。X 射线吸收近边结构谱表明，柳叶白菀富集硒以 C-Se-C 化合物为主（86%）（模型化合物为甲基硒半胱氨酸），这可能是其耐硒性的原因。艾菊叶翠菊中有 55% 的硒以 C-Se-C 化合物的形式积累，其余为无机硒。因此，在这项温室研究中，柳叶白菀表现出硒超积累植物的所有特征。而在硒超积累植物附近生长的柳叶白菀体型较大，除了是因为早先呈现的生态效益，也可能叠加有生理效益（El Mehdawi et al.，2014）。

（二）硒超积累植物中的硒形态与分布

Pickering 等（2003）早在 2003 年就利用锥形金属单毛细管光学系统（tapered metal monocapillary optics）产生的小 X 射线光束（small X-ray beams），对硒超积累植物双钩黄芪的纵向切面进行化学特异性成像观察（图 6-4），发现了硒在细胞水平上的分布。Lindblom 等（2013a）的一项研究首次表明，硒主要集中在超积累植物根的皮层，与叶片中的形态一样均为 C-Se-C，还存有少量亚硒酸盐，未发现硒酸盐。Cappa 等（2015）用 X 射线微探针分析（X-ray microprobe analysis）方法，结合系统发育、生理学研究了 *Stanleya*

属植物超积累特性的演变过程，结果表明，对硒的耐受性和合成有机硒的能力是 *Stanleya* 属植物超积累硒的前提。沙漠王羽和双钩黄芪根累积的硒由有机（C-Se-C）化合物组成（以甲基硒代半胱氨酸为模型化合物）（Prins et al.，2019）。双钩黄芪和沙漠王羽可富集高达植物干重 1% 的硒。在田间条件下，硒主要存在于这两种超积累植物的幼叶和生殖组织中。微聚焦扫描 X 射线荧光图谱（microfocused scanning X-ray fluorescence mapping）显示，在双钩黄芪幼叶的毛状体中超量积累了硒。显微 X 射线吸收光谱（micro X-ray absorption spectroscopy）和液相色谱-质谱联用（liquid chromatography-massspectrometry）分析表明，在幼叶中，除 70% 的 MeSeCys 外，还含有 30% 的无机硒（硒酸盐和亚硒酸盐），其毛状体中的硒主要以甲基硒代半胱氨酸（MeSeCys，53%）和 γ-glutamyl-MeSeCys（47%）的有机形式存在。同样的，X 射线能谱分析（energy-dispersive X-ray microanalysis）表明，硒主要分布在表皮细胞中，硒在沙漠王羽幼叶的叶缘和表面呈球状结构高度富集。液相色谱-质谱联用检测到叶缘中同时含有 MeSeCys（88%）和硒代胱硫醚（12%）。相反，富硒芥菜和非富硒拟南芥（*Arabidopsis thaliana*）叶片的硒积累在维管组织和叶肉细胞中。超积累植物中的硒在木质部和韧皮部中似乎都是可移动的，因为经硒处理的沙漠王羽对取食韧皮部的蚜虫有很高的毒性，而且 MeSeCys 存在于沙漠王羽幼叶叶柄的维管组织和导管液中。植物外围特定储存区对有机硒化合物的区隔作用似乎是硒超积累植物独有的特性。高浓度的硒分布于植物外周，可能有助于增强植物对硒的耐性，也可能是植物的一种基本防御机制（Freeman et al.，2006a）。

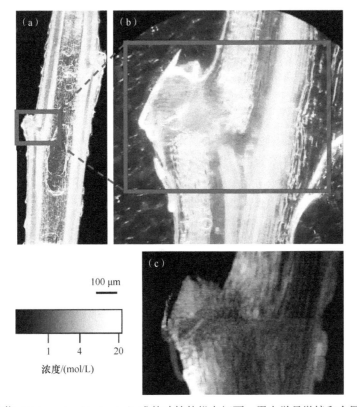

图 6-4 双钩黄芪（*Astragalus bisulcatus*）成熟叶枝的纵向切面，用光学显微镜和金属单毛细管光学系统成像（Pickering et al.，2003）

样品厚度约为 100 μm。（a）和（b）是光学显微照片；（c）是有机硒的分布图分析发现，玉蕊科（Lecythidaceae）初级积累植物小猴钵树（*Lecythis minor*）和次级积累植物巴西栗（*Bertholletia excelsa*）的硒形态存在差异，证实了超积累植物硒形态分布主要与积累机制有关，而与分类学无关的观点（Nemeth et al.，2013）。最丰富的硒化合物主要是硒代同型半胱氨酸（SeHCys）和硒代甲硫氨酸（SeMet）的衍生物，包括脂肪酸代谢相关的化合物。还有一系列含多个 Se 原子（＞2）的 SeHCy 衍生物，它们的结构可通过其 S-Se 类似物的合成验证（Nemeth et al.，2013）。植物主要吸收硒酸盐（SeO_4^{2-}）和亚硒酸盐（SeO_3^{2-}）两种形态，也吸收有机硒化合物，但不吸收元素硒或金属硒化物（Guignardi and Schiavon，2017；White and Broadley，2009）。硒通过根系中介导硫酸盐和磷酸盐离子内流的高亲和力膜转运蛋白进入细胞，进入硫（S）同化途径，从而被结合到硒代半胱氨酸（SeCys）和硒代甲硫氨酸（SeMet）中（Guignardi and Schiavon，2017）。

转基因研究表明（Bañuelos et al.，2005；Sors et al.，2005a；Sors et al.，2005b），在植物吸收、耐受和积累硒的过程中，某些酶对硒的吸收、耐受和积累具有限速作用。高浓度的硒对植物是有毒害作用的，这既是由于氧化压力，也是因为硒代氨基酸被非特异性地结合到蛋白质中，蛋白质因此失去折叠结构和功能（Brown and Shrift，1982；Terry et al.，2000；Kubachka et al.，2007）。因此，植物进化出不同的策略来应对硒毒害。它们通常涉及将硒代氨基酸转化为危害较小的挥发性化合物（Evans et al.，1968；Sors et al.，2005a；Sors et al.，2005b）。具体来说，不能积累高浓度硒的植物以 SeMet 为前体产生二甲基硒（DMeSe）；能够耐受和在组织中积累大量硒的超积累植物，从硒代半胱氨酸（SeCys）开始产生二甲基二硒（DMeDSe）。硒超积累植物具有防止 SeCys 误掺入蛋白质的额外机制（Guignardi and Schiavon，2017），如通过 SeCys 甲基转移酶（SMT）将 SeCys 甲基化为甲基硒代半胱氨酸（MeSeCys），以及将 SeCys 分解成元素硒和丙氨酸。

在壶瓶碎米荠积累的总硒中，有 70% 以上是以硒代胱氨酸（$SeCys_2$）的形式存在的，且随植物中总硒浓度的增加而增加，这与非富硒植物（如拟南芥）和次级积累植物（如芥菜）中的主形态硒代甲硫氨酸（SeMet），以及超积累植物（如沙漠王羽）中的主形态硒代甲基半胱氨酸（SeMeCys）不同（Yuan et al.，2013）。同样的，邵树勋等（2015）利用 HPLC-ICP-MS 及 HPLC-ESI-TOFMS 测得碎米荠幼苗期和半成熟期植物中含有硒代胱氨酸（$SeCys_2$）、硒代甲硫氨酸（SeMet）有机硒，以及无机硒 Se(Ⅳ)、Se(Ⅳ)；其中有机硒 $SeCys_2$、SeMet 的含量分别为 136.1 mg/kg、10.6 mg/kg 和 39.3 mg/kg、5.3 mg/kg，成熟期种子中富含高达 1081.4 mg/kg 的硒代胱氨酸（$SeCys_2$），占到总硒含量的 89%。与其他已知的硒超积累植物不同，在琼脂培养基和沙土中培养硒超积累植物堇叶碎米荠（*C. violifolia*）和同属非积累植物苦水芹（*C. pratensis*），均未表现出硒促进生长的迹象。硫酸盐可以抑制硒酸盐的吸收，但该研究中，两种不同硒积累能力的碎米荠属植物在这方面没有表现出上述吸收差异。此外，该研究中磷酸盐也没有抑制这两种碎米荠属植物对亚硒酸盐的吸收。采用 μ-XANES 分析在这两种植物的幼苗期和成熟期植株都发现了 C-Se-C，即有机硒。相反，添加硒酸盐培育的苦水芹（成熟期）含有大约一半的 C-Se-C 和一半的硒酸盐（Both et al.，2020）。SCX-ICP-MS 的分析数据表明，任何碎米荠属植物提取物中都没有硒代半胱氨酸（SeCys）的存在（Both et al.，2020）。因此，与苦水芹和其他硒超积累植物相比，堇叶碎米荠表现出明显的、与硒相关的生理生化差异。

对堇叶碎米荠施用三种外源硒（50～800 mg/kg）进行土壤培养，富硒效率为硒酸盐＞亚硒酸盐＞硒酵母，最高硒浓度可达 7000 mg/kg 以上，其中有机硒占 90% 左右。硒的主要形态为 SeCys$_2$，不同形态的硒所占比例受外源硒形态和浓度的影响（Wu et al.，2020）。对堇叶碎米荠叶片中的低分子含硒代谢产物研究发现，硒氨基酸（78 mg/kg 干重）占叶片总硒（261 mg/kg 干重）含量的 30%（Ouerdane et al.，2020）。恩施碎米荠 μ-XANES 原位分析表明，根维管组织含 16% SeO$_4^{2-}$、1.9% C-Se$^-$（以硒半胱氨酸为模型化合物）和 65% C-Se-C（以甲基硒代半胱氨酸为模型化合物）；地上部维管束含 10% SeO$_4^{2-}$、28% C-Se$^-$ 和 62% C-Se-C；叶片含 84% C-Se$^-$ 和 16% SeO$_3^{2-}$（Cui et al.，2018）。

三、植物解硒毒与硒的超积累

植物硒毒性是由于活性氧（ROS）物质和氧化应激导致的（Cabannes et al.，2011；Gomes-Junior et al.，2007；Tamaoki et al.，2008）。SOD、POD、CAT 和 GSH-Px 等抗氧化酶是抗氧化防御系统的主要组成部分，通过这些抗氧化防御系统清除自由基来保护细胞膜的结构和功能免受氧化损伤（Dai et al.，2019）。硒是谷胱甘肽过氧化物酶（GSH-Px）等一系列抗氧化酶的活性中心，在植物抗氧化防御系统中起着重要作用。通常情况下，组织中低浓度的硒可以增强不同植物的抗氧化防御机制，而过量的硒可以作为促氧化剂产生 ROS，导致膜脂过氧化和氧化应激（Gupta and Gupta，2016；Lima et al.，2018）。

组织中高浓度的无机硒可诱导活性氧（ROS）的产生，如超氧阴离子自由基（·O$_2^-$）、过氧化氢（H$_2$O$_2$）、羟基自由基（·OH）等导致氧化应激（Van Hoewyk，2013）。ROS 的生成量和氧化还原动态平衡是由抗氧化机制精确控制的。除了超氧化物歧化酶（SOD）、过氧化氢酶（CAT）和过氧化物酶（POD）等酶类成分外，抗坏血酸和谷胱甘肽（GSH）等非酶类抗氧化剂在防御氧化损伤方面起着至关重要的作用（Das and Roychoudhury，2014）。谷胱甘肽耗竭可能是硒诱导的活性氧积累的原因（Van Hoewyk，2013）。根据以往的数据，超积累植物更倾向于产生有机硒，可能是为了避免氧化应激（Freeman et al.，2006b；Van Hoewyk，2013；Kolbert et al.，2018）。

硒超积累植物具有较强的硫转运和同化基因的表达能力，具有比硫酸盐更高特异性的硒酸盐转运蛋白，同时能显著上调参与抗氧化反应和生物抗应激反应基因的表达（Freeman et al.，2010；Lima et al.，2018）。这些过程中的关键调节剂可能是生长调节剂茉莉酸、水杨酸和乙烯（Lima et al.，2018）。研究发现，超积累植物沙漠王羽（*Stanleya pinnata*）能完全耐受 20 mmol/L 亚硒酸盐，而次生富集植物 *Stanleya albescens* 在该浓度下表现出明显的叶片黄化坏死、活性氧（ROS）积累和光合性能下降（Freeman et al.，2010）。这些生长调节剂通常与植物的应激和防御反应有关，这些激素对硒耐受性的积极影响可能是由于上调了硫的吸收和同化，而硫同化的组成性上调可能使植物更有效地阻止硒取代蛋白质和其他硫化合物中的硫。因此，防御相关的植物激素可能在沙漠王羽过度积累硒过程中起着重要的信使作用（Freeman et al.，2010；Schiavon and Pilon-Smits，2017）。

砷（As）超积累植物蜈蚣草（*Pteris vittata*）也能大量积累硒，且没有明显的毒害症状，但生物量明显下降（Feng and Wei，2012）。研究发现，蜈蚣草的根比叶能积累更多的硒，根和叶中硒的最高浓度分别为 1536 mg/kg 和 242 mg/kg，表现出典型的富硒特征。加入 2 mg/L 硒可减少细胞丙二醛的生成，而 ≥5 mg/L 硒则增加丙二醛含量，说明低剂量硒具有抗氧化作用。过氧化氢酶（CAT）、抗坏血酸过氧化物酶（APx）和过氧化物酶（POD）只有在低硒浓度下才有抗氧化作用，表现为≤5 mg/L 时酶活性升高，＞5 mg/L 时酶活性降低。≥5 mg/L 能增加谷胱甘肽含量和增强谷胱甘肽还原酶（GR）活性，20 mg/L 的硒也能增强超氧化物歧化酶（SOD）活性。结果表明，SOD、GSH 和 GR 可能与蜈蚣草硒的积累有关，但 POD、APX 和 CAT 在蜈蚣草硒积累中的作用有限（Feng and Wei，2012）。

四、微生物与植物硒超积累

（一）硒超积累植物中的微生物

硒超积累植物在其天然富硒生境中蕴藏着一个丰富的内生细菌群落，与具有可比性的非富硒植物具有同等的多样性，这些内生菌具备较高的抗硒、产硒能力和促进植物生长的特性（Sura-de Jong et al.，2015；Cochran，2017）。另外，硒超积累植物的凋落物分解速率高于相关非超积累植物的凋落物，并且其中分解者丰度更高（Quinn et al.，2010a）。这种富硒生境中硒超积累植物凋落物的强化分解现象证明存在专性的抗硒微生物（Quinn et al.，2010a）。

芽孢杆菌属（Bacillus）是从硒超积累植物沙漠王羽和双钩黄芪中分离最多的一个属。其他种类丰富的细菌属有泛菌属（Pantoea）、假单胞菌属（Pseudomonas）和葡萄球菌属（Staphylococcus），另外还有类芽孢杆菌属（Paenibacillus）、小陌生菌属（Advenella）、节杆菌属（Arthrobacter）和贪噬菌属（Variovorax）（Sura-de Jong et al.，2015）。对分离纯化的菌株进行定性筛选，它们均能在硒酸盐和亚硒酸盐上生长，并能将亚硒酸盐还原为红色元素硒，但对硒酸盐均无此还原功能（Sura-de Jong et al.，2015）。

Li 等（2021）从硒超积累植物壶瓶碎米荠（Cardamine hupingshanensis）中分离到 11 个细菌属的 14 株可培养抗硒内生菌株，发现不同生境的植株内生菌都以变形菌门（Proteobacteria）为主（＞70%），其中 β 变形菌纲（Betaproteobacteria）、α 变形菌纲（Alphaproteobacteria）和 γ 变形菌纲（Gammaproteobacteria）的 OTU 比例较高。其中，海洋杆菌属（Oceanobacillus）和土杆菌属（Terribacillus）成员的耐硒特性是首次报道（Li et al.，2021）。

从富硒土壤中分离到的真菌比从非富硒土壤中分离到的真菌对硒的耐性更强（Wangeline et al.，2011）。来自富硒生境的根际真菌具有耐硒性，这可能是它们在富硒生境中的一种适应性优势。Wangeline 等（2011）从科罗拉多州和怀俄明州的 5 个地区（4 个富硒地点和 1 个非富硒地点）采集的富硒植物和非富硒植物中分离到根际真菌，对 259 株菌株进行了属、种鉴定，并对其耐硒性进行了评价。在 24 个代表性真菌属中，11 个属占总分离物的 86%。不考虑寄主植物种类（超积累与非积累植物）差异，富硒地点的大多数菌株不受硒（10 mg/L）的影响，而非富硒地点的根际真菌对 10 mg/L 的硒浓度高度敏感，总体耐性明显低于富硒地点的菌株（α=0.05）（Wangeline et al.，2011）。

（二）微生物对植物吸收硒的影响

关于根际微生物对植物生长和硒等元素积累的影响，已有较多报道。Cochran（2017）

的研究发现，根际微生物的存在对硒超积累和非超积累植物的生长及硒积累都有贡献（Cochran，2017）。微生物可以促进植物生长，强化植物对硒的耐受性和积累量（图6-5）（Sura-de Jong et al.，2015；Alford et al.，2014）。研究发现，根际细菌分泌物可以促进富硒植物印度芥菜的生长，并通过刺激根毛的形成解除部分吸收硒的速率限制，以及上调植物的S/Se同化来促进组织中硒的积累和挥发，进而增强植物的硒耐受性（de Souza et al.，1999；Lindblom et al.，2014；Freeman et al.，2010）。X射线探针分析表明，硒超积累植物的有机硒含量明显高于硒非超积累植物（Alford et al.，2014）。根瘤中的根瘤菌可为这些植物积累硒代氨基酸提供充足氮源，从而有助于这些植物超积累硒（Alford et al.，2012；Alford et al.，2014）。在Alford等（2014）的研究中，LC-MS硒形态分析结果表明，植物MeSeCys含量相近时，微生物的氮供应尤其有助于谷氨酸形成γ-谷氨酰甲基硒代半胱氨酸（γ-Glu-MeSeCys）。根部结瘤植株地上部32%的硒来自γ-Glu-MeSeCys，而在无根瘤的植株中，这一比例仅为2%（Alford et al.，2014）。

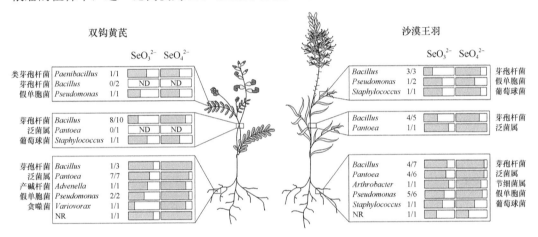

图6-5 内生菌在超积累植物双钩黄芪和沙漠王羽中的分布及其对亚硒酸盐和硒酸盐的抗性

（Sura-de Jong et al.，2015）

图中数值表示：硒抗性菌株数/分离得到的内生菌株数；ND表示未检测

另外，Lindblom等（2014）使用来自硒超积累植物沙漠王羽（*Stanleya pinnata*）的两种根际真菌硒链格孢菌（*Alternaria seleniiphila*）和兔粪曲霉（*Aspergillus leporis*）接种硒超积累植物沙漠王羽及其近缘种非超积累植物，研究了真菌对超积累植物中元素积累的影响，分析了植物生长和硒、硫的积累，同时利用X射线微探针法分析了元素分布和形态。结果发现，接种真菌同时降低了两种植物的硫转运，但却只降低了沙漠王羽的硒转运，而对两种植物根系硒的分布和形态无影响。两种植物主要积累有机硒（90%），而硫以有机和无机两种形式存在于根中。结果说明，根据寄主种类不同，这些根际真菌可以影响植物生长和硒或硫的积累（Lindblom et al.，2014）。

（三）微生物对植物硒转化以及抗硒胁迫的影响

土壤中，在缺氧环境下，亚砷酸硒菌（*Bacillusarseniciselenatis arseniciselenatis*）、亚硒酸亚硒菌（*Thauera selenatis*）和巴氏硫螺菌（*Sulfurospirillum barnesii*）等微生物可以将SeO_4^{2-}依次还原为SeO_3^{2-}和Se^0（Park et al.，2011；Kushwaha et al.，2021）。厌氧细菌

利用 SeO_4^{2-} 和 SeO_3^{2-} 作为末端电子受体，还原产生 Se^0 沉淀（Zawislanski et al., 2001）。有研究利用微聚焦 X 射线荧光光谱（μ-XRF）和 X 射线吸收近边结构（μ-XANES）光谱分别测定了硒的分布和形态，结果发现，单质硒在野外相当丰富，但在温室对照研究中仅在接种了可产单质硒的紫云英交链孢属真菌（*Alternaria astragali*）以及有固氮菌定居的根瘤中发现（Lindblom et al., 2013a）。除了将硒还原为 Se^0 外，从根际土壤中分离出来的根际细菌和真菌还能通过把 SeO_4^{2-} 或 SeO_3^{2-} 转化为二甲基硒（DMeSe）或二甲基二硒（DMeDSe）的形式挥发（de Souza et al., 1999）。与还原过程相比，氧化过程较为少见（Kushwaha et al., 2021）。氧化剂和土壤温度、水分等非生物因素可以导致一部分硒的氧化，但硒的氧化过程主要由微生物介导（Bassil et al., 2018）。据报道，硒的生物有效性会随着 SeO_3^{2-} 被氧化成 SeO_4^{2-} 而提高（Akiho et al., 2012）。对土壤成分更具有亲和力的 SeO_4^{2-} 是土壤中溶解度最大、生物有效性最高的一种硒形态（Peng et al., 2017）。

微生物可以产生元素硒（Se^0）并通过亚硒酸盐和硒酸盐（SeO_3^{2-} 和 SeO_4^{2-}）增加植物中硒的积累（Cochran, 2017）。硒超积累植物产生硒代氨基酸，这些硒代氨基酸在分解时为微生物提供了额外的（有机）硒来源（Cochran, 2017）。Di Gregorio 等（2005）从豆科硒超积累植物双钩黄芪根际土壤中分离到一株与嗜麦芽窄食单胞菌（*Stenotrophomonas maltophilia*）有关的细菌（SeITE02），该菌株对生长培养基中的亚硒酸盐 SeO_3^{2-} 的耐受性可达 50 mmol/L，能够迅速将有害的可溶性 SeO_3^{2-} 还原为不可溶和不可利用的 Se^0（Di Gregorio et al., 2005）。Sura-de Jong 等（2015）从超积累植物内生菌群落筛选了 7 个菌株进行植物接种，发现无论是在纯培养条件下，还是在与非积累作物芥菜或紫花苜蓿（*Medicago sativa*）共培养时，这些菌株都具有促进植物生长的特性，对植物硒的积累没有影响。而这些硒超积累植物内生菌都能将亚硒酸盐转化为单质硒，却不能转化硒酸盐（Sura-de Jong et al., 2015）。原因可能是细菌内生菌可以促进植物组织中的硒挥发，也有可能是接种的细菌也存在于根际，促进有效态硒向单质硒的转化，从而降低了植物对硒的生物有效性。

同样的，Staicu 等（2015）在筛选可用于废水生物修复和生产纳米硒颗粒的高耐硒性和还原能力细菌时，分离出超积累植物沙漠王羽中一种细菌内生菌——莫拉维亚假单胞菌亚种（*Pseudomonas moraviensis* subsp. Stanleyae）。这种菌可以耐受极端水平的硒酸盐和亚硒酸盐，在好氧条件下能有效地将亚硒酸盐还原为硒单质，对非积累作物芥菜的生长有 70% 的促进作用，同样对芥菜中硒的积累无明显影响（Staicu et al., 2015）。Alford 等（2012）研究了三种黄芪属植物 *Astragalus crotalariae*、*Astragalus praelongus* 和 *Astragalus preussii*，发现其根瘤中都含有硒，大部分以 C-Se-C 的形式储存。对硒超积累植物和非超积累植物的根瘤评价表明，由于植物的耐硒性，不存在结瘤抑制作用（Alford et al., 2012）。相反，在硒超积累植物中，较高水平的硒处理（高达 100 μmol/L）能够产生较多的根瘤数，这表明氮素固定在硒超积累中具有潜在的影响，而在非超积累植物中没有发生这种作用。Alford 等（2014）的研究进一步指出，根瘤互惠对黄芪体内硒的超积累有正向影响，一种可能的机制是由于植物根系结瘤后表面积增加，从而促进了根系对土壤元素的吸收。另一种互补的机制可能是，根瘤菌通过氮输入影响硒代氨基酸含量，从而改变植物的硒代谢，因为微生物的氮供应特别有助于谷氨酸转化形成 γ-谷氨酰甲基硒代半胱氨酸（γ-Glu-MeSeCys）（Alford et al., 2014）。

内生菌可以通过促进生长和影响元素积累来增强植物的抗逆性（Lindblom et al.，2018）。Lindblom 等（2018）的研究表明，极细链格孢菌（*Alternaria tenuissima*）在双钩黄芪根部形成单质硒，影响植株生长和硒的积累。细极链格孢菌对宿主组织中的硒含量敏感，但可以通过占据低硒部位（种皮、质外体）以及将植物组织中硒转化为无毒的单质硒来避免硒毒害（Lindblom et al.，2018）。Lindblom 等（2013b）研究两种真菌对黄芪属硒超积累植物总状黄芪（*Astragalus racemosus*）和硒非积累植物 *A. convallarius* 中硒的积累、运输和化学形态的影响，从黄芪属超积累植物根际中分离到两种真菌，即链格孢菌（*Alternaria astragali*）和锐顶镰刀菌（*Fusarium acuminatum*），接种链格孢菌可促进硒超积累植物的生长，但抑制非积累植物的生长。而硒处理消除了这些效应。锐顶镰刀菌降低了总状黄芪中硒从地上部到根部的转运（Lindblom et al.，2013b）。Li 等（2021）对硒超积累植物壶瓶碎米荠中的内生菌耐硒性和促进植物生长的能力进行比较，发现内生菌对不同浓度的亚硒酸盐和硒酸盐具有不同的抗性。进一步的接种试验结果表明，几种耐硒性强的内生菌可以促进白菜型油菜（*Brassica chinensis*）种子的萌发并缓解硒胁迫，具有促进高硒土壤植物生长的生物学潜力（Li et al.，2021）。

五、硒代谢与植物硒超积累

（一）乙烯、JA、SA 等防御相关激素调节的硒代谢机制

Wang 等（2018）研究发现，硒处理条件下，硒超积累植物沙漠王羽与其同属非富集植物落叶丹参（*Salvia elata*）根和地上部转录组基因表达水平有显著差异。研究还发现，与硒非超积累植物相比，在硒超积累植物沙漠王羽中，许多基因表现出组成性高表达，其中参与硫酸盐/硒酸盐转运和同化或参与抗氧化胁迫的相关基因（如谷胱甘肽和过氧化物酶相关基因）是这两种植物间表达差异较大的基因。另外，上调防御激素合成的上游信号基因在沙漠王羽中的表达高于落叶丹参，可能触发了这些硒介导的防御反应。因此，沙漠王羽中的硒超富集和强耐受性可能由组成性上调茉莉酸、水杨酸和乙烯调节的防御系统调节（Schiavon and Pilon-Smits，2017；Wang et al.，2018）。这与 Freeman 等（2010）的研究结果一致，在 20 mmol/L 亚硒酸盐处理下硒超积累植物沙漠王羽能完全耐受，而次生富集植物白花蛇舌草的生长受到毒害。转录组分析发现，与白花蛇舌草相比，沙漠王羽中硫同化、抗氧化活性、防御，以及与（甲基）茉莉酸、水杨酸或乙烯反应相关的基因组成性表达上调，激素的含量增加（Freeman et al.，2010）。此外，碱茅（*Puccinellia disans*）是一种具有极强的硼积累能力和耐高盐能力的植物，Kok 等（2020）对其富硒能力和耐盐性进行研究，认为硒同化和胁迫反应基因的上调可能是由于茉莉酸信号的诱导（Kok et al.，2020）。

（二）硫酸盐转运蛋白（SULTR）调控的硒超积累机制

Harris 等（2014）研究了不同浓度硫酸盐和硒酸盐对两种生态型沙漠王羽和非超积累芥菜硫、硒、钼的积累及转运的影响（Harris et al.，2014）。结果表明，与芥菜相比，当培养基中硫酸盐浓度从 0.5～5 mmol/L 增加 10 倍时，两种生态型沙漠王羽硒含量没有显著差异，而芥菜硒含量随着培养基中硫酸盐浓度的增加反而降低。此外，随着供硒

浓度的提高，沙漠王羽的硫含量略有下降，表明硫的吸收受到硒的竞争性抑制，而芥菜的硫水平则随着硒的增加而升高，这可能是由于调控硫酸盐转运蛋白的基因表达上调所致（Harris et al.，2014）。两种生态型沙漠王羽中的硒含量在幼叶中都要高得多，而芥菜在老叶中的硒含量略高于幼叶。这些结果表明，比之非超积累植物芥菜，硒超积累植物沙漠王羽可能含有修饰的硫酸盐转运蛋白，对硒酸盐比对硫酸盐有更高的特异性（Harris et al.，2014）。

Cabannes 等（2011）比较了硒超积累植物总状黄芪、双钩黄芪和其亲缘关系较近的非积累植物甜叶黄芪（*Astragalus glycyphyllos*）、鼓脉黄芪（*Astragalus drummondii*）对硫酸盐和硒酸盐的吸收能力，结果表明，硫饥饿培养增加了根和地上部组织中硒的积累，而增加硒供应则促进了根和地上部组织中硫酸盐的积累。在硒酸盐处理下，硫饥饿植株积累的硒是非硫饥饿植株的 2 倍以上，硫饥饿培养下不同植物品种的硒积累差异不显著（$P>0.1$），这在某种程度上可以解释为硒酸盐的吸收没有受到硫酸盐吸收的竞争性抑制（Cabannes et al.，2011）。从这些黄芪属植物中克隆了 1～4 组硫酸盐转运蛋白同源基因的 cDNA，研究其表达模式及其与硫酸盐和硒酸盐吸收的相互作用关系（Cabannes et al.，2011）。与对照相比，不考虑硫和硒的状态，无论是在富硒黄芪还是非富硒黄芪中，都观察到了丰富的硫酸盐转运蛋白基因表达。此外，基因表达的定量分析表明，在缺硫条件下，超积累黄芪的转录组水平与其他植物相当。硫酸盐转运蛋白在某些黄芪属植物中的高表达可能导致硒吸收和转运能力的增强，从而产生硒超积累特性。但这还不足以解释 S/Se 的选择机制（Cabannes et al.，2011）。另外有研究发现，从几个黄芪属硒超积累植物中，分离的 SULTR1 序列中含有丙氨酸残基，并没有非积累被子植物 SULTR1 亚型中发现的甘氨酸。这一结果说明这些黄芪属超积累植物优先吸收硒酸盐而不是硫酸盐（White，2016；Guignardi and Schiavon，2017；Cabannes et al.，2011）。这解释了超积累植物高 Se/S 比和硒超积累的特点（White，2016）。

Trippe 和 Pilon-Smits 报道（2021），除了 SULTR1;2 之外，硫酸盐转运蛋白还包括 SULTR2;1 和 SULTR3;5（主要负责硒酸盐从根到地上部的转运），以及 SULTRs 3;2-3;4（主要负责叶绿体硒酸盐转运）。ATPS 酶催化从硒酸到 SeCys 的代谢，CpNifS 酶催化从 SeCys 到 SeMys 的代谢，CγS 酶负责从 SeCys 到 SeMet 的代谢，MMT 酶负责从 SeMet 到 DMeSe 的代谢，SMT 和 CSl 参与从 SeCys 到 DMeDSe 的代谢（Trippe and Pilon-Smits，2021）。

El Mehdawi 等（2018）研究了外源硫酸盐对超积累和非超积累十字花科植物吸收硒动态及硫酸盐/硒转运蛋白表达的影响。研究发现，沙漠王羽不仅超量积累硒至其干重的 0.5%，而且组织中的 Se/S 比值也较高。在沙漠王羽中，硒的吸收和转运率很高，并且相对不依赖于硫酸盐。潜在的机制可能包括 *Sultr1;2* 和 *Sultr2;1* 的过表达，表明可能已进化出对硒酸盐比对硫酸盐更强的特异性（El Mehdawi et al.，2018）。Schiavon 等（2015）比较了超积累植物沙漠王羽和相关的非超积累植物芥菜对硫依赖的硒酸的吸收和转运，以及 3 种硫酸盐/硒酸盐转运体（SULTR）和 3 种 ATP 硫化酶（APS）的表达水平。结果表明，超积累植物的 Se/S 比值普遍高于非超积累植物，非超积累植物芥菜的组织 Se/S 比值基本对应了培养基中 Se/S 比值的变化，而超积累植物沙漠王羽对硒的富集度可达硫的 5 倍。沙漠王羽中 *Sultr1;2* 和 *2;1*，以及 *APS1*、*APS2* 和 *APS4* 的转录水平普遍高于芥菜，且不同植物对硫和硒供应的反应存在差异。此外，超积累植物具有较高的硫酸盐/硒酸盐转运蛋白和 APS 酶的表达水平，这可能与硒的超积累和超耐受表型有关（Schiavon et al.，2015）。

Rao 等（2021）对硒酸盐处理的硒超积累植物堇叶碎米荠幼苗进行了研究，发现在其他硒超积累植物中未报道过的新机制。结果表明，硫酸盐转运蛋白（SULTR）基因（*Sultr1;1*，*Sultr1;2* 和 *Sultr2;1*）和硫同化酶基因对硒酸盐的响应表现出高表达水平。多个钙蛋白和富含半胱氨酸激酶的基因在硒酸盐诱导下表达下调，而硒结合蛋白 1（SBP1）和缺硫诱导蛋白 2（SDI2）在硒酸盐诱导下表达上调。钙蛋白和富含半胱氨酸激酶的表达下调以及 SBP1 和 SDI2 的表达上调是堇叶碎米荠耐硒性增强的重要因素（Rao et al.，2021）。

（三）硒代半胱氨酸甲基转移酶（SMT）介导的硒超积累机制

在硒超积累黄芪和非积累黄芪中，都发现了一种硒代半胱氨酸甲基转移酶（SMT），可使 SeCys 甲基化。不同黄芪品种的硒积累能力与 SMT 的酶活性直接相关，活性 SMT 有助于转移硒进入蛋白质，从而减少硒毒性（Sors et al.，2005b）；硒超积累黄芪中的 SMT 活性是非积累黄芪物种的 17.5 倍（Sors et al.，2009；Cabannes et al.，2011）。SMT 转基因植株中硒的总积累量远低于超积累植物中的总积累量，这表明 SMT 是一种重要的超积累和耐性酶，也可能还参与了对硒超积累有协同作用的其他过程。Sors 等（2009）发现 SMT 活性与硒的超积累有关，但没有证据表明酶减少或吸收硫能力的差异对黄芪中硒的超积累有重要作用。此外，由于非积累型 SMT 同系物在体外会丧失活性，因此在非积累型物种中几乎没有发现可检测到的 MeSeCys（Freeman et al.，2010；Sors et al.，2005b；Sors et al.，2009）。McKenzie 等（2009）的研究也指出，富硒植物对高硒土壤的耐性与其生物合成甲基硒代半胱氨酸（MeSeCys）的能力有关，这与 SMT 的活性有关。然而，在非硒积累植物拟南芥中，SMT 的过表达没有导致硒酸盐生物代谢成 MeSeCys，因为 ATP 硫酰化酶将硒酸盐还原为亚硒酸盐的速率很低，在其他能够产生 MeSeCys 的十字花科超积累植物中却几乎没有这一限制。与单独过量表达 SMT 相比，同时高表达 ATP 硫酰化酶和 SMT 的植物品系并没有表现出进一步的总硒积累增加或叶片毒害症状加剧，而是促进了更大比例的硒转化为 MeSeCys 形态（McKenzie et al.，2009）。

LeDuc 等（2004）从硒超积累植物双钩黄芪中发现，过表达的 SMT 基因会增加硒代半胱氨酸甲基转移酶的表达和活性，使硒代半胱氨酸甲基化形成甲基硒代半胱氨酸（MeSeCys）（LeDuc et al.，2004）。这是一种非蛋白组成氨基酸，可将硒从硒代半胱氨酸（SeCys）转移到随后的硒代甲硫氨酸（SeMet）。这两种硒氨基酸都可以错误地结合到蛋白质中，改变蛋白质的天然结构和功能（Brown and Shrift，1982；Terry et al.，2000；Kubachka et al.，2007）。然而，当硒以硒酸盐形式供应时，SMT 酶的作用要小得多。为了提高硒酸盐的植物修复效率，除 SMT 外，还可以培育高表达 ATP 硫酰化酶（APS）基因的双转基因植株，即 APS×SMT。结果表明，在供应硒酸盐条件下，同时表达 APS 和 SMT 的转基因植物中硒的积累有显著的增强（增加 4～9 倍）（LeDuc et al.，2006）。

（四）调控植物超积累硒的其他相关物质

硒酸盐是硒代谢的起始化合物（Zhou et al., 2018）。硒的化学结构与硫相似，植物通过相同的代谢途径吸收硒（Van Hoewyk et al., 2008; Cakir et al., 2016b）。前已述及，大多数植物通过硫酸盐转运蛋白从环境中非特异性地吸收硒，并通过硫代谢途径将硒同化成有机形式的硒。硒酸盐转化为亚硒酸盐需要两种酶的连续作用：ATP 硫酰化酶（APS）介导硒酸盐与 ATP 的结合，形成腺苷磷酸硒酸酯（APSe），这种化合物通过 APS 还原酶（APR16）被还原为亚硒酸盐。在壶瓶碎米荠幼苗的根组织中，亚硒酸盐可能首先被转化为硒酸盐，然后被 APS 结合到 ATP 中，再被 APR 还原为亚硒酸盐，然后被还原为硒化物，最后被结合到 SeCys 中（Zhou et al., 2018）。利用拟南芥 DNA 微阵列对十字花科超积累植物沙漠王羽和非积累植物 *Stanleyaalbency albency* 进行了比较分析，结果表明，在沙漠王羽中，许多重要的硫同化基因在叶片和根中都有组成性上调。在叶片中，硒和硫积累的潜力似乎与三个 Cys 合成酶基因的组成性上调有关（Freeman et al., 2010; Cabannes et al., 2011）。

Rao 等（2020）通过 PacBio SMRT-Seq 和 Illumina RNA-Seq 的元分析揭示了硒酸钠处理的硒超积累植物堇叶碎米荠中与硒代谢相关的候选基因和通路。该研究从 134 个转录本中鉴定出 51 个与硒代谢有关的基因，包括转运蛋白、同化酶和几个特异基因，进一步揭示了硒代谢的分子机制（Rao et al., 2020）。对 Illumina RNA-Seq 数据的分析表明，Na_2SeO_4 处理组共筛选出 948 个差异表达基因（DEG），其中 11 个 DEG 与硒代谢相关。对所有 DEG 的 KEGG 途径的富集分析表明，它们在激素信号转导途径和植物病原物相互作用途径等 5 条途径中显著富集。4 个与硒代谢相关的基因均受 lncRNA 的调控，它们分别是：腺苷三磷酸硫酶 1、腺苷 5′-磷酸硫酸盐还原酶 3、半胱氨酸脱硫酶 1 和丝氨酸乙酰转移酶 2。通过加权基因共表达网络分析，筛选并鉴定了 20 个潜在的关键基因（如硫酸盐转运蛋白 Sultr1;1、甲硫氨酸 γ 裂解酶 MGL 和硒结合蛋白 SBP1 的基因），这些基因在堇叶碎米荠的硒积累和耐性中起重要作用（Rao et al., 2020）。

高浓度硒显著影响植物体内部氧化还原动态平衡，主要通过调节硫氧还蛋白（TRX）和谷胱甘肽（GRX）对活性氧（ROS）的清除和蛋白质的硫醇/二硫状态影响氧化还原状态，并通过感知和还原大量的靶蛋白（如还原酶、过氧化物酶、转录因子、糖酵解代谢酶等）来维持细胞氧化还原动态平衡（Zhou et al., 2018）。

在 Hung 等（2012）的研究中应用荧光差异显示技术分析了 20 μmol/L 硒酸盐（K_2SeO_4）处理 2 周的总状黄芪的转录谱。在 125 个确定的候选应答基因中，有 9 个基因在两个独立的实验中被硒酸盐处理诱导或抑制了 2 倍以上的表达水平，而在亚硒酸盐处理下，有 14 个基因的表达水平出现了这样的变化。其中的 6 个基因被发现对硒酸盐和亚硒酸盐处理都有反应，一个新基因 *CEJ367* 被硒酸盐（1920 倍）和亚硒酸盐（579 倍）高度诱导（Hung et al., 2012）。

植物中小 RNA（miRNA）在转录后调节基因表达，是许多生物和代谢过程中的关键分子，也可能在植物中硒的积累中发挥重要作用。Cakir 等（2016a）利用深度测序技术（deep sequencing strategy）鉴定黄芪属植物 *Astragalus chrysochlorus* 愈伤组织中对硒响应的 miRNA，结果显示 miR1507a、miR1869 和 miR2867-3p 表达上调，miR1507-5p

和 miR8781b 表达明显下调（Cakir et al.，2016a）。

第四节　植物硒超积累的生态效应

一、富硒土壤上植物群落组成和植被特征

超积累植物可能对硒具有生态依赖性，以保护其免受各种生物胁迫。超积累植物可以影响邻近生物对硒的积累，而土壤硒可影响植物之间的竞争和互利共生（El Mehdawi et al.，2012）。超积累植物体内及其周边极端硒含量水平不利于硒敏感的生物，而有利于耐硒物种。这种超积累植物产生的双重效应，可能会影响其邻近环境中的物种组成及硒的循环（Reynolds and Pilon-Smits，2018）。

研究发现，菊科（Asteraceae）硒超积累植物柳叶白菀在富硒和非富硒土壤中的硒积累能力具有显著差异。柳叶白菀种群在富硒地区松岭（Pine Ridge，Colorado，USA）能够超积累硒，而在附近非富硒地区云关（Cloudy Pass，Colorado，USA）没有超量积累硒，说明柳叶白菀对硒的超积累能力与土壤硒含量有关（El Mehdawi et al.，2015b）。

Reynolds 等（2019）调查了两个自然富硒地点的 54 种超积累植物及其 5 个相近物种，并将这些物种在超积累植物周围的相对丰度与整个植被中的相对丰度进行了比较，确定了一些物种与超积累植物呈正或负相关关系。与负共生物种相比，两个正向共生物种中的有机硒形态相对更多。硒超积累植物周围土壤硒水平较高，与生长在该地区不同地方的同一物种相比，硒超积累植物的附近植物组织中硒含量也较高。硒超积累植物周围土壤硒含量较多可能是由于凋落物沉积造成的，会显著影响当地植物群落，即在促进耐硒植物群落成员发展的同时，降低硒敏感成员的生态适应性（Reynolds et al.，2019）。

高硒植物凋落物的分解可能是由一种特殊的耐硒微生物群落介导的，是硒循环中的一个重要因素。Quinn 等（2010a）采用凋落物袋法，比较了超积累植物双钩黄芪及其近缘非积累种鼓状黄芪和紫花苜蓿凋落物的分解、硒损失、分解者群落和凋落物土壤中的硒浓度。高硒凋落物分解速率较高、分解物丰度较高，表明在这一含硒生境中存在耐硒分解者，这可能是导致高硒凋落物分解速率增加的原因之一（Quinn et al.，2010a）。El Mehdawi 等（2011a；2011b）用 X 射线荧光图谱和 X 射线吸收近边结构谱研究了生长在超积累植物和非超积累植物旁的银叶艾（*Artemisia ludoviciana*）、柳叶白菀、藜草（*Chenopodium album*）叶片中硒的分布和形态。结果表明，在富硒的土壤中，硒的分布与形态存在明显的差异，与其他植被相邻的土壤相比，硒超积累植物附近的土壤中的生物可利用硒水平高出 2.5～3 倍（El Mehdawi et al.，2011a；2011b）。因此，超积累植物凋落物可能是土壤有机硒的来源之一，而土壤微生物也可能起到一定作用（El Mehdawi et al.，2015a）。

Reynolds 等（2020）调查了科罗拉多州三个富硒地带植被物种的组成特征和土壤硒的含量与分布，发现当地超积累植物会至少在直径 3m 的区域或 4 倍于其冠层的范围内显著影响土壤中的硒浓度和植被。在整个采样区内，有超积累植物的样地土壤硒含量范围为 1.7～34.7 μg/g，没有超积累植物的样地土壤硒含量范围为 0.8～16.1 μg/g，超积累植物生境土壤平均硒含量是非超积累植物生境土壤硒水平的 2 倍左右（Reynolds et al.，2020）。该研究在较小的尺度上取得了极显著的结果：超积累植物周边土壤比离它们更远

的土壤有更高的硒水平，超积累植物生存的样地有更多的裸地，冠层覆盖更少，环境异质性增加，植被物种丰富度更大。超积累植物生存的土壤中硒含量较高有两种可能的原因：一是超积累植物通过重新分配硒而使土壤富硒；二是超积累植物更适合在富硒的小区生境中存活，在这种情况下，是植物直接选择了那些小区生境，而不是植物导致了土壤硒的富集。但也有可能是这两种机制发生共同作用（Reynolds et al.，2020）。

进一步的研究表明，超积累植物可以通过提高周围的土壤硒水平减少植食者取食和促进生长来促进耐硒植物的生长，同时抑制硒敏感植物的竞争（El Mehdawi et al.，2011a）。通过这些竞争和促进效应，硒超积累植物可能会影响植物群落组成，从而影响更高的营养级（El Mehdawi et al.，2011a）。换言之，超积累植物通过将硒以高度生物有效性的有机形式聚集在植物体内和周围，促进群落耐硒植物成员发展，同时降低硒敏感成员的生态适应性，从而对当地植物群落产生影响。这些对当地植被的影响可能将反过来影响其他营养级，以及当地生态系统中整体的硒循环（Reynolds et al.，2019）。

二、硒超积累植物对硒循环的影响

硒以有机和无机形式广泛存在于海洋和淡水系统、土壤、生物和大气中。植物组织中硒的积累和挥发性甲基化硒形态的产生都是环境中硒循环的关键（Lima et al.，2018；Winkel et al.，2015）。研究发现，某些低硒水平陆地环境导致人类缺硒，而水和土壤中过高的硒含量可能是有毒害的，会造成水生野生动物和其他动物的死亡。人类膳食硒的摄入量在很大程度上受植物中硒含量的控制，而植物根对硒的吸收作用可影响土壤硒浓度、形态和生物有效性。此外，植物和微生物可以将硒甲基化，这导致硒逸散流失到大气中（Winkel et al.，2015）。硒的浓度达到有毒水平后，再加上硒迁移到更多生物可接触到的位置，可能会对生物环境产生显著影响。有毒浓度的硒在超积累植物组织中的存在可能会极大地影响这些植物与食草动物、传粉者、其他植物和微生物（无论是在植物内部还是在其外部影响区域）之间的生物相互作用。这些生物的相互作用有利于耐硒的生物类群，可能会改变群落物种组成，对超积累植物生长的生态系统产生较大的影响（Reynolds and Pilon-Smits，2018）。

硒超积累植物对硒再分配和形态的影响可能会创造出一个更加异质的环境，创造出独特的生态位，增加物种丰富度和景观异质性（Reynolds and Pilon-Smits，2018）。土壤是超积累植物中硒的来源。超积累植物可能通过将硒同化为有机形式，进而影响高浓度硒在根和地上部组织的渗出及转运，从而影响土壤中硒的形态（Sors et al.，2005a；Schiavon and Pilon-Smits，2017；Reynolds et al.，2020）。生态适合度的变化会通过增加或减少物种丰富度、相对物种丰富度和改变群落组成来影响当地的物种多样性。这些局部效应将对它们所属区域的物种多样性产生影响。随着时间推移，通过局部和区域生态过程之间的循环作用，代表着强大选择压力的硒超积累植物可能会在空间和时间上对当地生态系统产生不成比例的巨大影响（Reynolds and Pilon-Smits，2018）。

三、其他物种对硒超积累植物的反应

一方面，硒可以保护超积累植物免受普通食草动物的侵害；另一方面，在自然选择过程中，生物间相互作用发生协同进化也产生了抗硒或耐硒性食草动物。有研究表明，具有硒耐受性的草食昆虫 *Bruchophagus mexicanus* 和甲虫 *Acanthoscelides fraterculus* 能够取食超积累植物的种子，这些草食动物可将硒排出体外，从而大大减少了组织硒积累，而耐硒小菜蛾（*Plutella xylostella*）将硒隔离在腹部不同的区域，可能参与了排毒或幼虫对捕食者的防御（Freeman et al., 2006b; Freeman et al., 2012）。这解释了它们食用高硒植物而不遭受毒害的能力，从而使其能够占据硒超积累植物所提供的独特生态位。这些抗性植食者与超积累植物又可能继续发生协同进化，从而促进植物硒富集能力的增强。

第五节 硒污染土壤的植物修复及案例研究

植物对微量元素的超积累是一个很有吸引力的研究领域，在污染场地的植物修复、食品中必需营养的生物强化和生物燃料的生产等方面具有潜在的应用前景（Zhu et al., 2009; Barceló and Poschenrieder, 2010）。植物硒的积累和挥发可以应用于作物的生物强化和植物修复（Padmavathiamma and Li, 2007; Schiavon and Pilon-Smits, 2017; 王晓东, 2019）。硒生物强化是一种农业过程，通过植物育种、基因工程或使用硒肥来增加作物中的硒积累。有关硒的生物强化技术已经在第五章中进行了较为系统的讨论，本节主要讨论硒污染的植物修复问题。硒污染植物修复是一种主要通过植物提取和植物挥发来净化受硒污染环境的绿色生物技术。通过将硒植物修复和生物强化技术相结合，从植物修复中获得的高硒含量植物材料可以作为富硒绿肥或其他的硒补充来源，用于生产硒生物强化农产品（宋成祖, 1995; Wu et al., 2015）。

一、硒超积累植物的污染修复应用

植物修复是指利用植物去除、控制环境污染物或将环境污染物无害化（Garbisu and Alkorta, 2001; Berken et al., 2010），对去除环境有毒金属是一种生态友好、经济有效的技术（Girdhar et al., 2014）。植物修复包括植物提取、植物稳定、根际过滤和植物挥发（Salt et al., 1995; Salt et al., 1998; Garbisu and Alkorta, 2001; Dhillon and Dhillon, 2003; Usmani et al., 2019）。与传统的物理和化学修复方法相比，植物修复是一种可选择的、可持续的修复技术（Wu et al., 2015）。植物修复技术有可能为修复土壤、沉积物、尾矿、固体废物和水体等低到中度污染的基质提供一种具有成本效益的就地替代技术（Mench et al., 2010）。

（一）植物提取

植物提取是利用植物从污染或矿化的土壤中吸收有毒元素，并将其运输和富集到地上部的技术。可以通过收割植物地上部来去除土壤中的污染元素，在某些情况下，根也可以收获（Garbisu and Alkorta, 2001; Chaney et al., 2007; Chaney and Baklanov, 2017）。植物提取的优点之一是，含有污染物的植物材料可以作为一种资源来利用（Prasad and Freitas, 2003）。通过种植植物来清除一个地区的过量硒，同时为其他缺硒地区提供必需的硒来源，这是一种引人入胜的技术（Valdez Barillas et al., 2011）。植物提取还是减轻土壤高硒风险的一种低成本替代方法。由于多数含硒过高的土壤中还存在着高水平的硫酸盐，植物组织对硒的高选择性吸收和对硫酸盐的选择性积累，再加上植物组织中高效的硒代谢解毒，是硒植物提取和土壤修复的关键特征（Chaney et al.，2007）。

Freeman 和 Banuelos（2011）筛选和鉴定的耐盐（Na_2SO_4、NaCl）/耐硼（B）基因型沙漠王羽为成功修复高盐/硼和富硒农业排水底泥中的硒提供了新方法。Freeman 和 Banuelos 等（2011）的研究表明，经过两个生长季节（约 18 个月），耐受基因型硒超积累植物及其伴生微生物对硒的超积累和挥发，可以去除 0～30cm 沉积物中约 30% 的土壤硒（Freeman and Banuelos，2011）。El-Ramady 等（2015）的研究表明，利用芦荻（*Arundo donax*）从自然环境中提取硒是一种新兴的绿色植物修复技术。巨型芦苇可以在富硒土壤上茁壮成长，在遭受林火等自然灾害后，其恢复土壤生态系统的效果好、能力强。因此，巨型芦苇在硒污染土壤植物修复中具备一定潜力（El-Ramady et al.，2015）。白菜型油菜（*Brassica chinensis*）也可用于污染土壤中硒的植物提取。有研究表明，在亚硒酸盐和硒酸盐处理的土壤中，小白菜的提取效率分别为 4.91% 和 31.90%，说明在这两种无机硒形态中，小白菜更容易吸收积累硒酸盐（Li et al.，2015）。一些研究认为芥菜是一种富硒植物；除了芥菜，还有可食用的十字花科植物芜菁（*Brassica rapa* var. *hakabura* 和 *Brassica rapa* var. *peruviridis*）（Yawata et al.，2010；Ogra et al.，2010）。对这些独特的十字花科植物的富硒能力进行评价发现，3 种植物根和叶中硒含量差异不显著，积累的硒形态是相同的。生长最快、生物量最高的叶用芜菁对硒的积累效率最高。此外，叶用芜菁含有铁、铜和锌等易提取的必需矿物质。叶用芜菁具有作为矿物质来源以及修复硒污染土壤和水体的潜力（Yawata et al.，2010）。Ogra 等（2010）评估了富硒植物芥菜在植物修复碲（Te）和硒方面的潜在用途。将印度芥菜植株暴露于硒酸盐和碲酸盐中，测定了植株中硒、碲的含量和化学形态，印度芥菜对碲的积累量小于硒，约为硒的 1/69。印度芥菜能严格区分碲酸盐和硒酸盐，其具备作为硒植物修复的潜力，但不适合碲的植物修复（Ogra et al.，2010）。

利用基因工程培育具有更强的耐硒、积累和挥发硒能力的速生植物可提高硒的修复效率（LeDuc et al.，2006；Merkle，2006；Zhang et al.，2006）。这首先需要鉴定出具有增强硫酸盐或硒酸盐特异性的转运蛋白，并克隆相关基因，这些转运蛋白可能会在作物生产、植物修复和生物强化方面得到应用。硫酸盐特异性转运蛋白的表达可以用于生物工程作物，特别是存在氧离子竞争的情况下，植物对硫饥饿的敏感性较低。编码硒酸转运蛋白的基因可用于富硒植物生物工程，特别是用于在高硫酸盐环境中的生物强化或植物修复（DeTar et al.，2015）。当利用转基因植物通过植物提取进行环境修复时，在植物达到较高生物量并仍具有较高的硒含量时收获地上部材料，可使修复效率最大化（Galeas et al.，2007）。

（二）植物挥发

植物挥发（phytovolatilization）是利用植物从土壤中提取挥发性污染物（如硒、砷、汞），并从枝叶中挥发的过程（Garbisu and Alkorta，2001）。由于硒容易挥发，硒的生物挥发是从受污染的水和土壤中向大气中去除硒的重要途径（Lin et al.，2002）。硒的氧化态（亚硒酸盐和硒酸盐）是高度可溶的，因此具有生物有效性和潜在的毒性（Hansen et al.，1998）。植物和微生物都可以通过生物甲基化作用就地挥发硒，其挥发速率与硒浓度、硒形态、温度、有机碳等环境因素有关（Pilon-Smits et al.，1999）。根据硒的同化途径（Terry et al.，2000），供应硒代甲硫氨酸（SeMet）的植物比供应硒代半胱氨酸（SeCys）的植物更易挥发硒，其次是亚硒酸盐，然后是硒酸盐。植被对硒挥发通常很重要，不仅因为植物可以直接挥发硒，还因为植物创造的根际环境支持特定的土壤微生物，从而可能对挥发作用有重大贡献（Lin et al.，2002）。从底泥中额外释放的挥发性硒表明，在无植被的裸地中观察到的硒和硫挥发是由于微生物活动产生的，植物和与植物相关的土壤微生物在硒挥发中起着重要作用，这解释了无植被的裸地中硒挥发的平均速率要高于盐草（*Distichlis spicata*）植被区的现象（Lin et al.，2002；Bañuelos and Lin，2010）。

通过植物的吸收和积累，将无机硒代谢为相对无毒的挥发性甲基硒化物挥发到大气中，以及还原为不溶性形式固定在沉积物中，可实现人工湿地废水中硒的去除（Pilon-Smits et al.，1999）。挥发性硒主要以二甲基硒（DMeSe）的形式存在（Karlson and Frankenberger，1988；Lin et al.，2000），其毒性仅为硒酸盐或亚硒酸盐的0.17%（Wilber，1980）。另外还有少量的二甲基二硒（DMeDSe）和其他硒有机化合物（Lin et al.，2002）。有研究表明，在美国西部半干旱地区的高温、干旱、高盐和高硼环境中，硒超积累植物沙漠王羽可以用于提取硒以及强化甲基硒化物挥发（Parker et al.，2003）。尽管土壤中高浓度硫酸盐对土壤硒的吸收和挥发有竞争性抑制作用，会显著降低植物修复的效率，但沙漠王羽对硒酸盐的亲和性明显高于硫酸盐（Parker et al.，2003）。

Lin 等（2002）在一项为期12个月的研究中观察到，北美海蓬子（*Salicornia bigelovii*）通过生物挥发到大气中的硒为 (62.0 ± 3.6) mg/($m^2 \cdot a$)，这比其组织中积累的硒高5.5倍。在淹滨草地块中，植物挥发除硒比例占年总硒输入量[957.7 mg/($m^2 \cdot a$)]的6.5%（Lin et al.，2002）。Hansen 等（1998）研究了位于旧金山湾附近的36 hm^2 人工湿地中生物硒挥发的作用。人工湿地植被系统对亚硒酸盐污染的炼油废水中的硒有很好的去除效果，去除率可达89%。出水硒浓度由 20~30 μg/L 降至 <5 μg/L。生物挥发占硒去除总量的10%~30%，另外大部分硒被固定到底泥和植物组织中，硒浓度分别达到5 mg/kg 和 15 mg/kg 左右。平均挥发率最高的是长芒棒头草（*Polypogon monspeliensis*）、香蒲（*Typha angustifolia*、*Typha domingensis*、*Typha latifolia*）和荆三棱（*Scirpus maritimus*），分别为 190 μg/($m^2 \cdot d$)、180 μg/($m^2 \cdot d$) 和 150 μg/($m^2 \cdot d$)。最具优势的盐沼芦苇在一年中的速率不同：2月、6月和10月的平均速率分别为 150 μg/($m^2 \cdot d$)、70 μg/($m^2 \cdot d$) 和 25 μg/($m^2 \cdot d$)。因此，生物硒挥发是湿地除硒的重要途径（Hansen et al.，1998）。

（三）根际过滤

根际过滤（rhizofiltration）是利用植物根部吸收废水中的污染物，主要是金属（Salt et al.，1995；Salt et al.，1998），例如，通过在曝气水中生长的植物根系或幼苗吸收、沉淀和浓缩污染环境中的有毒金属（Garbisu and Alkorta，2001）。理想的根滤植物应该有快速生长的根，能够在较长时间内去除溶液中的有毒金属。对于污染物浓度低的大型水体，根际过滤修复技术尤为经济高效（Salt et al.，1995）。该技术作为一种可持续、生态友好、经济实惠的技术，在污染场地的修复方面受到了世界各国的重视。根滤的主要优点是既可以原位利用，也可以易位利用，还可以考虑超积累植物以外的其他物种（Malik et al., 2015）。它涉及使用水生和陆地植物来处理工业排放、农业径流或矿山酸性废水。为了有效地去除现场的污染物，植物必须拥有庞大、高效和快速生长的根系，能够清除这些有毒物质，从而减轻它们的有害影响。低维护成本、易于处理和植物对重金属的高度抗性是从特定区域有效去除重金属的其他需要考虑的重要标准（Malik et al., 2015）。

湿地植物对硒的根际过滤研究较少（Dhillon and Dhillon, 2003）。Zayed 等（1998）的调查研究表明，浮萍（*Lemna minor*）是一种漂浮的水生植物，其富集硒的浓度与其他已知的富硒物种相当（Zayed et al., 1998）。在一项类似的研究中，水葫芦（*Eichhornia crassipes*）被证明是硒的积累者。这两种植物通常用于人工湿地进行废水处理（Zhu et al., 1999）。为评价人工湿地处理系统的除硒效果及其潜在的生态风险，Zhou 等（2019）构建了一个由三个主要营养级组成的浮叶植物系统，实施处理 21d 后，水体硒浓度降低了 40.40%，其中的 24.03% 由植物去除，74.41% 进入底泥沉积。

Garousi 等（2016）以水培条件下培养的向日葵（*Helianthus annuus*）作为根际过滤系统模型，测定了不同浓度硒酸钠和亚硒酸钠处理的向日葵植株的含硒量和叶绿素参数，对其耐硒和富硒能力进行评价。结果表明，随着施硒量的增加，向日葵植株的硒含量显著增加。此外，在 3 mg/L 硒酸钠和亚硒酸盐处理下暴露 3 周后，叶绿素 a 和 b 均未受到损害，说明可用于水体净化的向日葵植株具有很高的耐硒性。在植物修复过程中，向日葵根部的硒酸盐向地上部的转运比亚硒酸盐更容易（Garousi et al., 2016）。

Srivastava 等（2005）在 20 mg/L 硒酸盐和亚硒酸钠处理水培条件下，研究了 11 种蕨类植物的富硒潜力。结果表明，植物根中硒的浓度分别达到 245～731 mg/kg 和 516～1082 mg/kg，叶片中的硒含量分别为 153～745 mg/kg 和 74～1028 mg/kg，无明显毒性症状。其中 3 种蕨类植物的叶部比根部积累了更多的硒，这 3 种蕨类植物分别是硒酸盐处理的杯盖阴石蕨（*Davallia griffithiana*）、亚硒酸盐处理的蜈蚣草和两种无机硒处理的眉刷蕨（*Actiniopteris radiata*）。富硒蕨类植物硒积累高、受硒毒性很小，这表明利用蕨类植物修复受硒污染的水体和土壤是具有可行性的（Srivastava et al., 2005）。

二、硒污染土壤修复的典型案例研究

美国加利福尼亚州帕洛阿尔托（Palo Alto）的圣华金河谷西部广泛分布着高硒土壤。20 世纪 80 年代，由于农业灌溉，大量硒元素随径流和圣路易斯排水渠（San Luis Drain）输送并积累在凯斯特森（Kesterson）水库中。在 4 年时间内，高浓度硒的毒性导致多达 64% 的孵化野生水鸟和候鸟死亡或畸形（Marshall, 1985）。一夜之间，凯斯特森水库变成了鱼类和候鸟的死亡陷阱，硒也被认为是一种环境污染物（Presser, 1994）。

为了控制硒的环境毒害效应，水库被排干，并被土壤填充物覆盖（Wu, 2004）。然而，这种硒的释放机制仍然威胁着加州和世界各地（Lemly, 2004）。

植物修复作为一种低成本、环境友好的方法来控制土壤和水环境中的硒，得到了越来越多人的认可。但是，植物修复硒污染技术的成功取决于技术的潜在经济效益和种植者接受的程度。实践表明，如果采用的修复植物能同时生产有经济价值的产品，则更有助于技术的推广应用（Bañuelos, 2006）。有价值的产品包括富硒农产品、饲料、有机肥和可用作生物燃料的油料（Bañuelos, 2006）。

基于 Bañuelos 和 Meek（1990）等研究制定的硒污染植物修复技术策略，美国科学家发现硒超积累植物双钩黄芪和沙漠王羽硒含量高于 0.5%，是比较理想的修复植物（Zhang et al.，2007）。一些研究人员开发了在不同条件下利用这些植物对硒污染的修复治理技术，并进行了大面积示范（Frankenberger and Karlson，1995；Wu et al.，2000；Bañuelos et al.，2002a；Lin et al.，2002）。主要的经验是，为了使硒污染植物修复技术具有实用性和可持续性，要求所选择的植物可以与农作物进行轮作，且不应该给土地所有者造成经济损失。

植物修复是一种有效的除硒技术，但富硒植物的应用受到植物生长缓慢、栽培技术要求不明确或生物量低的限制（Dorado et al.，2003），因此，选择合适的富硒植物种类具有重要意义（He et al.，2018）。Bañuelos（2002）的报道指出，加利福尼亚州中部用于植物提取硒的潜在作物包括芸薹属的富硒植物，如西兰花（*Brassica oleracea*）和油菜（*Brassica napus*）等。经研究，油菜和西兰花不仅能从土壤中去除可溶性硒，收获的富硒作物还可生产出对种植者具有潜在重要经济价值的产品，例如，使用富硒植物材料作为补充动物饲料或食用蔬菜、将榨出的菜籽油与柴油混合用于生产柴油发动机的生物燃料，以及将榨油后的种子副产品作为动物饲料。油菜、向日葵（*Helianthus annuus*）、红花菜（*Helianthus annuus*）、大豆（*Glycine max*）、棉花（*Gossypium hirsutum*）和花生（*Arachis hypogaea*）等含油性作物的油被认为是潜在的柴油发动机替代燃料，这些燃料性能良好，具有与商业级柴油相似的化学性质和物理性质（Dorado et al.，2003）。

Bañuelos（2006）还在红岩农场（Red Rock Ranch）进行了一项为期 3 年的田间轮作修复试验，分别在两个 20 hm^2 的场地上种植油菜（*Brassica napus* var. Hyola）和西兰花（*Brassica oleracea* var. Marathon）。前期使用硒含量 < 0.01 mg/L 的运河水灌溉，直至作物长出 2~3 片真叶后改用硒含量 0.100~0.150 mg/L 的农场沟渠排水进行日常灌溉。结果表明，两地的油菜籽产量相当，而加利福尼亚州中部种植的西兰花产量较高（Bañuelos et al.，2002b）。由于硫酸盐和硒酸盐是土壤中的生物地球化学类似物，灌溉水中高浓度的硫酸盐可能会抑制植物对硒酸盐的吸收，并将植物的硒浓度保持在 7 mg/kg 以下。硒也是人类必需的微量营养素，美国食品药品监督管理局（FDA）建议每天摄入约 200 mg 硒（Clark，1996），欧洲食品安全局（EFSA）推荐每天摄入硒 50~200 μg（EFSA，2014）。在西兰花中测得的硒浓度 < 5 mg/kg，日常食用不会超出安全的剂量范围。上述两个试验表明，油菜和西兰花对土壤硒的植物提取和生物挥发不仅可以去除由富硒排水灌溉后土壤中积累的硒，而且收获的富硒作物为种植者带来了经济效益（Bañuelos，2002）。

三、超积累植物硒生物强化与人类健康

硒是人类、动物和微生物必需的营养物质（Ellis and Salt，2003；Kaur et al.，2014）。一些有机硒，如甲基硒代半胱氨酸（MeSeCys）是膳食硒的最好来源。硒以硒代半胱氨酸（SeCys）的形式结合在一系列硒蛋白的活性位点上，这些硒蛋白参与一些关键代谢途径，如甲状腺激素代谢、抗氧化防御和免疫功能等。硒对某些癌症、哮喘和心肌病具有保护作用（Barceló and Poschenrieder，2010）。有研究报道表明，硒对 COVID-19 具有良好的抑制效果（Zhang et al.，2020）。然而，饮食中硒摄入量过低可能导致一系列疾病，包括心脏病、甲状腺功能减退、男性生育力下降、免疫系统减弱，以及感染和癌变的易

感性增加（Malagoli et al., 2015）。哺乳动物和许多其他动物，以及许多原核生物和某些藻类，需要硒作为硒蛋白的结构成分来进行必需的新陈代谢，硒蛋白具有多种氧化还原功能，对免疫、甲状腺和生殖健康至关重要。在植物中，必需的硒代谢似乎已在进化中丢失。硒虽然不是植物必需的，但它可以对植物产生生理益处，特别是在胁迫条件下，硒可能上调植物抗氧化剂代谢产物和酶的活性，从而提高植物清除破坏性活性氧（ROS）的能力（Pilon-Smits, 2019）。

甲基硒代半胱氨酸（MeSeCys）是一种氨基酸衍生物，在动物体内具有很强的抗癌活性。在高硒土壤上，硒耐受植物，或硒超积累植物可通过硒代半胱氨酸甲基转移酶（SMT）将无机硒转化为 MeSeCys（Brummell et al., 2011）。有研究报道，在番茄（*Solanum lycopersicum*）中过表达编码硒超积累植株 SMT 的 cDNA，在果实发育过程中向根部添加亚硒酸盐或硒酸盐，结果表明，MeSeCys 在果实中积累，而在叶片中不积累。形态分析表明，果实中高达 16% 的总硒以 MeSeCys 的形式存在。按百分比转化率计算，施用亚硒酸盐可以更有效地产生 MeSeCys，但由于硒酸盐更容易从根部转运，施用硒酸盐可以获得更多的 MeSeCys 积累量。MeSeCys 热稳定性好，能在加工番茄汁的过程中保留活性（Brummell et al., 2011）。

前已述及，硒不是植物的必需元素，但植物积累硒并将其转化为生物活性化合物的能力对人类的营养和健康以及环境保护具有重要意义（Ellis and Salt, 2003）。利用超积累植物进行植物提取硒可以降低高硒土壤的硒负担，而富硒植物材料作为肥料可以用于低硒土壤的硒生物强化（Barceló and Poschenrieder, 2010）。Bañuelos 等（2015）把硒超积累植物沙漠王羽生物质施入土壤，种植西兰花和胡萝卜。结果发现，西兰花和胡萝卜中积累了大量的有机硒。这说明沙漠王羽不仅可以作为硒污染修复植物，还可以作为有机富硒肥料生产富硒农产品（Banuelos et al., 2015）。

微生物可以促进土壤硒的转化，提高土壤硒的有效性（El Mehdawi et al., 2015a）。此外，有益微生物具有促进作物生长、提高植物吸收硒的能力，因此利用植物-微生物的相互作用提高植物硒的生物强化效果是一个新的研究领域（Malagoli et al., 2015）。Yasin 等（2014）研究发现，印度芥菜可以从富硒土壤中提取硒用于生产富硒植物材料，而接种菌群可促进植物生长，提高硒的吸收，从而进一步提高硒的强化效果。

综上所述，植物修复技术富有潜力。目前已确认了一些典型的硒超积累植物，但由于它们可能存在生长缓慢、栽培技术要求不明确或生物量低的短板，鲜有超积累植物修复土壤硒污染大规模应用的案例。对于应用超积累植物进行硒污染修复，从目前的实验室小尺度研究阶段迈入到大规模应用还需要进一步发展。

对于已知的硒超积累植物，未来可以进一步深入研究其代谢通路和分子机制，同时可以充分探讨其治疗、抑制相关疾病的效果及对人体和动物的营养健康效应；研究和筛选最佳的硒超积累植物栽培技术条件，寻找食品补充剂或功能性食品中的新硒源，开发具有更高水平的抗癌硒化合物的粮食作物，为合理商业价值的开发提供参考。对于新发现的硒超积累植物，其对硒毒害的耐受机制尚不清楚，限制了在硒植物修复中的应用。一方面，应加强对新型硒超积累植物种质资源的调查收集；另一方面，继续深入研究不同超积累植物的硒积累代谢机制，为硒的植物修复策略提供依据。通过过表达转运蛋白来增加对硒的吸收，以及过表达硒代谢途径中的酶来增加植物中无毒硒化合物的比例，是提高硒积累和耐性的一个有效途径。与硒吸收和耐性相关的基因将为今后利用基因工程技术进行硒植物修复提供丰富的资源。

参 考 文 献

樊俊, 王瑞, 向必坤, 等. 2014. 不同硒源对堇叶碎米荠吸收、转化硒的效应研究. 中国土壤与肥料, (1): 6.
刘亚峰, 龙胜桥, 邵树勋. 2018. 碎米荠对硒、镉超富集特性研究. 地球与环境, 46(2): 173-177.
邵树勋, Mihaly Dernovics, 邓国栋, 等. 2015. 湖北恩施地区富硒植物中硒的形态研究. 吉林大学学报 (地球科学版), (S1): 457-458.
沈振国. 2002. 污染土壤的植物修复. 见: 陈怀满等. 土壤中化学物质的行为与环境质量. 北京: 科学出版社: 601-635.
宋成祖. 1995. 鄂西南渔塘坝硒矿区硒污染成因探讨. 地质论评, 41(2): 6.
王晓东. 2019. 水土环境硒污染生物修复研究. 环境与发展, 31(4): 2.
Akiho H, Yamamoto T, Tochihara Y, et al. 2012. Speciation and oxidation reaction analysis of selenium in aqueous solution using X-ray absorption spectroscopy for management of trace element in FGD liquor. Fuel, 102: 156-161.
Alford E R, Lindblom S D, Pittarello M, et al. 2014. Roles of rhizobial symbionts in selenium hyperaccumulation in *Astragalus* (Fabaceae). Am J Bot, 101(11): 1895-1905.
Alford E R, Pilon-Smits E A, Fakra S C, et al. 2012. Selenium hyperaccumulation by *Astragalus* (Fabaceae) does not inhibit root nodule symbiosis. Am J Bot, 99(12): 1930-1941.
Baker A J M, Proctor J, Reeves R D. 1992. The Vegetation of Ultramafic (Serpentine) Soils. In: Reeves R D. Hyperaccumulation of Nickel by Serpentine Plants. Andover: Intercept Ltd: 253-277.
Baker A J M, Whiting S N. 2002. In search of the Holy Grail—a further step in understanding metal hyperaccumulation? New Phytol, 155(1): 1-4.
Bañuelos G S, Arroyo I, Pickering I J, et al. 2015. Selenium biofortification of broccoli and carrots grown in soil amended with Se-enriched hyperaccumulator *Stanleya pinnata*. Food Chem, 166: 603-608.
Bañuelos G S, Bryla D R, Cook C G. 2002b. Vegetative production of kenaf and canola under irrigation in central California. Industrial Crops and Products, 15(3): 237-245.
Bañuelos G S, Lin Z Q, Wu L, et al. 2002a. Phytoremediation of selenium-contaminated soils and waters: fundamentals and future prospects. Rev Environ Health, 17(4): 291-306.
Bañuelos G S, Lin Z Q. 2010. Cultivation of the Indian fig Opuntia in selenium-rich drainage sediments under field conditions. Soil Use and Management, 26(2): 167-175.
Banuelos G S, Meek D W. 1989. Selenium accumulation in selected vegetables. Journal of Plant Nutrition, 12(10): 1255-1272.
Bañuelos G S, Meek D W. 1990. Accumulation of selenium in plants grown on selenium-treated soil. Journal of Environmental Quality, 19(4): 772-777.
Bañuelos G S, Terry N Leduc D L, et al. 2005. Field trial of transgenic Indian mustard plants shows enhanced phytoremediation of selenium-contaminated sediment. Environ Sci Technol, 39(6): 1771-1777.
Bañuelos G S. 2002. Irrigation of broccoli and canola with boron- and selenium-laden effluent. J Environ Qual, 31(6): 1802-1808.
Bañuelos G S. 2006. Phyto-products may be essential for sustainability and implementation of phytoremediation. Environ Pollut, 144(1): 19-23.
Barceló J, Poschenrieder C. 2010. Hyperaccumulation of trace elements: from uptake and tolerance mechanisms to litter decomposition; selenium as an example. Plant and Soil, 341(1-2): 31-35.
Bassil J, Naveau A, Bueno M, et al. 2018. Leaching behavior of selenium from the karst infillings of the

hydrogeological experimental site of Poitiers. Chemical Geology, 483: 141-150.

Beath O A, Eppson H F, Gilbert C S. 1912. Selenium distribution in and seasonal variation of type vegetation occurring on seleniferous soils. The Journal of the American Pharmaceutical Association, 26(5): 394-405.

Beath O A, Gilbert C S, Eppson H F. 1939a. The use of indicator plants in locating seleniferous areas in western United States. Ⅰ. General. American Journal of Botany, 26(4): 257-269.

Beath O A, Gilbert C S, Eppson H F. 1939b. The use of indicator plants in locating seleniferous areas in western United States. Ⅱ. Correlation studies by states. American Journal of Botany, 26(5): 296-315.

Beath O A, Gilbert C S, Eppson H F. 1940. The use of indicator plants in locating seleniferous areas in Western United States Ⅲ Further studies. American Journal of Botany, 27(7): 564-573.

Beath O A, Gilbert C S, Eppson H F. 1941. The use of indicator plants in locating seleniferous areas in Western United States Ⅳ Progress report. American Journal of Botany, 28(10): 887-900.

Beath O A. 1943. Toxic vegetation growing on the salt wash sandstone member of the Morrison formation. American Journal of Botany, 30(9): 698-707.

Berken A, Mulholland M M, LeDuc D L, et al. 2010. Genetic engineering of plants to enhance selenium phytoremediation. Critical Reviews in Plant Sciences, 21(6): 567-582.

Bitterli C, Banuelos G S, Schulin R. 2010. Use of transfer factors to characterize uptake of selenium by plants. Journal of Geochemical Exploration, 107(2): 206-216.

Both E B, Shao S, Xiang J, et al. 2018. Selenolanthionine is the major water-soluble selenium compound in the selenium tolerant plant Cardamine violifolia. Biochim Biophys Acta Gen Subj, 1862(11): 2354-2362.

Both E B, Stonehouse G C, Lima L W, et al. 2020. Selenium tolerance, accumulation, localization and speciation in a *Cardamine* hyperaccumulator and a non-hyperaccumulator. Sci Total Environ, 703: 135041.

Boyd R S. 2007. The defense hypothesis of elemental hyperaccumulation: status, challenges and new directions. Plant and Soil, 293(1-2): 153-176.

Boyd R S. 2013. Exploring tradeoffs in hyperaccumulator ecology and evolution. New Phytol, 199(4): 871-872.

Brown T A, Shrift A. 1982. Selenium - toxicity and tolerance in higher-plants. Biological Reviews, 57(Feb): 59-84.

Brummell D A, Watson L M, Pathirana R, et al. 2011. Biofortification of tomato (*Solanum lycopersicum*) fruit with the anticancer compound methylselenocysteine using a selenocysteine methyltransferase from a selenium hyperaccumulator. J Agric Food Chem, 59(20): 10987-10994.

Cabannes E, Buchner P, Broadley M R, et al. 2011. A comparison of sulfate and selenium accumulation in relation to the expression of sulfate transporter genes in *Astragalus* species. Plant Physiol, 157(4): 2227-2239.

Cakir O, Candar-Cakir B, Zhang B. 2016a. Small RNA and degradome sequencing reveals important microRNA function in *Astragalus chrysochlorus* response to selenium stimuli. Plant Biotechnol J, 14(2): 543-556.

Cakir O, Turgut-Kara N, Ari Ş. 2016b. Selenium induced selenocysteine methyltransferase gene expression and antioxidant enzyme activities in *Astragalus chrysochlorus*. Acta Botanica Croatica, 75(1): 11-16.

Cappa J J, Cappa P J, El Mehdawi A F, et al. 2014. Characterization of selenium and sulfur accumulation across the genus Stanleya (Brassicaceae): A field survey and common-garden experiment. Am J Bot, 101(5): 830-839.

Cappa J J, Pilon-Smits E A. 2014. Evolutionary aspects of elemental hyperaccumulation. Planta, 239(2): 267-275.

Cappa J J, Yetter C, Fakra S, et al. 2015. Evolution of selenium hyperaccumulation in Stanleya (Brassicaceae) as inferred from phylogeny, physiology and X-ray microprobe analysis. New Phytol, 205(2): 583-595.

Chaney R L, Angle J S, Broadhurst C L, et al. 2007. Improved understanding of hyperaccumulation yields commercial phytoextraction and phytomining technologies. J Environ Qual, 36(5): 1429-1443.

Chaney R L, Baklanov I A. 2017. Phytoremediation and Phytomining: Status and Promise. In: Cuypers A, Vangronsveld J. Advances in Botanical Research, Volume 83 Phytoremediation. New York: Academic Press: 189-221.

Chen X, Yang G, Chen J, et al. 1980. Studies on the relations of selenium and Keshan disease. Biol Trace Elem Res, 2(2): 91-107.

Clark L C. 1996. Effects of selenium supplementation for cancer prevention in patients with carcinoma of the skin. Jama, 276(24): 1957-1963.

Cochran A T. 2017. Selenium and the Plant Microbiome. In: Elizabeth A H, Pilon-Smits, Lenny H E. Selenium in plants: Molecular, Physiological, Ecological and Evolutionary Aspects. Berlin: Springer: 109-121.

Combs G F Jr. 2001. Selenium in global food systems. Br J Nutr, 85(5): 517-547.

Cui L W, Zhao J T, Chen J Y, et al. 2018. Translocation and transformation of selenium in hyperaccumulator plant *Cardamine enshiensis* from Enshi, Hubei, China. Plant and Soil, 425(1-2): 577-588.

Dai Z, Imtiaz M, Rizwan M, et al. 2019. Dynamics of Selenium uptake, speciation, and antioxidant response in rice at different panicle initiation stages. Sci Total Environ, 691: 827-834.

Das K, Roychoudhury A. 2014. Reactive oxygen species (ROS) and response of antioxidants as ROS-scavengers during environmental stress in plants. Frontiers in Environmental Science, 53: 2.

de Souza M P, Chu D, Zhao M, et al. 1999. Rhizosphere bacteria enhance selenium accumulation and volatilization by indian mustard. Plant Physiol, 119(2): 565-574.

DeTar R A, Alford E R, Pilon-Smits E A. 2015. Molybdenum accumulation, tolerance and molybdenum-selenium-sulfur interactions in *Astragalus* selenium hyperaccumulator and nonaccumulator species. J Plant Physiol, 183: 32-40.

Dhillon K S, Dhillon S K. 2003. Distribution and management of seleniferous soils. Advances in Agronomy Volume, 79: 119-184.

Di Gregorio S, Lampis S, Vallini G. 2005. Selenite precipitation by a rhizospheric strain of *Stenotrophomonas* sp. isolated from the root system of *Astragalus bisulcatus*: a biotechnological perspective. Environ Int, 31(2): 233-241.

Dorado M P, Ballesteros E, López F J, et al. 2003. Optimization of alkali-catalyzed transesterification of *Brassica carinata* oil for biodiesel production. Energy & Fuels, 18(1): 77-83.

EFSA (European Food Safety Authority). 2014. Scientific Opinion on the safety and efficacy of DL-selenomethionine as a feed additive for all animal species. EFSA Journal, 12(2): 3567.

El Mehdawi A F, Cappa J J, Fakra S C, et al. 2012. Interactions of selenium hyperaccumulators and nonaccumulators during cocultivation on seleniferous or nonseleniferous soil--the importance of having good neighbors. New Phytol, 194(1): 264-277.

El Mehdawi A F, Jiang Y, Guignardi Z S, et al. 2018. Influence of sulfate supply on selenium uptake dynamics and expression of sulfate/selenate transporters in selenium hyperaccumulator and nonhyperaccumulator Brassicaceae. New Phytol, 217(1): 194-205.

El Mehdawi A F, Lindblom S D, Cappa J J, et al. 2015a. Do selenium hyperaccumulators affect selenium speciation in neighboring plants and soil? An X-Ray Microprobe Analysis. Int J Phytoremediation, 17(8): 753-765.

El Mehdawi A F, Paschke M W, Pilon-Smits E A H. 2015b. *Symphyotrichum ericoides* populations from seleniferous and nonseleniferous soil display striking variation in selenium accumulation. New Phytol, 206(1): 231-242.

El Mehdawi A F, Pilon-Smits E A H. 2012. Ecological aspects of plant selenium hyperaccumulation. Plant

Biol (Stuttg), 14(1): 1-10.

El Mehdawi A F, Quinn C F, Pilon-Smits E A H. 2011a. Effects of selenium hyperaccumulation on plant-plant interactions: evidence for elemental allelopathy? New Phytol, 191(1): 120-131.

El Mehdawi A F, Quinn C F, Pilon-Smits E A H. 2011b. Selenium hyperaccumulators facilitate selenium-tolerant neighbors via phytoenrichment and reduced herbivory. Curr Biol, 21(17): 1440-1449.

El Mehdawi A F, Reynolds R J, Prins C N, et al. 2014. Analysis of selenium accumulation, speciation and tolerance of potential selenium hyperaccumulator *Symphyotrichum ericoides*. Physiol Plant, 152(1): 70-83.

Ellis D R, Salt D E. 2003. Plants, selenium and human health. Curr Opin Plant Biol, 6(3): 273-279.

El-Ramady H R, Abdalla N, Alshaal T, et al. 2015. Giant reed for selenium phytoremediation under changing climate. Environmental Chemistry Letters, 13(4): 359-380.

Evans C S, Asher C J, Johnson C M. 1968. Isolation of dimethyl diselenide and other volatile selenium compounds from *Astragalus racemosus* (Pursh). Australian Journal of Biological Sciences, 21(1): 13-20.

Favorito J E, Grossl P R, Davis T Z, et al. 2020. Soil-plant-animal relationships and geochemistry of selenium in the western phosphate resource area (United States): A review. Chemosphere, 266: 128959.

Feng R W, Wei C Y. 2012. Antioxidative mechanisms on selenium accumulation in *Pteris vittata* L., a potential selenium phytoremediation plant. Plant, Soil and Environment, 58(No. 3): 105-110.

Frankenberger W T, Karlson U. 1995. Volatilization of selenium from a dewatered seleniferous sediment: A field study. Journal of Industrial Microbiology, 14(3-4): 226-232.

Freeman J L, Banuelos G S. 2011. Selection of salt and boron tolerant selenium hyperaccumulator *Stanleya pinnata* genotypes and characterization of Se phytoremediation from agricultural drainage sediments. Environ Sci Technol, 45(22): 9703-9710.

Freeman J L, Lindblom S D, Quinn C F, et al. 2007. Selenium accumulation protects plants from herbivory by Orthoptera via toxicity and deterrence. New Phytol, 175(3): 490-500.

Freeman J L, Marcus M A, Fakra S C, et al. 2012. Selenium hyperaccumulator plants *Stanleya pinnata* and *Astragalus bisulcatus* are colonized by Se-resistant, Se-excluding wasp and beetle seed herbivores. PLoS One, 7(12): e50516.

Freeman J L, Quinn C F, Lindblom S D, et al. 2009. Selenium protects the hyperaccumulator *Stanleya pinnata* against black-tailed prairie dog herbivory in native seleniferous habitats. Am J Bot, 96(6): 1075-1085.

Freeman J L, Quinn C F, Marcus M A, et al. 2006b. Selenium-tolerant diamondback moth disarms hyperaccumulator plant defense. Curr Biol, 16(22): 2181-2192.

Freeman J L, Tamaoki M, Stushnoff C, et al. 2010. Molecular mechanisms of selenium tolerance and hyperaccumulation in *Stanleya pinnata*. Plant Physiol, 153(4): 1630-1652.

Freeman J L, Zhang L H, Marcus M A, et al. 2006a. Spatial imaging, speciation, and quantification of selenium in the hyperaccumulator plants *Astragalus bisulcatus* and *Stanleya pinnata*. Plant Physiol, 142(1): 124-134.

Galeas M L, Zhang L H, Freeman J L, et al. 2007. Seasonal fluctuations of selenium and sulfur accumulation in selenium hyperaccumulators and related nonaccumulators. New Phytol, 173(3): 517-525.

Garbisu C, Alkorta I. 2001. Phytoextraction: a cost-effective plant-based technology for the removal of metals from the environment. Bioresource Technology, 77(3): 229-236.

Garousi F, Kovács B, Andrási D, et al. 2016. Selenium phytoaccumulation by sunflower plants under hydroponic conditions. Water, Air & Soil Pollution, 227(10): 382.

Girdhar M, Sharma N R, Rehman H, et al. 2014. Comparative assessment for hyperaccumulatory and phytoremediation capability of three wild weeds. 3 Biotech, 4(6): 579-589.

Girling C A. 1984. Selenium in agriculture and the environment. Agriculture Ecosystems & Environment, 11(1): 37-65.

Gomes-Junior R A, Gratao P L, Gaziola S A, et al. 2007. Selenium-induced oxidative stress in coffee cell suspension cultures. Funct Plant Biol, 34(5): 449-456.

Guignardi Z, Schiavon M. 2017. Biochemistry of plant selenium uptake and metabolism. Selenium in Plants: Molecular, Physiological, Ecological and Evolutionary Aspects. AG: Springer International Publishing: 21-34.

Gupta M, Gupta S. 2016. An overview of selenium uptake, metabolism, and toxicity in plants. Front Plant Sci, 7: 2074.

Hansen D, Duda P J, Zayed A, et al. 1998. Selenium removal by constructed wetlands: Role of biological volatilization. Environmental Science & Technology, 32(5): 591-597.

Hanson B Garifullina G F, Lindblom S D, et al. 2003. Selenium accumulation protects *Brassica juncea* from invertebrate herbivory and fungal infection. New Phytol, 159(2): 461-469.

Hanson B, Lindblom S D, Loeffler M L, et al. 2004. Selenium protects plants from phloem-feeding aphids due to both deterrence and toxicity. New Phytol, 162(3): 655-662.

Harris J, Schneberg K A, Pilon-Smits E A. 2014. Sulfur-selenium-molybdenum interactions distinguish selenium hyperaccumulator *Stanleya pinnata* from non-hyperaccumulator *Brassica juncea* (Brassicaceae). Planta, 239(2): 479-491.

Hasanuzzaman M, Bhuyan M, Raza A, et al. 2020. Selenium toxicity in plants and environment: biogeochemistry and remediation possibilities. Plants (Basel), 9(12): 710-718.

Hatfield D L, Tsuji P A, Carlson B A, et al. 2014. Selenium and selenocysteine: roles in cancer, health, and development. Trends Biochem Sci, 39(3): 112-120.

He Y, Xiang Y, Zhou Y, et al. 2018. Selenium contamination, consequences and remediation techniques in water and soils: a review. Environ Res, 164: 288-301.

Hung C Y, Holliday B M, Kaur H, et al. 2012. Identification and characterization of selenate- and selenite-responsive genes in a Se-hyperaccumulator *Astragalus racemosus*. Mol Biol Rep, 39(7): 7635-7646.

Jaiswal S K, Prakash R, Acharya R, et al. 2012. Selenium content in seed, oil and oil cake of Se hyperaccumulated *Brassica juncea* (Indian mustard) cultivated in a seleniferous region of India. Food Chemistry, 134(1): 401-404.

Karlson U, Frankenberger W T. 1988. Determination of gaseous selenium-75 evolved from soil. Soil Science Society of America Journal, 52(3): 678-681.

Kaur N, Sharma S, Kaur S, et al. 2014. Selenium in agriculture: a nutrient or contaminant for crops? Archives of Agronomy and Soil Science, 60(12): 1593-1624.

Knott S G, McCray C W R. 1959. Two naturally occurring outbreaks of selenosis in queensland. Australian Veterinary Journal, 35(4): 161-165.

Kok A B, Mungan M D, Doganlar S, et al. 2020. Transcriptomic analysis of selenium accumulation in *Puccinellia distans* (Jacq.) Parl, a boron hyperaccumulator. Chemosphere, 245: 125665.

Kolbert Z, Molni R I, Szollosi R K, et al. 2018. Nitro-oxidative stress correlates with Se tolerance of *Astragalus* species. Plant Cell Physiol, 59(9): 1827-1843.

Krzciuk K, Galuszka A. 2014. Prospecting for hyperaccumulators of trace elements: a review. Crit Rev Biotechnol, 35(4): 522-532.

Kubachka K M, Meija J, LeDuc D L, et al. 2007. Selenium volatiles as proxy to the metabolic pathways of selenium in genetically modified *Brassica juncea*. Environ Sci Technol, 41(6): 1863-1869.

Kushwaha A, Goswami L, Lee J, et al. 2021. Selenium in soil-microbe-plant systems: Sources, distribution, toxicity, tolerance, and detoxification. Critical Reviews in Environmental Science and Technology, 52(13): 1-42.

Lakin H W. 1972. Selenium accumulation in soils and its absorption by plants and animals. Geological

Society of America Bulletin, 83(1): 181-189.

Lakin, H W, Byers, H G. 1941. Selenium occurrence in certain soils in the United States with a discussion of related topics, sixth report. U.S. Dept Agric Tech.

LeDuc D L, AbdelSamie M, Montes-Bayon M, et al. 2006. Overexpressing both ATP sulfurylase and selenocysteine methyltransferase enhances selenium phytoremediation traits in Indian mustard. Environ Pollut, 144(1): 70-76.

LeDuc D L, Tarun A S, Montes-Bayon M, et al. 2004. Overexpression of selenocysteine methyltransferase in *Arabidopsis* and Indian mustard increases selenium tolerance and accumulation. Plant Physiol, 135(1): 377-383.

Lemly A D. 2004. Aquatic selenium pollution is a global environmental safety issue. Ecotoxicology and Environmental Safety, 59(1): 44-56.

Li J T, Gurajala H K, Wu L H, et al. 2018. Hyperaccumulator plants from China: a synthesis of the current state of knowledge. Environ Sci Technol, 52(21): 11980-11994.

Li J, Liang D, Qin S, et al. 2015. Effects of selenite and selenate application on growth and shoot selenium accumulation of pak choi (*Brassica chinensis* L.) during successive planting conditions. Environ Sci Pollut Res Int, 22(14): 11076-11086.

Li Q, Zhou S, Liu N. 2021. Diversity of Endophytic Bacteria in *Cardamine hupingshanensis* and potential of culturable selenium-resistant endophytes to enhance seed germination under selenate stress. Curr Microbiol, 78(5): 2091-2103.

Lima L W, Pilon-Smits E A H, Schiavon M. 2018. Mechanisms of selenium hyperaccumulation in plants: A survey of molecular, biochemical and ecological cues. Biochim Biophys Acta Gen Subj, 1862(11): 2343-2353.

Lin Z Q, Cervinka V, Pickering I J, et al. 2002. Managing selenium-contaminated agricultural drainage water by the integrated on-farm drainage management system: role of selenium volatilization. Water Res, 36(12): 3150-3160.

Lin Z Q, Schemenauer R S, Cervinka V, et al. 2000. Selenium volatilization from a soil-plant system for the remediation of contaminated water and soil in the San Joaquin Valley. Journal of Environmental Quality, 29(4): 1048-1056.

Lindblom S D, Fakra S C, Landon J, et al. 2013b. Inoculation of *Astragalus racemosus* and *Astragalus convallarius* with selenium-hyperaccumulator rhizosphere fungi affects growth and selenium accumulation. Planta, 237(3): 717-729.

Lindblom S D, Fakra S C, Landon J, et al. 2014. Inoculation of selenium hyperaccumulator *Stanleya pinnata* and related non-accumulator *Stanleya elata* with hyperaccumulator rhizosphere fungi-investigation of effects on Se accumulation and speciation. Physiol Plant, 150(1): 107-118.

Lindblom S D, Valdez-Barillas J R, Fakra S C, et al. 2013a. Influence of microbial associations on selenium localization and speciation in roots of *Astragalus* and *Stanleya* hyperaccumulators. Environmental and Experimental Botany, 88: 33-42.

Lindblom S D, Wangeline A L, Valdez Barillas J R, et al. 2018. Fungal endophyte *Alternaria tenuissima* can affect growth and selenium accumulation in its hyperaccumulator host *Astragalus bisulcatus*. Front Plant Sci, 9: 1213.

Malagoli M, Schiavon M, dall'Acqua S, et al. 2015. Effects of selenium biofortification on crop nutritional quality. Front Plant Sci, 6: 280.

Malik B, Pirzadah T B, Tahir I, et al. 2015. Recent trends and approaches in phytoremediation. In: Hakeem K R. Soil Remediation and Plants: Prospects and Challenges. New York: Academic Press: 131-146.

Marshall E. 1985. Selenium Poisons Refuge, California Politics: Drainage from the San Joaquin Valley has dumped selenium into a wildlife refuge and pitted two federal agencies against each other. Science,

229(4709): 144-146.

McGrath S P, Zhao F J. 2003. Phytoextraction of metals and metalloids from contaminated soils. Curr Opin Biotechnol, 14(3): 277-282.

McKenzie M J, Hunter D A, Pathirana R, et al. 2009. Accumulation of an organic anticancer selenium compound in a transgenic Solanaceous species shows wider applicability of the selenocysteine methyltransferase transgene from selenium hyperaccumulators. Transgenic Res, 18(3): 407-424.

Mehdi Y, Hornick J L, Istasse L, et al. 2013. Selenium in the environment, metabolism and involvement in body functions. Molecules, 18(3): 3292-3311.

Mench M, Lepp N, Bert V, et al. 2010. Successes and limitations of phytotechnologies at field scale: outcomes, assessment and outlook from COST Action 859. Journal of Soils and Sediments, 10(6): 1039-1070.

Merkle S A. 2006. Engineering forest trees with heavy metal resistance genes. Silvae Genetica, 55(6): 263-268.

Mohiley A, Tielborger K, Seifan M, et al. 2020. The role of biotic interactions in determining metal hyperaccumulation in plants. Functional Ecology, 34(3): 658-668.

Nemeth A, Garcia Reyes J F, Kosary J, et al. 2013. The relationship of selenium tolerance and speciation in *Lecythidaceae* species. Metallomics, 5(12): 1663-1673.

Ogra Y, Okubo E, Takahira M. 2010. Distinct uptake of tellurate from selenate in a selenium accumulator, Indian mustard (*Brassica juncea*). Metallomics, 2(5): 328-333.

Ouerdane L, Both E B, Xiang J, et al. 2020. Water soluble selenometabolome of *Cardamine violifolia*. Metallomics, 12(12): 2032-2048.

Padmavathiamma P K, Li L Y. 2007. Phytoremediation technology: Hyper-accumulation metals in plants. Water Air and Soil Pollution, 184(1-4): 105-126.

Park J H, Lamb D, Paneerselvam P, et al. 2011. Role of organic amendments on enhanced bioremediation of heavy metal(loid) contaminated soils. J Hazard Mater, 185(2-3): 549-574.

Parker D R, Feist L J, Varvel T W, et al. 2003. Selenium phytoremediation potential of *Stanleya pinnata*. Plant and Soil, 249(1): 157-165.

Peng Q, Wang M, Cui Z, et al. 2017. Assessment of bioavailability of selenium in different plant-soil systems by diffusive gradients in thin-films (DGT). Environ Pollut, 225: 637-643.

Pfister J A, Davis T Z, Hall J O. 2013. Effect of selenium concentration on feed preferences by cattle and sheep. J Anim Sci, 91(12): 5970-5980.

Pickering I J, Hirsch G, Prince R C, et al. 2003. Imaging of selenium in plants using tapered metal monocapillary optics. Journal of Synchrotron Radiation, 10(Pt 3): 289-290.

Pilon-Smits E A H, Souza M P, Hong G, et al. 1999. Selenium volatilization and accumulation by twenty aquatic plant species. Journal of Environmental Quality, 28(3): 1011-1018.

Pilon-Smits E A H. 2019. On the ecology of selenium accumulation in plants. Plants (Basel), 8(7): 197.

Prasad M N V, Freitas H M D. 2003. Metal hyperaccumulation in plants - Biodiversity prospecting for phytoremediation technology. Electronic Journal of Biotechnology, 6(3): 285-321.

Presser T S. 1994. The Kesterson Effect. Environmental Management, 18(3): 437-454.

Prins C N, Hantzis L J, Quinn C F, et al. 2011. Effects of selenium accumulation on reproductive functions in *Brassica juncea* and *Stanleya pinnata*. J Exp Bot, 62(15): 5633-5640.

Prins C N, Hantzis L J, Valdez-Barillas J R, et al. 2019. Getting to the root of selenium hyperaccumulation-localization and speciation of root selenium and its effects on nematodes. Soil Systems, 3(3): 47.

Quinn C F, Freeman J L, Galeas M L, et al. 2008. The role of selenium in protecting plants against prairie dog herbivory: implications for the evolution of selenium hyperaccumulation. Oecologia, 155(2): 267-275.

Quinn C F, Freeman J L, Reynolds R J, et al. 2010b. Selenium hyperaccumulation offers protection from cell disruptor herbivores. BMC Ecol, 10: 19.

Quinn C F, Prins C N, Freeman J L, et al. 2011. Selenium accumulation in flowers and its effects on pollination. New Phytol, 192(3): 727-737.

Quinn C F, Wyant K A, Wangeline A L, et al. 2010a. Enhanced decomposition of selenium hyperaccumulator litter in a seleniferous habitat—evidence for specialist decomposers? Plant and Soil, 341(1-2): 51-61.

Rao S, Yu T, Cong X, et al. 2020. Integration analysis of PacBio SMRT- and Illumina RNA-seq reveals candidate genes and pathway involved in selenium metabolism in hyperaccumulator *Cardamine violifolia*. BMC Plant Biol, 20(1): 492.

Rao S, Yu T, Cong X, et al. 2021. Transcriptome, proteome, and metabolome reveal the mechanism of tolerance to selenate toxicity in *Cardamine violifolia*. J Hazard Mater, 406: 124283.

Rascio N, Navari-Izzo F. 2011. Heavy metal hyperaccumulating plants: how and why do they do it? And what makes them so interesting? Plant Sci, 180(2): 169-181.

Rayman M P. 2000. The importance of selenium to human health. Lancet, 356(9225): 233-241.

Rayman M P. 2012. Selenium and human health. Lancet, 379(9822): 1256-1268.

Reynolds R J B, Jones R R, Heiner J, et al. 2020. Effects of selenium hyperaccumulators on soil selenium distribution and vegetation properties. Am J Bot, 107(7): 970-982.

Reynolds R J B, Jones R R, Stonehouse G C, et al. 2019. Identification and physiological comparison of plant species that show positive or negative co-occurrence with selenium hyperaccumulators. Metallomics, 12(1): 133-143.

Reynolds R J B, Pilon-Smits E A H. 2018. Plant selenium hyperaccumulation- Ecological effects and potential implications for selenium cycling and community structure. Biochim Biophys Acta Gen Subj, 1862(11): 2372-2382.

Salt D E, Blaylock M, Kumar N P B A, et al. 1995. Phytoremediation: A novel strategy for the removal of toxic metals from the environment using plants. Nature Biotechnology, 13(5): 468-474.

Salt D E, Smith R D, Raskin I. 1998. Phytoremediation. Annu Rev Plant Physiol Plant Mol Biol, 49: 643-668.

Schiavon M, Pilon M, Malagoli M, et al. 2015. Exploring the importance of sulfate transporters and ATP sulphurylases for selenium hyperaccumulation-a comparison of *Stanleya pinnata* and *Brassica juncea* (Brassicaceae). Front Plant Sci, 6: 2.

Schiavon M, Pilon-Smits E A H. 2017. The fascinating facets of plant selenium accumulation - biochemistry, physiology, evolution and ecology. New Phytol, 213(4): 1582-1596.

Schwarz K, Foltz C M. 2002. Selenium as an Integral Part of Factor 3 against Dietary Necrotic Liver Degeneration. Journal of the American Chemical Society, 79(12): 3292-3293.

Shao S, Deng G, Mi X, et al. 2014. Accumulation and speciation of selenium in Cardamine sp. in Yutangba Se Mining Field, Enshi, China. Chinese Journal of Geochemistry, 33(4): 357-364.

Shier W T. 1994. Metals as toxins in plants. Journal of Toxicology-Toxin Reviews, 13(2): 205-216.

Sors T G, Ellis D R, Na G N, et al. 2005b. Analysis of sulfur and selenium assimilation in *Astragalus* plants with varying capacities to accumulate selenium. Plant J, 42(6): 785-797.

Sors T G, Ellis D R, Salt D E. 2005a. Selenium uptake, translocation, assimilation and metabolic fate in plants. Photosynth Res, 86(3): 373-389.

Sors T G, Martin C P, Salt D E. 2009. Characterization of selenocysteine methyltransferases from *Astragalus* species with contrasting selenium accumulation capacity. Plant J, 59(1): 110-122.

Spallholz J E. 1994. On the nature of selenium toxicity and carcinostatic activity. Free Radic Biol Med, 17(1): 45-64.

Srivastava M, Ma L Q, Cotruvo J A. 2005. Uptake and distribution of selenium in different fern species. Int J Phytoremediation, 7(1): 33-42.

Staicu L C, Ackerson C J, Cornelis P, et al. 2015. *Pseudomonas moraviensis* subsp. stanleyae, a bacterial endophyte of hyperaccumulator *Stanleya pinnata*, is capable of efficient selenite reduction to elemental selenium under aerobic conditions. J Appl Microbiol, 119(2): 400-410.

Statwick J, Majestic B J, Sher A A. 2016. Characterization and benefits of selenium uptake by an *Astragalus* hyperaccumulator and a non-accumulator. Plant and Soil, 404(1-2): 345-359.

Statwick J, Sher A A. 2017. Selenium in soils of western Colorado. Journal of Arid Environments, 137: 1-6.

Steven J C, Culver A. 2019. The defensive benefit and flower number cost of selenium accumulation in *Brassica juncea*. AoB Plants, 11(5): plz053.

Sura-de Jong M, Reynolds R J, Richterova K, et al. 2015. Selenium hyperaccumulators harbor a diverse endophytic bacterial community characterized by high selenium resistance and plant growth promoting properties. Front Plant Sci, 6: 113.

Tamaoki M, Freeman J L, Pilon-Smits E. A. 2008. Cooperative ethylene and jasmonic acid signaling regulates selenite resistance in Arabidopsis. Plant Physiol, 146(3): 1219-1230.

Terry N, Zayed A M, De Souza M P, et al. 2000. Selenium in higher plants. Annu Rev Plant Physiol Plant Mol Biol, 51: 401-432.

Tinggi U. 2003. Essentiality and toxicity of selenium and its status in Australia: a review. Toxicol Lett, 137(1-2): 103-110.

Trippe R C, Pilon-Smits E A H. 2021. Selenium transport and metabolism in plants: Phytoremediation and biofortification implications. J Hazard Mater, 404(Pt B): 124178.

Tunnicliffe F W. 1901. Selenium compounds as factors in the recent beer-poisoning epidemic. The Lancet, 157: 4041.

Usmani Z, Kumar A, Ahirwal J, et al. 2019. Scope for Applying Transgenic Plant Technology for Remediation and Fortification of Selenium. In: Prasad M N V. Transgenic Plant Technology for Remediation of Toxic Metals and Metalloids. Chennai: Academic Press: 429-461.

Valdez Barillas J R, Quinn C F, Freeman J L, et al. 2012. Selenium distribution and speciation in the hyperaccumulator *Astragalus bisulcatus* and associated ecological partners. Plant Physiol, 159(4): 1834-1844.

Valdez Barillas J R, Quinn C F, Pilon-Smits E A. 2011. Selenium accumulation in plants-phytotechnological applications and ecological implications. Int J Phytoremediation, 13 Suppl 1: 166-178.

Van der Ent A, Baker A J M, Reeves R D, et al. 2013. Hyperaccumulators of metal and metalloid trace elements: Facts and fiction. Plant and Soil, 362(1-2): 319-334.

Van Hoewyk D, Takahashi H, Inoue E, et al. 2008. Transcriptome analyses give insights into selenium-stress responses and selenium tolerance mechanisms in *Arabidopsis*. Physiol Plant, 132(2): 236-253.

Van Hoewyk D. 2013. A tale of two toxicities: malformed selenoproteins and oxidative stress both contribute to selenium stress in plants. Ann Bot, 112(6): 965-972.

Vesk P A, Reichman S M. 2009. Hyperaccumulators and herbivores-a Bayesian meta-analysis of feeding choice trials. J Chem Ecol, 35(3): 289-296.

Wang J, Cappa J J, Harris J P, et al. 2018. Transcriptome-wide comparison of selenium hyperaccumulator and nonaccumulator *Stanleya* species provides new insight into key processes mediating the hyperaccumulation syndrome. Plant Biotechnol J, 16: 1582-1594.

Wangeline A L, Valdez J R, Lindblom S D, et al. 2011. Characterization of rhizosphere fungi from selenium hyperaccumulator and nonhyperaccumulator plants along the eastern Rocky Mountain Front Range. Am J Bot, 98(7): 1139-1147.

White P J, Broadley M R. 2009. Biofortification of crops with seven mineral elements often lacking in human diets—iron, zinc, copper, calcium, magnesium, selenium and iodine. New Phytol, 182(1): 49-84.

White P J. 2016. Selenium accumulation by plants. Ann Bot, 117(2): 217-235.

Winkel L H, Vriens B, Jones G D, et al. 2015. Selenium cycling across soil-plant-atmosphere interfaces: a critical review. Nutrients, 7(6): 4199-4239.

Wu L, Banuelos G, Guo X. 2000. Changes of soil and plant tissue selenium status in an upland grassland contaminated by selenium-rich agricultural drainage sediment after ten years transformed from a wetland habitat. Ecotoxicol Environ Saf, 47(2): 201-209.

Wu L. 2004. Review of 15 years of research on ecotoxicology and remediation of land contaminated by agricultural drainage sediment rich in selenium. Ecotoxicol Environ Saf, 57(3): 257-269.

Wu M, Cong X, Li M, et al. 2020. Effects of different exogenous selenium on Se accumulation, nutrition quality, elements uptake, and antioxidant response in the hyperaccumulation plant *Cardamine violifolia*. Ecotoxicol Environ Saf, 204: 111045.

Wu Z, Banuelos G S, Lin Z Q, et al. 2015. Biofortification and phytoremediation of selenium in China. Front Plant Sci, 6: 136.

Xiang J, Ming J, Yin H, et al. 2019. Anatomy and histochemistry of the roots and shoots in the aquatic selenium hyperaccumulator *Cardamine hupingshanensis* (Brassicaceae). Open Life Sciences, 14(1): 318-326.

Xiao K C, Lu L F, Tang J J, et al. 2020. Parent material modulates land use effects on soil selenium bioavailability in a selenium-enriched region of southwest China. Geoderma, 376: 114554.

Yasin M, El Mehdawi A F, Jahn C E, et al. 2014. Seleniferous soils as a source for production of selenium-enriched foods and potential of bacteria to enhance plant selenium uptake. Plant and Soil, 386(1-2): 385-394.

Yawata A, Oishi Y, Anan Y, et al. 2010. Comparison of selenium metabolism in three Brassicaceae plants. Journal of Health Science, 56(6): 699-704.

Yuan L, Zhu Y, Lin Z Q, et al. 2013. A novel selenocystine-accumulating plant in selenium-mine drainage area in Enshi, China. PLoS One, 8(6): e65615.

Zawislanski P T, Mountford H S, Gabet E J, et al. 2001. Selenium distribution and fluxes in intertidal wetlands, San Francisco Bay, California. J Environ Qual, 30(3): 1080-1091.

Zayed A, Gowthaman S, Terry N. 1998. Phytoaccumulation of trace elements by wetland plants: I. Duckweed. Journal of Environmental Quality, 27(3): 715-721.

Zhang J, Taylor E W, Bennett K, et al. 2020. Association between regional selenium status and reported outcome of COVID-19 cases in China. Am J Clin Nutr, 111(6): 1297-1299.

Zhang L, Ackley A R, Pilon-Smits E A. 2007. Variation in selenium tolerance and accumulation among 19 *Arabidopsis thaliana* accessions. J Plant Physiol, 164(3): 327-336.

Zhang R Q, Tang C F, Wen S Z, et al. 2006. Advances in research on genetically engineered plants for metal resistance. Journal of Integrative Plant Biology, 48(11): 1257-1265.

Zhou C, Huang J C, Liu F, et al. 2019. Selenium removal and biotransformation in a floating-leaved macrophyte system. Environ Pollut, 245: 941-949.

Zhou Y, Tang Q, Wu M, et al. 2018. Comparative transcriptomics provides novel insights into the mechanisms of selenium tolerance in the hyperaccumulator plant *Cardamine hupingshanensis*. Sci Rep, 8(1): 2789.

Zhu Y G, Pilon-Smits E A H, Zhao F J, et al. 2009. Selenium in higher plants: understanding mechanisms for biofortification and phytoremediation. Trends Plant Sci, 14(8): 436-442.

Zhu Y L, Zayed A M, Qian J H, et al. 1999. Phytoaccumulation of trace elements by wetland plants: II. water hyacinth. Journal of Environmental Quality, 28(1): 339-344.

第七章　硒与重金属的交互作用

硒（Se）是一种人和动物必需的微量有益元素，参与了哺乳动物体内多种蛋白质与酶的合成，在抗氧化、抗癌、提高机体免疫力等方面具有重要的作用。虽然硒不是植物必需的微量元素，但是，适量的硒也可以缓解植物的氧化胁迫，促进植物的生长和发育，提高作物产量和品质，是一种植物的有益元素。同时，硒能够与多种重金属元素产生拮抗效应，缓解重金属对植物的毒害，降低植物对重金属的累积，这就是交互作用的表现之一。

交互作用是指在一定条件下两个或多个元素的结合效应小于或超过它们各自效应之和，即拮抗或者协同作用。就土壤-植物体系而言，拮抗作用是指一种离子阻挡或抑制另一离子从土壤进入植物的生理效应；而协同作用则是指某一离子促进另一离子"释放"或增强植物体对某一离子的吸收。因而，当以植物效应来制定环境标准时，交互作用影响应给予足够的关注。这就将交互作用的研究从单纯的营养学角度扩展到维持生态平衡和环境保护与人类健康等方面，从而赋予交互作用在环境科学和生命科学中新的含义。所以，交互作用的研究无论在理论上还是实践上都有重要的意义（郑春荣，2002）。土壤-植物系统中的交互作用十分复杂，其作用发生的部位有可能在土壤中、植物体中，或者两者兼而有之。

自 1960 年发现亚硒酸钠能预防氯化镉对大鼠睾丸的损伤后，人们对硒与重金属的交互作用展开了大量的研究，发现硒对铅、镉、汞、砷等重金属在水果、蔬菜、谷物中的富集有拮抗作用，可减缓其对植物生长的抑制；而硒与锌的交互作用，亦可有效降低第三者，如镉在水稻中的累积，缓解镉对水稻生长发育的毒害作用（呼艳姣等，2022）。本章主要论述硒与重金属的交互作用。

第一节　硒对镉、铅吸收和累积的影响

镉、铅等阳离子与硒酸盐的带电性不同，SeO_4^{2-} 或 SeO_3^{2-} 在转运蛋白上与包括镉离子在内的重金属阳离子不存在吸收竞争。镉主要是通过自然抗性相关巨噬细胞蛋白（natural resistance associated macrophage protein，Nramp）、锌铁调控转运蛋白（IRT-like proteins，ZRT）家族及阳离子/氢离子反向转运蛋白（antitransporter）的吸收通道吸收，而硒酸盐主要是通过硫酸盐转运子吸收，亚硒酸盐主要是通过磷酸盐和硅酸盐转运子吸收。但是，很多研究表明硒可以影响植物对重金属阳离子的吸收和累积。

一、硒与镉铅的拮抗作用

硒肥施用、富硒作物品种选育等生物强化措施显著提高了植物体内的硒含量，同时适量硒的添加能够显著降低重金属的积累。多项研究表明，营养液或土壤（盆栽和大田）培养条件下，基施或喷施适量浓度的硒，具有降低植物镉、铅吸收和累积的作用（表 7-1）。

表 7-1 不同形态和浓度的硒对不同植物吸收转运镉和铅的影响汇总

Se 形态	Se 浓度	重金属	Cd/Pb 浓度	植物	条件	处理时间/d	结果	作用	参考文献
Na_2SeO_3	1.27~63.6 mmol/L	Cd	8.9~178 μmol/L	水稻	水培	14	Cd 浓度≤35.6 μmol/L 时，Se 降低水稻 Cd 含量；Cd 浓度≥89 μmol/L 时，Se 增加水稻 Cd 含量	拮抗/协同	Ding et al., 2014; Feng et al., 2013
Na_2SeO_3	3 μmol/L	Cd	50 μmol/L	水稻	水培	10, 15	Se 降低根系和地上部 Cd 含量	拮抗	Lin et al., 2012
Na_2SeO_3	0.1 μmol/L	Cd	0.5, 5μmol/L	水稻	水培	9	Cd 为 0.5 μmol/L 时，Se 对水稻 Cd 吸收无显著影响；Cd 为 5 μmol/L 时，Se 显著降低水稻对 Cd 的吸收和转运	拮抗	庞晓辰等, 2014
Na_2SeO_3/ Na_2SeO_4	0.1~5 μmol/L	Cd	0.1~5 μmol/L	水稻/小麦	水培	0.1~6	短时间培养时，添加 Se 对水稻 Cd 吸收无显著影响；当培养时间增加时，添加 Se 可以降低水稻对 Cd 的吸收及向地上部的转运，且 Na_2SeO_3 的降低程度大于 Na_2SeO_4	拮抗	Wan et al., 2016; 2019; Wang et al., 2021
Na_2SeO_3/ Na_2SeO_4/ SeNP/SeMet	1 μmol/L	Cd	20 μmol/L	水稻	水培	25	Na_2SeO_3、SeNP 和 SeMet 的添加降低了根系和地上部 Cd 含量；Na_2SeO_4 降低了地上部 Cd 含量，但对根系 Cd 含量无显著影响	拮抗	Xu et al., 2021
Na_2SeO_3	2.5 μmol/L	Cd	10 μmol/L, 50 μmol/L	樱桃番茄	水培	—	Se 降低了植株各部位的 Cd 含量，提高了果实品质	拮抗	Xie et al., 2021
Na_2SeO_4	2 μmol/L	Cd	400 μmol/L, 600μmol/L	油菜	水培	14	Se 降低植株 Cd 含量	拮抗	Filek et al., 2008
Na_2SeO_3	3 μmol/L, 7 μmol/L	Cd	250 μmol/L, 500 μmol/L	辣椒	水培	—	Se 减少可食部位 Cd 含量，7 μmol/L 的效果＞3 μmol/L	拮抗	Mozafariyan et al., 2014

续表

Se 形态	Se 浓度	重金属	Cd/Pb 浓度	植物	条件	处理时间/d	结果	作用	参考文献
Na$_2$SeO$_3$/ Na$_2$SeO$_4$	0.1 μmol/L, 1 μmol/L, 5 μmol/L	Cd	1 μmol/L, 5 μmol/L	生菜	水培	3	Na$_2$SeO$_3$ 降低了生菜中的 Cd 含量，降低效果随着硒添加量增加而增加；Na$_2$SeO$_4$ 影响效果不稳定	拮抗	齐田田，2017
Na$_2$SeO$_4$	5 μmol/L, 10 μmol/L, 20 μmol/L	Cd	20 μmol/L	向日葵	水培	4	Se 降低了根系地上部 Cd 含量，效果：5 μmol/L > 10 μmol/L > 20 μmol/L	拮抗	Saidi et al.，2014
Na$_2$SeO$_3$/ Na$_2$SeO$_4$	10 μmol/L	Cd	10 μmol/L	青菜	水培	3	Na$_2$SeO$_3$ 显著降低了杭州油冬儿 40 d 苗龄时地上部 Cd 含量和苏州青菜 19 d 苗龄地上部 Cd 含量；Na$_2$SeO$_4$ 提高了地上部的 Cd 含量	拮抗/协同	Yu et al.，2019
Na$_2$SeO$_3$/ Na$_2$SeO$_4$	63.3 μmol/L	Cd	11 μmol/L, 22 μmol/L, 44 μmol/L	白磨	水培	21	Cd 为 11 μmol/L，22 μmol/L 时，Se 降低了 Cd 含量；Cd 为 44 μmol/L 时，Na$_2$SeO$_3$ 降低了 Cd 含量，Na$_2$SeO$_4$ 增加了 Cd 含量	拮抗/协同	Muñoz et al.，2007
Na$_2$SeO$_3$	12.7 μmol/L	Cd	8.9 μmol/L	西兰花	水培	10, 40	Se 降低了茎、果实 Cd 含量，增加了根 Cd 含量	拮抗/协同	Pedrero et al.，2008
Na$_2$SeO$_3$	38 μmol/L	Pb	483 μmol/L	白芥	水培	8	Se 显著降低了根系和地上部的 Pb 含量	拮抗	Fargašová et al.，2006
Na$_2$SeO$_3$	1.6 μmol/L, 6 μmol/L	Pb	50 μmol/L	蚕豆	水培	14	Se 的添加具有降低根系 Pb 吸收的趋势	拮抗	Mroczek-Zdyrska and Wójcik，2012
Na$_2$SeO$_3$	0.5 mg/kg, 1 mg/kg, 8 mg/kg	Cd	1.5 mg/kg, 5.12 mg/kg	水稻	盆栽基施	—	随着 Se 添加量的升高，水稻各部位 Cd 含量明显下降	拮抗	Chang et al.，2013; Wan et al.，2018b;
Na$_2$SeO$_3$/ Na$_2$SeO$_4$	1 mg/kg	Cd	1.75 mg/kg	水稻	盆栽基施	—	Se 降低了水稻各部位 Cd 含量，Na$_2$SeO$_3$ 效果优于 Na$_2$SeO$_4$	拮抗	Huang et al.，2018;

续表

Se形态	Se浓度	重金属	Cd/Pb浓度	植物	条件	处理时间/d	结果	作用	参考文献
Na₂SeO₃/SeNPs	0.5 mg/kg, 2.5 mg/kg	Cd	1.6 mg/kg	水稻	盆栽基施	—	Na₂SeO₃ 和 SeNP 均降低了水稻对 Cd 的吸收、运输和累积，2.5 mg/kg SeNP 处理效果最佳	拮抗	徐境懋等, 2021
Na₂SeO₃	0.5 mg/kg, 1.0 mg/kg	Cd/Pb	Cd: 1.03 mg/kg Pb: 98.1 mg/kg	水稻	盆栽基施	—	Se 减少了水稻地上部的 Cd、Pb 含量	拮抗	Hu et al., 2014
Na₂SeO₃	0.5 mg/kg, 1.0 mg/kg	Pb	300.9 mg/kg	水稻	盆栽基施	—	Se 减少了水稻对 Pb 的吸收	拮抗	万亚男等, 2018
Na₂SeO₃	0.5 mg/L, 2 mg/L, 4 mg/L	Cd	0.5 mg/kg, 2.0 mg/kg	菠菜	盆栽基施	—	Se 降低了菠菜 Cd 含量；Cd 为 0.5 mg/kg 时，Se 为 2.0 mg/L 最有效；Cd 为 2.0 mg/kg 时，Se 为 4 mg/L 最有效	拮抗	郭峰等, 2014
Na₂SeO₃	1.0～15 mg/kg	Cd	0.5～5 mg/kg	樱桃小萝卜	盆栽基施	—	Se 能显著降低萝卜对 Cd 的吸收，且这种作用表现为地下部分强于地上部分	拮抗	铁梅等, 2014
Na₂SeO₃	1 mg/kg 338 g/hm²	Cd/Pb	Cd: 1 mg/kg Pb: 10 mg/kg	白菜/生菜	盆栽/大田基施	—	Se 降低了地上部的 Cd 和 Pb 含量	拮抗	He et al., 2004
Na₂SeO₃	10 mg/kg, 20 mg/kg, 30 mg/kg	Cd	4 mg/kg, 8 mg/kg, 12 mg/kg	油菜	盆栽基施	—	土壤 Cd≤8 mg/kg 时，添加 Se 抑制根吸收，但 Se 为 30 mg/kg 时略促进根吸收；土壤 Cd 为 12 mg/kg 时，Se 促进根吸收；Se 降低了叶片 Cd 含量	拮抗/协同	李荣林等, 2008
Na₂SeO₃/Na₂SeO₄	0.5 mg/kg, 1 mg/kg, 2.5 mg/kg	Cd	5.36 mg/kg	小白菜	盆栽基施	—	Se 降低了小白菜对 Cd 的吸收，Na₂SeO₃ 效果优于 Na₂SeO₄；高浓度的 Se 促进了 Cd 的转运	拮抗/协同	刘杨等, 2020
Na₂SeO₃/氨基酸螯合硒	15～100 g/hm²	Cd/Pb	—	水稻	大田喷施	—	施 Se 15 g/hm² 可降低水稻籽粒 Cd 含量；喷施 75 g/hm² 和 100 g/hm² Se 肥对 Cd 和 Pb 的降低效果最好	拮抗	Gao et al., 2018; 方勇等, 2013; 谭俊等, 2018; 黄青青等, 2020

续表

Se 形态	Se 浓度	重金属	Cd/Pb 浓度	植物	条件	处理时间/d	结果	作用	参考文献
SeNPs	2.3mg/pot, 4.7mg/pot, 9.4 mg/pot	Cd/Pb	Cd: 3 mg/kg, Pb: 300 mg/kg	水稻	盆栽喷施	—	喷施 4.7 和 9.4 mg/pot Se 降低了籽粒 Cd 含量; 喷施 9.4 mg/pot Se 降低了籽粒 Pb 含量	拮抗	Wang et al., 2021
Na_2SeO_3	2~8 mg/pot	Cd	—	小麦	盆栽喷施	—	Se 降低了小麦各部位 Cd 含量, 降低程度随 Se 添加量的增加而增加	拮抗	Wu et al., 2020
Na_2SeO_3	0.5 mg/kg, 1.0 mg/kg, 1.95 mg/kg, 3.9 mg/pot	Cd	1.12 mg/kg	小麦	盆栽基施/喷施	—	基施 Se 降低了小麦籽粒的 Cd 含量, 效果优于喷施; 喷施 Se 降低了高镉积累型软质小麦籽粒 Cd 含量, 对低镉积累型硬质小麦籽粒 Cd 含量无显著影响	拮抗	Zhou et al., 2021
Na_2SeO_3	39 mg/L	Cd	20 mg/kg	生菜	盆栽喷施	—	Se 减少了 Cd 吸收	拮抗	吕选忠等, 2006
Na_2SeO_3	0.5 mg/L, 1 mg/L	Cd	20 mg/kg	番茄	盆栽喷施	—	使根、茎、叶、果实 Cd 含量分别降低了 24.5%~48.5%、9.8%~30.9%、7.3%~41.9% 和 14.4%~37.1%	拮抗	Chi et al., 2017
Na_2SeO_3	0.5~8 mg/L	Cd/Pb	Cd: 0.195 mg/kg, Pb: 24.9 mg/kg	西瓜	大田喷施	—	喷施 2.0~4.0 mg/L Se 时, 对 Cd、Pb 的抑制效果最好	拮抗	张翠萍等, 2013
Na_2SeO_3	0.5~10 mg/L	Cd/Pb	Cd: 0.153 mg/kg, Pb: 28.9 mg/kg	草莓	大田喷施	—	喷施 2.5~5.0 mg/L Se 时, 可有效抑制草莓叶片和果实中 Cd 和 Pb 的含量	拮抗	张海英等, 2011
Na_2SeO_3	2.3 mg/棵, 0.56~9.02 mg/棵	Cd/Pb	Cd: 0.27 mg/kg, Pb: 25.6 mg/kg	猕猴桃	穴施/喷施	—	2.25 mg/L 叶面喷 Se 处理下对果实 Cd 和 Pb 含量的抑制效果最好, 叶面喷施处理较土壤基施效果好	拮抗	龙友华等, 2016
硒矿粉/含硒叶面肥	30 g/hm²	Cd/Pb	2.1 mg/kg	油菜	大田基施喷施	—	单独基施、单独喷施和基施+喷施分别使油菜籽粒 Cd 含量分别降低 9.24%、5.88%、12.61%	拮抗	陈火云等, 2019

对水稻（*Oryza sativa*）、小麦（*Triticum aestivum*）、蔬菜等多种作物的水培试验表明，在一定范围镉浓度暴露下，硒的添加降低了植物对镉的吸收和向地上部的转运，且这个范围因不同种类或同种植物不同基因而异。水稻在镉暴露浓度≤50 μmol/L 时，添加 10 μmol/L 以下的硒酸盐或亚硒酸盐降低了水稻根系和地上部的镉含量，且对镉含量的降低程度随着硒添加量的增加呈现增加的趋势，亚硒酸盐的降低效果优于硒酸盐（Ding et al.，2014；Feng et al.，2013；Lin et al.，2012；Wan et al.，2016；2019；Wang et al.，2021）。对于镉耐性较强的植物，如油菜（*Brassica napus*）和辣椒（*Capsicum annuum*）等，在较高镉浓度（250～600 μmol/L）暴露下，添加硒也可以降低可食部位的镉含量（Filek et al.，2008；Mozafariyan et al.，2014）。对莴苣（*Lactuca sativa*）、辣椒等的研究发现，在一定浓度范围内，亚硒酸盐对地上部镉累积的降低程度随添加量的增加而增加；而在 20 μmol/L 镉处理下，添加不同浓度的硒酸盐对向日葵（*Helianthus annuus*）地上部镉累积的降低效果为 5 μmol/L＞10 μmol/L＞20 μmol/L（Saidi et al.，2014），这可能与不同植物的镉耐受机制以及硒的不同添加形态有关。此外，同种植物不同基因型以及植物的不同生长时期对硒、镉的耐受性不同，也会导致硒对镉积累影响的差异。Yu 等（2019）研究了青菜（*Brassica chinensis*）不同苗龄对硒、镉的响应，结果发现，在苏州青根 19 d 苗龄时，添加亚硒酸盐降低了其地上部镉含量，该效果随着苗龄增加而逐渐减弱；而杭州油冬儿幼苗前期时添加亚硒酸盐对地上部镉含量无显著影响，在 40 d 时受其影响而显著降低。硒能够降低铅在植物体中的积累，以白芥（*Sinapis alba*）和蚕豆（*Vicia faba*）为供试作物的研究发现，添加亚硒酸盐降低了铅在植物体中的积累，减弱了重金属对植物的毒性作用（Fargašová et al.，2006；Mroczek-Zdyrska and Wójcik，2012）。

土壤中基施硒肥可以降低水稻植株对镉和铅的累积，且在一定范围内（土壤硒≤8 mg/kg）对镉的降低效果随着硒添加量的增加而增强（万亚男等，2018；Chang et al.，2013；Hu et al.，2014；Wan et al.，2018b）。镉污染土壤（pH 7.26，Cd 5.12 mg/kg）淹水和非淹水下加亚硒酸盐（0.5 mg/kg 和 1.0 mg/kg）分别使水稻籽粒的镉含量降低了 67.7%～77.4% 和 45.2%（Wan et al.，2018b）；土壤基施硒酸盐也可以降低水稻各部位的镉含量，且降低程度较亚硒酸盐小（Huang et al.，2018）。对其他蔬菜水果等进行硒肥基施也可以降低可食部位的镉和铅含量，且在不同土壤镉条件下，降低效果最佳时的施硒量也存在差异，如土壤镉含量为 0.5 mg/kg 时，土壤添加 2.0 mg/L 亚硒酸盐对菠菜（*Spinacia oleracea*）可食部位的镉含量降低最多；而在土壤镉为 2.0 mg/kg 时，添加 4.0 mg/L 硒时降低效果最好（郭锋等，2014）；镉污染土壤（5.36 mg/kg）添加 1 mg/kg 硒时，对小白菜（*Brassica chinensis*）可食部位镉含量的降低效果最好，亚硒酸盐和硒酸盐处理分别使镉含量降低了 40.0% 和 20.5%（刘杨等，2020）。He 等（2004）的盆栽试验结果表明，给镉和铅胁迫的白菜（*Brassica rapa*）和生菜（*Lactuca sativa*）施用亚硒酸盐，显著降低了镉和铅在植物地上部的含量；对生菜的大田试验也发现硒具有降低镉和铅累积、增加矿质元素吸收的作用。

叶面喷施硒肥可显著降低稻米籽粒中铅和镉的含量，且提高了水稻籽粒的营养品质，在水稻分蘖期和齐穗期分两次喷施 15 g/hm^2 的亚硒酸盐或者在水稻孕穗期喷施 15 g/hm^2 的氨基酸螯合硒，可分别使稻米籽粒镉含量降低 13.0% 和 9.0%（黄青青等，2018；谭俊等，2020）；喷施 75 g/hm^2 以有机物质为载体的亚硒酸盐，可使水稻籽粒镉和铅含量分别降低 11.8% 和 34.8%（方勇等，2013）；硒对水稻籽粒镉累积的影响与水稻品种有关，Gao 等（2018）研究发现，在水稻分蘖期和孕穗期喷施亚硒酸盐可使 WYHZ 品种籽粒镉含量降低 61.6%，略微降低了南粳 5055 籽粒的镉含量，而对 ZF1Y 无显著影响。在小麦分蘖期、拔节期和孕穗期进行叶面喷硒，也可以降低小麦籽粒对镉的累积，并且促进了小麦的生长（Wu et al.，2020）。喷施亚硒酸盐同样抑制了生菜、番茄（Lycopersicon esculentum）、西瓜（Citrullus lanatus）、草莓（Fragaria ananassa）、猕猴桃（Actinidia chinensis）等不同蔬菜水果可食部位对镉、铅的积累（龙友华等，2016；吕选忠等，2006；张翠翠等，2013；张海英等，2011；Chi et al.，2017），且最佳施硒量随土壤重金属含量和植物类型的不同而不同。龙友华等（2016）对猕猴桃的研究发现，高质量浓度的硒肥处理对镉和铅的拮抗效果较相对低质量浓度处理差，且叶面喷施处理较土壤基施效果好。而 Zhou 等（2021）研究发现，基施硒肥对小麦籽粒镉含量的降低效果优于喷施处理，且喷施处理只降低了高镉积累型软质小麦籽粒镉含量，对低镉积累型硬质小麦籽粒镉含量无显著影响。陈火云等（2019）对油菜（Brassica campestris）的研究发现，播前底施活化硒矿粉结合始花期追施硒叶面肥，可以有效提高油菜籽粒硒含量且降低籽粒镉含量。除了亚硒酸盐和硒酸盐，有机硒（SeMet）和纳米硒（SeNP）的添加也可以降低植物对镉的吸收和向地上部的转运，缓解镉对植物的胁迫（Wan et al.，2016；Xu et al.，2021）。纳米硒作为一种新型元素态硒，具有高活性和低毒性的特点，在农业环境领域的研究和应用也越来越多。研究发现，土壤基施和叶面喷施纳米硒均可以降低水稻籽粒对镉或铅的累积（徐境懋等，2021；Wang et al.，2021）。

关于硒与镉、铅吸收累积的拮抗作用机制，主要包括降低土壤中镉、铅的有效性，以及降低镉向地上部的转运（图 7-1）。有研究表明，还原条件下，根际的 SeO_4^{2-} 和 SeO_3^{2-} 可以发生还原作用转化为 Se^{2-}，从而与重金属离子形成难溶的复合物如 Cd-Se、Pb-Se，降低金属离子的有效性（Shanker et al.，1995；Plant et al.，2006）。亚硒酸盐对植物镉吸收的降低程度大于硒酸盐，这可能是由于亚硒酸盐不易被植物利用，在土壤中的移动性较弱，因而更容易直接还原成 Se^{2-} 而与镉形成 Cd-Se 复合物（Shanker et al.，1995）。万亚男（2018）研究发现，重金属污染土壤上添加硒可降低土壤溶液中的镉含量和铅含量。Huang 等（2018）和刘达等（2016）也发现，镉污染土壤上添加硒降低了交换态镉含量和占土壤总镉的比例，增加了碳酸盐结合态、铁锰氧化物结合态、有机结合态的比例。

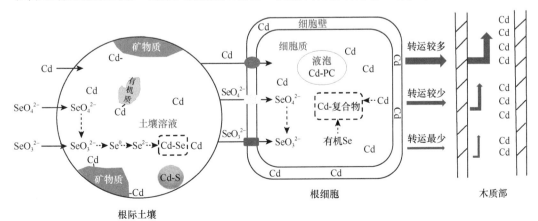

图 7-1 硒对植物吸收转运镉的影响过程示意图（万亚男等，2017）

硒和镉、铅等阳离子型重金属在水稻根表不存在离子间的竞争作用，水稻对其吸收是独立的过程。Wan等（2016）对水稻的吸收动力学试验发现，硒和镉共同处理水稻根系60 min时，硒的添加没有降低水稻根系对镉的吸收速率，也证明了这一点。硒可能主要是通过降低镉、铅由植物根系向地上部的转运从而减少重金属积累的，硒与镉、铅在细胞内可能存在相互作用。植物受到重金属胁迫时，可通过自身的耐受缓解镉的毒害。植物细胞壁中一些蛋白质和多糖含有大量的氨基、羧基等配位基，可将金属固持在细胞壁上，从而减少重金属进入细胞。而重金属穿过细胞壁和细胞膜进入细胞质后，细胞质内含有大量巯基的PC可与金属离子发生配合作用，将其转化为毒性较小的结合态。金属-PC螯合物多储存在液泡中，而不是积累在生物代谢活动的细胞器上，从而减少对细胞器的伤害和重金属向木质部的运输（Ismael et al., 2019）。研究发现，加硒可促进PC的形成，并提高PC合成酶的活性，而且添加硒能够促进PC合成前体GSH的形成（Ismael et al., 2019；Li et al., 2016；Wan et al., 2019）。Wan等（2019）研究发现，硒的添加增加了细胞可溶组分的分配率，减少了细胞器的分配率，可能就是增加了PC与镉的螯合作用而将较多的镉储存在液泡中。此外，庞晓辰等（2014）和李虹颖等（2016）研究还发现，添加硒增加了镉在根系细胞壁的分配率。Cui等（2018）也发现，加硒提高了水稻悬浮细胞壁的厚度和机械强度。镉由根系细胞吸收后，通过共质体途径和质外体途径进入木质部并向地上部运输。余垚（2019）研究发现，亚硒酸盐提高了植物根部木栓质和木质素合成以及细胞壁过敏性增强的相关基因表达，而增强的木质素合成可以促进细胞壁的强化，进而促进镉的质外体固定；硒酸盐削弱了镉的共质体固定，从而促进了镉在地上部的累积。但是，硒酸盐代谢速率慢于亚硒酸盐，所以它与镉的拮抗可能需要更长的时间才能表现出来（图7-2）。

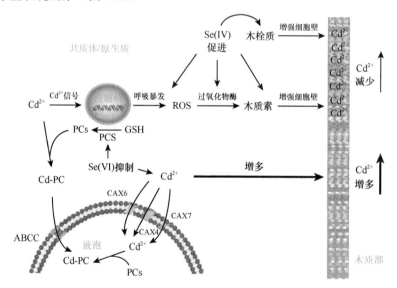

图7-2　硒影响镉在植物根部质外体和共质体的固定过程示意图（余垚，2019）

研究表明，硒也可能通过影响镉相关转运蛋白的表达影响植物对镉的积累，如与镉跨质膜运输有关的自然抗性相关巨噬细胞蛋白（OsNramp5）、与镉在液泡中隔离有关的重金属ATP酶（OsHMA3）等（Cui et al., 2018）。无机硒进入植物体内，可转化为有机态硒［硒代甲硫氨酸、硒代半胱氨酸（MeSeCys）］，而 MeSeCys 是谷胱甘肽过氧化物酶（GSH-Px）的活性中心。GSH-Px 能将有害的过氧化物还原为无害的羟基化合物，并且可以清除机体内的氧化产物如自由基，保护细胞膜的正常功能（Yadav，2010）。因此，硒还可能通过影响 GSH-Px 的形成及活性从而减少重金属毒性。亚硒酸盐对镉吸收转运的降低程度大于硒酸盐，这也可能是由于亚硒酸在植物体内更容易转化为 MeSeCys，且亚硒酸盐进入体内后主要累积在根部，从而在根系与重金属发生相互作用而减少重金属向地上部的转运，硒酸盐进入植物体后快速向地上部转运（Li et al.，2008）。此外，硒能有效调节抗氧化酶（超氧化物歧化酶、过氧化氢酶和过氧化物酶）的活性，在一定程度上缓解镉胁迫对植物生理生长的毒害作用，减少镉的累积。

二、硒与镉的协同作用

硒对植物中重金属累积的作用在很大程度上依赖于硒的剂量。通常认为，在较低的适宜浓度范围内，硒可以抑制植物中镉的累积；而超出这个范围，即在较高的暴露剂量时，硒反而会促进镉的累积，并且这个范围也因不同植物或同种植物不同基因而异（表 7-1）。

研究发现，当水稻镉暴露浓度和硒添加量相对较低时，亚硒酸盐减少了水稻根系和地上部的镉含量，而在镉浓度≥89 µmol/L 时，硒的添加增加了根系对镉的吸收（Ding et al.，2014；Feng et al.，2013）。当土壤镉含量为 4 mg/kg 和 8 mg/kg 时，施加 10 mg/kg 和 20 mg/kg 的硒降低了油菜根系镉含量，而施硒量为 30 mg/kg 时略增加了根系的镉含量；在高镉（12 mg/kg）土壤中，施硒增加了根系镉含量（李荣林等，2008）。对小白菜（*Brassica chinensis*）的盆栽试验也发现，低浓度的硒降低了镉由根系向地上部的转运，而高浓度的硒促进了镉的转运（Liu et al.，2015）。这可能是因为在高浓度条件下，硒与镉同时施加对植物根系的毒性高于单独施镉，对植物造成了协同胁迫。对白蘑（*Pleurotus ostreatus*）的研究发现，镉含量为 11 µmol/L 和 22 µmol/L 时，亚硒酸盐或硒酸盐的添加均降低了植物体内的镉；而当镉含量为 44 µmol/L 时，亚硒酸盐降低了镉含量，而硒酸盐增加了镉含量（Muñoz et al.，2007），与青菜的研究结果相似（Yu et al.，2019）。余垚（2019）的研究发现，处理 3d 时，硒酸盐对青菜抗氧化系统的促进效果小于亚硒酸盐，并且下调了抗病基因的表达；硒酸盐削弱了镉的共质体固定，从而促进了镉在地上部的累积。但是在暴露 5～7d 时，硒酸盐的添加降低了青菜地上部的镉含量。对生菜的研究结果也表明，硒酸盐对植物吸收累积镉的影响效果不稳定，在不同浓度和不同处理时间下产生的效果不同（齐田田，2017），可能是因为硒酸盐代谢速率慢于亚硒酸盐，其与重金属的作用需要更长的时间才能稳定。对甘蓝（*Brassica oleracea*）和油菜的研究发现，在一定浓度下，硒的添加增加了植物根系的镉含量，却降低了地上部的镉含量，降低了镉由根系向地上部的转运（李荣林等，2008；Pedrero et al.，2008）。

由于硒对植物重金属吸收和累积的双重作用，且最佳施硒量随土壤镉含量和植物类型的不同而不同，因此在利用硒缓解植物镉积累时，要根据不同植物对镉的耐性选择合适的硒添加量和添加形态。

第二节 硒对砷铬吸收和累积的影响

自然条件下，砷通常以阴离子的形式存在（AsO_3^{3-}、AsO_4^{3-}），硒通常也是以阴离子的形式存在（SeO_3^{2-}、SeO_4^{2-}），它们具有相似的结构，可能竞争相似的吸收通道。水稻对亚砷酸盐的吸收主要依靠根细胞膜上的水通道蛋白，可以在硅酸盐转运子的作用下进入水稻根系（Meharg and Jardine，2003；Ma et al.，2006；Ma et al.，2008），而亚硒酸盐也可以通过硅的转运蛋白进入水稻根系（Zhao et al.，2010）。砷酸盐的化学性质与磷酸盐离子非常相似，因此，砷酸盐可以在磷酸盐转运子的作用下进入植物根系（Meharg and Macnair，1992；Meharg et al.，1994；Meharg and Hartley-Whitaker，2002），而水稻根系的磷酸盐转运子在亚硒酸盐的主动吸收过程中也起着重要作用（Li et al.，2008；Zhang et al.，2012）。亚硒酸盐与亚砷酸盐和砷酸盐有相似的吸收通道，在吸收过程中可能存在竞争。

有研究表明，硒可以影响植物对砷的吸收、转运及形态转化（Kramárová et al.，2012；Han et al.，2015；Camara et al.，2018）。但是，由于硒与砷的形态、供试作物不同、试验条件等不同，硒与砷之间的相互作用有所差异，即硒与砷之间既可能表现为拮抗作用，又可能表现为协同作用。Khattak等（1991）通过营养液砂培试验发现，营养液中添加0.05 mg/L和0.1 mg/L的硒酸盐能够抑制紫花苜蓿（*Medicago sativa*）对砷酸盐的吸收，降低苜蓿地上部砷的含量。在5 mg/L砷处理下，亚硒酸盐的添加（0.1 mg/L、1 mg/L和5 mg/L）可以降低烟草根系和地上部的砷含量（Han et al.，2015）。Malik等（2012）的研究也发现，在2.5 μmol/L、5.0 μmol/L和10 μmol/L砷处理下的绿豆（*Phaseolus aureus*）添加2.5 μmol/L和5 μmol/L硒后，显著降低了绿豆地上部的砷含量。Feng等（2009）通过对蜈蚣草（*Pteris vittata*）的研究发现，在一定浓度砷、硒处理下（0~5 mg/L），亚硒酸盐对砷的吸收转运具有抑制竞争作用。然而Srivastava等（2009）发现硒酸盐（5 μmol/L）的添加增加了蜈蚣草对砷（Na_2HAsO_4，浓度150μmol/L、300 μmol/L）的吸收，两者之间表现为协同作用。

理论上来讲，亚硒酸盐会与亚砷酸盐竞争硅酸转运蛋白而影响水稻对亚砷酸盐的吸收。植物可通过磷酸盐转运子吸收砷酸盐（Meharg and Macnair，1992；Meharg and Hartley-Whitaker，2002），并且砷酸盐的向上运输也受控于磷酸盐的调控系统（Wu et al.，2011）；而植物根系对亚硒酸盐的吸收同样受生长介质中磷酸盐的影响，细胞膜上的磷酸盐转运子参与调控植物根系对亚硒酸盐的吸收过程（Li et al.，2008；Zhang et al.，2014b）。因此，亚硒酸盐会与砷酸盐竞争磷酸盐转运子而影响水稻对砷酸盐的吸收和转运。但是很多研究表明硒对砷的吸收无显著影响，砷污染土壤中亚硒酸盐的添加对水稻根系砷吸收无降低作用（Lv et al.，2020；Zhou et al.，2017）。Camara等（2018）和Wang等（2021）的研究发现，水培条件下，亚砷酸盐胁迫时，亚硒酸盐的添加增加了水稻对砷的吸收，降低了砷向地上部的转运，而硒酸盐的添加有降低砷吸收的趋势，对砷转运无显著影响；砷酸盐处理下，亚硒酸盐的添加对水稻砷吸收有增加的趋势，降低了砷向地上部的转运，而硒酸盐的添加降低了水稻根系对砷吸收的趋势，却有促进砷向地上部转运的趋势。此外，Wang等（2021）的研究还发现，纳米硒的添加显著降低了亚砷酸盐处理下水稻地上部的砷含量，而对砷酸盐处理下水稻地上部砷含量影响较小，略有降低。因此，硒对砷吸收转运的作用因硒形态的不同而不同：硒酸盐主要是通过抑制砷的吸收以降低砷在水稻植株中的积累，而亚硒酸盐主要是通过抑制砷的向上运输以降低地上部的砷含量来实现，纳米硒对砷转运的抑制效果弱于亚硒酸盐；亚硒酸盐的添加增加了砷在水稻根部不活跃部位（细胞壁和液泡）的分配，从而抑制了砷向地上部的运输（图7-3）。Wan等（2018a）通过土壤盆栽试验研究了不同水分管理（即淹水和非淹水）条件下，亚硒酸盐对水稻吸收、积累砷的影响。研究发现，在非淹水条件下，添加亚硒酸盐降低了水稻各部位的砷含量；而在淹水条件下，添加亚硒酸盐降低了水稻茎和叶中的砷含量，却增加了籽粒和颖壳中的砷含量，所以硒还可能影响砷的韧皮部再运输过程。

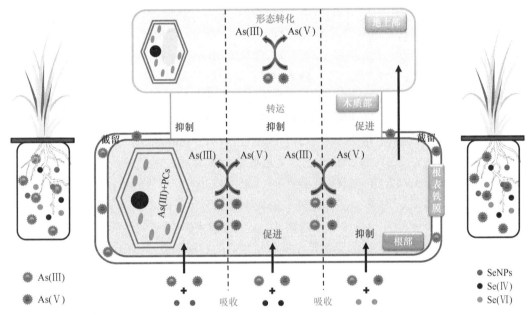

图7-3　不同形态硒对植物砷的吸收、转运及形态转化影响的示意图（王康，2021）

无论是水稻茎叶还是水稻籽粒，无机砷均是砷的主要存在形态（Norton et al., 2010b；Batista et al., 2011）。Batista等（2011）也发现，无机砷和DMA是稻米中砷的主要存在形态，且无机砷占总砷含量的78.4%。而且，硒对植物体吸收、积累砷的影响也表现在砷的形态转化上。例如，在同一砷处理条件下，虽然添加亚硒酸盐对烟草（Nicotiana tabacum）根部As(Ⅲ)和As(Ⅴ)各自所占的百分比无显著影响，但其地上部As(Ⅲ)所占百分比随亚硒酸盐添加浓度的增加而降低，而As(Ⅴ)所占百分比却随之增加（Han et al., 2015）。Wang等（2021）对水稻的研究也发现，添加亚硒酸盐增加了根系中As(Ⅲ)占根系总砷的比例，降低了地上部As(Ⅲ)的比例，说明亚硒酸盐促进了根系砷向As(Ⅲ)的转化，抑制了地上部砷向As(Ⅲ)的转化。

综上，硒对植物吸收和累积砷的作用主要与硒的形态有关，表现为硒酸盐抑制了根系对砷的吸收而亚硒酸盐抑制了砷由根部向地上部的转运；同时也通过影响砷的形态转化和韧皮部的再运输过程而影响砷的活性和地上部（籽粒）对砷的累积。

铬是人体不可缺少的微量元素，但过量的摄入会对机体产生严重危害。Cr(Ⅲ)毒性较小，是人体和动物必需微量元素，而Cr(Ⅵ)对动、植物有较大的毒性，接触Cr(Ⅵ)有致癌、致畸等潜在危害。Xu等（2021）研究发现，硫酸盐转运蛋白Sultr1;2是植物

Cr(Ⅵ)吸收的主要转运蛋白，而 Se(Ⅳ)也可以通过硫酸盐转运蛋白进入植物体内（Sors et al.，2005），因此硒与铬在吸收上可能存在竞争关系。然而，Srivastava 等（1998）研究了硒酸盐及亚硒酸盐对菠菜不同形态铬吸收的影响，结果表明硒对植物 Cr(Ⅲ)吸收的抑制作用显著强于硒对 Cr(Ⅵ)的抑制，但不同形态的硒对植物铬吸收的拮抗效应相近。Zhao 等（2019）对紫菜薹（*Brassica campestris* ssp. Pekinensis）的水培试验结果表明，亚硒酸盐（0.1 mg/L）的添加缓解了铬胁迫[1 mg/L Cr(Ⅵ)]导致的根系损伤，减少了根系对铬的吸收，且促进了铬向活性较低的细胞组分的分配；但是没有降低地上部的铬含量，却缓解了铬对地上部的氧化损伤（Qing et al.，2015）。对芥菜（*Brassica juncea*）的研究也发现，在铬胁迫下[Cr(Ⅵ)]添加硒酸盐，可以提高植物的抗氧化能力和金属螯合剂的含量（Handa et al.，2017；2018）。刘晓娟等（2019）对茄子（*Solanum melongena*）的盆栽试验结果发现，在土壤铬[Cr(Ⅵ)]含量为 10 mg/kg 和 20 mg/kg 时，植株各部位的铬含量随亚硒酸盐添加量（0~48 mg/kg）的增加呈先降低后增加的趋势，土壤铬含量为 40 mg/kg 和 80 mg/kg 时，植株铬含量随硒添加量的增加而降低，硒缓解茄子铬毒害的最佳浓度也因土壤铬含量的不同而不同。Cai 等（2019）研究发现，铬以 Cr(Ⅵ)或 Cr(Ⅲ)的形式添加到土壤中，当铬含量较低[Cr(Ⅵ)：60 mg/kg，Cr(Ⅲ)：100 mg/kg]时，亚硒酸盐（5 mg/kg）的添加对小白菜植株铬含量有增加的趋势；而当铬含量较高[Cr(Ⅵ)：120 mg/kg，Cr(Ⅲ)：200 mg/kg]时，硒的添加降低了 Cr(Ⅵ)处理下小白菜根系和地上部的铬含量，而对 Cr(Ⅲ)处理下植物铬含量无显著影响。除了在植物体内的影响作用外，硒与铬可能在土壤中竞争吸附位点，导致土壤中铬有效性的增加（Cai et al.，2021），所以硒对植物累积铬的影响是土壤和植物两个过程动态平衡的结果。此外，硒对铬吸收和累积的作用与元素的形态和剂量有关，硒与铬之间也是拮抗或协同的双重作用。

第三节 硒对汞吸收和累积的影响

一、硒对水稻吸收和累积汞的影响

汞（Hg）是一种高毒性、全球性的持久性污染物，广泛存在于食物链和各类环境介质中，不同形态的汞具有不同的毒性，各种形态中甲基汞的毒性较高。随着工业化进程的发展，土壤汞污染问题愈发严重。由于稻米极易富集甲基汞（MeHg），同时全球 40%以上人口以稻米为主食，因此，稻田土壤汞（Hg）污染，尤其是稻米甲基汞浓度超标问题越来越受到人们的关注。硒与金属有较强的亲和力，能够与多种有毒重金属元素产生拮抗作用并降低其毒性。硒与重金属的拮抗作用为使用硒降低作物中甲基汞含量提供了科学理论依据。Zhang 等（2012）的野外调查研究结果表明，在贵州万山汞矿区，随土壤中硒含量的增加稻米中，无机汞和甲基汞的含量不断降低。Zhao 等（2014）的土壤盆栽试验结果发现，0.5 mg/kg、1 mg/kg、5 mg/kg 亚硒酸盐的添加，降低了水稻根、秸秆、籽粒中无机汞的含量，且硒对无机汞的抑制作用比对甲基汞的抑制作用更强。Wang 等（2016a）的盆栽试验表明，土壤中施加不同剂量的硒（3.0~6.0 mg/kg），稻米中甲基汞的含量降低最大可达 72%，同时土壤中甲基汞的含量也显著降低。赵家印等（2019）的田间试验进一步证实了土壤施硒（1.5 mg/kg）后可以有效降低稻米中甲基汞的含量，稻米中甲基汞的含量降低了 35%；同时，在水稻生长前中期（10～80 d），硒肥的施用使土壤中甲基汞的含量降低了 55%～69%。

稻米中甲基汞含量的降低与甲基汞和硒的植物竞争吸收关系不大，其主要原因可能是硒的添加导致土壤中无机汞的生物可利用性降低，进而降低土壤甲基汞的生成。在水稻生长期，土壤处于还原状态，土壤中高价态的硒（SeO_4^{2-} 或 SeO_3^{2-}）在还原条件下可转化为低价态硒（Se^0 或 Se^{2-}）（Jayaweera and Biggar，1996；Li et al.，2010）；同时，二价汞离子（Hg^{2+}）易被还原为元素汞（Hg^0）（Björnberg et al.，1988）。Se^{2-} 与无机汞结合生成惰性硒化汞（Zhang et al.，2014a；Wang et al.，2016a），进而降低土壤甲基汞的生成（赵家印等，2019），同样硒也有可能与汞形成硒化汞沉淀。硒与汞相互作用可能的反应如下：

$$SeO_4^{2-}/SeO_3^{2-} \rightleftharpoons Se^0/Se^{2-}$$
$$Hg^{2+}Cl_2/Hg^{2+}(OH)_2 \rightleftharpoons Hg^0$$
$$Hg^{2+} + Se^{2-} \rightleftharpoons HgSe$$
$$Hg^0 + Se^0 \rightleftharpoons HgSe$$

硒与汞具有极强的结合力（HgSe 的稳定常数 $\log K_{sp}=-58$），HgSe 难溶于水（Björnberg et al.，1988），亦难被微生物转化形成甲基汞，因此，稻田土壤添加硒后降低了土壤中汞的生物有效性。土壤模拟试验结果表明，即使土壤添加高浓度硒和汞（Se 150 mg/kg，Hg 100 mg/kg），硒仍能抑制汞的甲基化，X 射线吸收近边结构（XANES）进一步分析发现，施硒土壤中汞的形态主要为硒化汞（HgSe），同时透射电镜-能谱分析（TEM-EDX）结果发现在厌氧环境下施硒土壤中形成了 HgSe 纳米颗粒（Wang et al.，2016a；2016b）。

在植物体内也可能发生上述反应，从而降低植物体内汞的转运。Zhao 等（2014）的营养液培养试验结果表明，在低浓度汞处理下（0、0.1 mg/L），亚硒酸钠的加入增加了水稻根系对无机汞吸收；而在高浓度汞处理下（1 mg/L、10 mg/L），随着硒添加浓度的增加，水稻对无机汞的吸收逐渐降低，同时降低了汞由根部向地上部的转运；利用同步辐射 X 射线荧光（SRXRF）分析技术发现水稻根部硒和汞分布基本一致，认为根部"硒-汞作用"可能是抑制无机汞由根部向地上部的运输原因。Li 等（2015）在贵州清镇汞矿区稻田喷施 0.01 μg/mL、0.1 μg/mL、0.5 μg/mL、1 μg/mL、5 μg/mL 不同浓度亚硒酸钠，结果发现 0.5 μg/mL 亚硒酸钠处理后的稻米总汞含量降低了 30%，但是 1 μg/mL 和 5 μg/mL 亚硒酸钠的处理却增加了稻米中总汞的含量；0.5 μg/mL 亚硒酸钠处理的水稻根系 X 射线荧光光谱（XRF）分析结果表明，硒显著降低了根系中汞的吸收，原位 X 射线吸收光谱（XAS）形态分析表明汞和硒形成了复合物。田间叶面喷施 75 g/hm^2 和 100 g/hm^2 硒肥可显著降低稻米籽粒中汞含量（方勇等，2013）。

硒与汞在土壤-水稻系统中的拮抗作用主要表现在以下 4 个方面（Li et al.，2015，图 7-4）：①在土壤中形成 Hg-Se 难溶性复合物，降低土壤中汞的生物有效性（Zhang et al.，2012）；②硒诱导水稻根表铁膜的形成，阻碍了水稻对汞的吸收（高阿祥等，2017）；③硒诱导凯氏带和木栓层等生物屏障的形成，降低了汞的转运；④在植物根部形成 Hg-Se 复合物，不能通过凯氏带，从而减少汞向地上部的运输（Zhao et al.，2014）；强化植物体内半胱氨酸对无机汞离子的络合，降低其向籽粒的转运（Meng et al.，2014）。

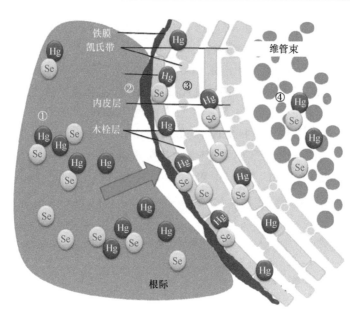

图 7-4　硒降低水稻吸收和累积汞的影响机制示意图（引自 Li et al., 2015）

二、硒对其他植物吸收和累积汞的影响

硒-汞作用机理的研究最早始于 Shanker 等研究土壤施硒对番茄（*Lycopersicum esculentum*）和萝卜（*Raphanus sativus*）吸收富集汞的影响，发现随着土壤中添加亚硒酸盐和硒酸盐浓度的增加，番茄对汞的吸收显著降低，降低了根和果实中汞的含量，但却促进了地上部的汞含量，并且亚硒酸盐和硒酸盐对汞吸收的影响效果相似（Shanker et al., 1996a）；在萝卜的研究中得到了相似的结果，土壤中亚硒酸盐和硒酸盐的添加显著降低了萝卜根和地上部汞的含量（Shanker et al., 1996b）。

分子排阻色谱和蛋白酶解分析结果表明，水溶性的汞与硒形成高分子物质后积累在植物的根部，限制了汞的转运和代谢。Mounicou 等（2006）对印度芥菜（*Brassica juncea*）的研究表明，芥菜的根部积累了大量的以硒-汞高分子聚合物（≥70 kDa）形式存在的物质，推测汞可能和根部的一些生物分子，如含硒蛋白质发生反应，限制了汞由根向茎叶的运输。Yathavakilla 和 Caruso（2007）对大豆（*Glycine max*）的研究发现，大豆中的汞绝大部分积累在根中，而硒通过与汞形成高分子物质（>600 kDa）将汞固定在植物根部。Afton 和 Caruso（2009）通过应用高效液相色谱-电感耦合等离子体-质谱（HPLC-ICP-MS）对大葱（*Allium fistulosum*）的研究发现，大葱根部可能存在有机硒-汞高聚物。McNear 等（2012）应用 HPLC-ICP-MS、μ-XANES 和 X 射线荧光分析进一步发现，大葱根中的汞可能与细胞壁的巯基、质膜蛋白结合，在根际与还原的硒反应生成稳定的 Hg-Se 复合物，从而限制汞从根部向地上部的转运。

在不同浓度汞、硒和不同硒形态条件下，硒的影响效果也不同。在高浓度汞（100 mg/L）处理的大蒜（*Allium sativum*）培养液中添加高浓度的硒（>1 mg/L）可以显著抑制大蒜对汞的吸收、转运及积累，且亚硒酸盐和硒酸盐在减少汞积累方面的作用效果相似；但当汞的处理浓度<1 mg/L 时，高浓度硒的添加反而促进了大蒜对汞的吸收；同步辐射微束 X 射线荧光（μ-SRXRF）分析结果表明，加硒能够显著降低汞在大蒜（*Allium sativum*）根尖中柱鞘中的含量，表明硒能够拮抗汞从根部向地上部转运，从而减少汞在地上部的积累（Zhao et al.，2013a）。Tran 等（2018）的土培试验发现，2.5 mg/kg 亚硒酸盐的添加可以使 3 mg/kg 汞处理的青菜（*Brassica chinensis*）地上部汞含量显著降低 59%，而等量的硒酸盐却使其增加了 3 倍以上。在田间施加含硒肥可使水蜜桃（*Prunus persica*）中汞含量减少 23.9%（张志元等，2011）。硒对重金属积累的这种拮抗效应受硒和重金属浓度及形态等多种因素的影响。

第四节 重金属对硒吸收和累积的影响

硒对植物吸收转运重金属有一定的影响，反过来，重金属也可以改变植物对硒的吸收、转运及代谢。Khattak 等（1991）通过营养液砂培试验发现，营养液中添加 0.05 mg/L 和 0.1 mg/L 的砷酸盐能够增加紫花苜蓿（*Medicago sativa*）对硒酸盐的吸收，增加苜蓿地上部硒的含量。Ebbs 和 Weinstein（2001）进行的砷酸盐和硒酸盐添加试验结果表明，营养液中添加 1.0 mg/L 的砷酸盐能够显著增加大麦（*Hordeum vulgare*）地上部和根中硒的含量，但降低了挥发态硒的含量；营养液中添加 0.01 mg/L 的砷酸盐时，对大麦吸收和转运硒没有显著影响。Feng 等（2009）的研究发现，当外加的硒浓度较低时，砷能够增加蜈蚣草对硒的吸收及转运，而当硒浓度较高时，砷则拮抗植物对硒的吸收和积累。Malik 等（2012）的研究也发现，在 2.5 μmol/L 和 10 μmol/L 硒处理下的绿豆中分别添加 2.5 μmol/L、5.0 μmol/L、10 μmol/L 砷后，可显著增加绿豆对硒的吸收。胡莹等（2013）的水培研究结果表明亚砷酸盐处理显著提高了水稻根系对亚硒酸盐的吸收，但却显著降低了水稻茎叶对硒的积累，而砷酸盐处理除了显著降低水稻茎叶对硒的吸收积累外，对水稻根系积累硒没有影响。砷除了影响硒的吸收和转运外，还影响了硒的形态转化。Hu 等（2014）的研究表明砷酸盐或亚砷酸盐可以显著提高亚硒酸盐处理的水稻根部硒代甲硫氨酸含量，却降低了地上部硒代甲硫氨酸的含量；而对于硒酸盐处理的水稻，砷酸盐或亚砷酸盐并没有影响硒代甲硫氨酸的含量。

除了砷会影响植物对硒的吸收外，汞对植物吸收硒也有一定的影响。Zhao 等（2013b）的研究发现，在亚硒酸盐处理浓度较低时（<0.1 mg/L），随着汞添加量（0～50 mg/L）的增加，大蒜根和地上部硒的含量显著增加；但在较高亚硒酸盐浓度处理下，汞的添加则减少了大蒜根对硒的吸收。同步辐射 X 荧光分析（μ-SRXRF）结果表明，当亚硒酸钠和汞的处理浓度为 50 mg/L 时，汞的添加减少了硒在大蒜根部维管束组织的含量，降低了硒在大蒜球茎中的积累；X 射线吸收近边结构（XANES）的硒形态分析表明，汞的添加促进了大蒜各组织中无机硒向有机硒的转化。

土壤盆栽的试验结果表明，当土壤中亚硒酸盐的添加浓度<5.0 mg/kg 时，5.0 mg/kg 镉的添加不影响萝卜对硒的吸收；而当土壤中硒浓度≥10.0 mg/kg 时，镉的存在显著抑制了萝卜对硒的吸收，且这种作用表现为地上部分强于地下部分；当土壤中亚硒酸盐的添加浓度介于 5.0～10.0 mg/kg 时，镉虽然降低了萝卜对硒的吸收，但萝卜中硒的含量趋于稳定（铁梅等，2014）。Williams 等（2009）的土壤盆栽试验也发现，当土壤中添加 10 mg/kg 和 20 mg/kg 的砷酸盐时，显著降低了水稻籽粒中硒的含量。

在孟加拉国调查的 230 个水稻样品分析结果发现，水稻地上部和籽粒硒含量随着砷含量的增加而明显降低（Williams et al.，2009）。Norton 等（2010a）对来自孟加拉国 2 个采样点的 76 个水稻品种以及印度 2 个采样点的 89 个印度品种中的硒和砷含量进行了分析，结果表明砷含量高的作物品种硒含量较低。这种负相关可能受作物品种等遗传因素调控，也可能是由土壤环境条件导致的。重金属对植物硒吸收、转运及代谢的影响非常复杂，其可能因重金属的浓度、植物的类型及部位、硒的浓度及形态不同而与硒产生协同效应或拮抗效应。

第五节　硒对重金属胁迫的缓解作用

虽然硒不是植物生长所必需的微量元素，但适量的硒有利于植物的生长发育，参与调控植物的光合作用和呼吸作用（Xue et al.，2001），同时也能够提高植物的抗氧化能力、降低脂质过氧化反应（Malik et al.，2012；Kumar et al.，2014；Singh et al.，2018）。而且，添加适量的硒还能够降低重金属在植物体内的富集和积累，缓解有毒重金属对植物机体的损伤（Zhang et al.，2012；Saidi et al.，2014；Hussain et al.，2020；Feng et al.，2021）。由于硒的非必需性，其只能影响植物本身已有的防御机制来缓解重金属胁迫，而不会产生全新的生理过程。重金属的植物毒性主要表现在抑制生长、养分吸收、光合作用、呼吸作用，而硒的作用就在于保护这些过程，恢复植物的正常状态。硒缓解重金属胁迫的机制主要表现在：硒的抗氧化作用（调节生物体内的活性氧和抗氧化剂）、硒对细胞的修复作用（重建细胞膜和叶绿体组织）、硒与植物螯合肽的形成（改变重金属的存在形态，使重金属离子失去活性）、硒与重金属复合物的形成（阻碍重金属的吸收和转运）等几个方面（图 7-5）。

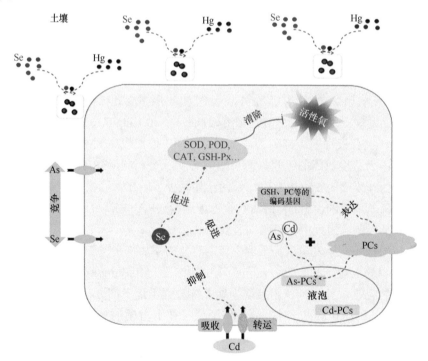

图 7-5　硒缓解重金属胁迫的机制示意图（Jiang et al.，2021）

一、硒的抗氧化作用

当植物暴露于重金属如镉、砷、铅、汞、铬等胁迫时，植物体内的自由基等氧化性物质增多，导致膜脂发生过氧化作用，破坏了生物膜结构和功能的完整性。有研究表明，硒具有抗氧化作用，能够缓解重金属对植物的胁迫，适量的硒可以缓解重金属对植物生长的抑制。Saidi 等（2014）通过对种子进行硒酸盐的预处理，使得 20 μmol/L 镉处理 4 d 的向日葵地上部和根部生物量显著提高，最大增幅分别可达 93% 和 150%。Sun 等（2016a）发现，3 μmol/L 亚硒酸盐显著提高了 50 μmol/L 镉处理黄瓜的株高、根长和生物量，使之回到对照水平。1 μmol/L 亚硒酸盐显著提高了以 44 μmol/L 镉处理的苎麻（*Boehmeria nivea*）根部和地上部干重（Tang et al.，2015）。然而，过量的硒同样会抑制植物生长，因为高浓度的硒对植物也有毒害作用。Feng 等（2013）发现，1.27 μmol/L 和 12.7 μmol/L 亚硒酸盐可以使 35.6 μmol/L 镉处理的水稻地上部生物量提高约 30%，但 63.5 μmol/L 亚硒酸盐则显著抑制了其生长，而且当镉处理浓度更高时，硒的阈值则更低。

很多研究已经证明了硒的抗氧化功能，在小白菜研究中发现适当浓度的外源 Se(Ⅳ) 和 Se(Ⅵ) 均提高了小白菜的抗氧化作用，促进了叶绿素的合成和生长（薛瑞玲等，2010）；在莴苣的研究中也发现这两种形态硒的抗氧化及促生长作用（Ramos et al.，2010）；对水稻的研究也发现，植物体内膜脂过氧化产物丙二醛（MDA）的含量及 $O_2^{-} \cdot$ 的产生速率随着硒添加浓度的增加而降低（吴永尧等，2000）。

适量的硒可以清除重金属胁迫下植物体内的活性氧（reactive oxygen species，ROS）。硒的添加可以减少植物体内 ROS 的含量，从而缓解铅、镉等对植物的毒害（Kumar et al.，2012；Mroczek-Zdyrska and Wójcik，2012）。Liu 等（2015）发现，3 μmol/L 亚硒酸盐显著抑制了烟草叶片中由 50 μmol/L 镉诱导的 H_2O_2 和 $\cdot O_2^{-}$ 的产生。Sun 等（2016b）发现，以 100 μmol/L 镉处理 20 d 的黄瓜叶片、茎和根中的 $\cdot OH$ 和 H_2O_2 含量显著增加了 30%~40%；加入 6 μmol/L 亚硒酸盐后，两种 ROS 的含量回落到对照水平。相应地，脂质过氧化程度的指示物丙二醛（malondialdehyde，MDA）的含量也显著下降，ROS 导致的脂质过氧化被硒缓解（Saidi et al.，2014；Sun et al.，2016a）。关于硒降低 $\cdot O_2^{-}$ 的机制，有学者认为可能有三个方面：$\cdot O_2^{-}$ 发生自发歧化作用生成 H_2O_2；硒化合物直接清除 $\cdot O_2^{-}$ 和 $\cdot OH$；调节抗氧化保护酶的活性（Feng et al.，2013）。

硒可以缓解镉对植物幼苗生长的抑制作用，使得保护性酶（超氧化物歧化酶 SOD、过氧化氢酶 CAT 和过氧化物酶 POD）的活性得到改善，减少植物的氧化损伤，降低叶片中镉含量，缓解植物的氧化胁迫，降低脂质过氧化反应，促进植物的生长（陈平等，2006；Filek et al.，2008；Pedrero et al.，2008；Filek et al.，2010）。例如，添加硒增加了绿豆苗中抗氧化剂（SOD、CAT、POD、谷胱甘肽还原酶）的含量，减少了砷对其的氧化损坏作用（Malik et al.，2012）。在受镉毒害的水稻幼苗体内过氧化作用强烈，POD 的活性得到显著提高，SOD 和 CAT 活性下降，而硒的添加则抑制了水稻幼苗剧烈的过氧化作用，降低了 POD 活性，提高了 SOD 和 CAT 活性（陈平等，2002）。硒对于抑制植物体内的过氧化作用发挥着非常重要的作用，这些保护酶类的活性与胁迫程度、硒的添加量及自身的含量有关。

在植物体内，硒可以清除由重金属胁迫产生的有害氧化性物质，维持机体的氧化还原物质含量正常，保护膜结构和功能的完整性，从而提高植物对重金属胁迫的抗性。硒在植物体内发挥的抗氧化性是硒缓解重金属胁迫的机制之一。

二、硒对细胞的修复作用

细胞膜是环境物质进入细胞的屏障，它的完整性对于防止原生质体电解质外泄和降低重金属吸收有重要意义，而硒可以修复被重金属破坏的植物细胞膜系统。Filek 等（2008）发现，2 μmol/L 硒酸盐显著降低了 400 μmol/L 和 600 μmol/L 镉处理下甘蓝型油菜地上部的细胞脂肪酸不饱和度，而磷脂中的不饱和脂肪酸会降低细胞膜的厚度并提高其通透性。这种效果同样可以作用于细胞内膜系统，Filek 等（2010）用同样浓度的硒、镉处理离体培养的甘蓝型油菜叶绿体时，发现硒酸盐不仅显著增加了叶绿体的尺寸，还降低了外膜脂肪酸的不饱和度。同时，硒酸盐还缓解了镉导致的叶绿体膜脂质极化减弱，而较低的极化度也会阻碍镉穿过叶绿体膜，因为金属阳离子的跨膜运输很大程度上受膜表面电位的阻碍。

植物体内叶绿素含量的降低是重金属对植物毒害作用的普遍现象。硒同样能缓解重金属对光合作用的抑制，其作用主要体现在：提高植物叶片中叶绿素和类胡萝卜素等的含量，增加叶片的干物质积累量（张驰等，2002；陈平等，2006；许自成等，2011）；修复受损的叶绿体，修复膜系统并重建内囊体和基质（Filek et al.，2010），进而提高相应的光合参数。重金属胁迫下，叶绿体的双层膜结构遭到了破坏，而施加硒后重新活化了细胞膜酶，恢复了叶绿体的代谢运输功能（郭锋等，2014）。此外，加入硒后可增加植物对其他营养物质如磷、钾、钙、镁、锌等的吸收，而这些元素对于叶绿素酸酯还原酶的合成具有促进作用，并且可以促进叶绿体双层膜结构的修复（陈铭和刘更另，1996；陈平等，2006）。光合作用是能量代谢的重要过程，硒的意义不仅仅在于光合效率的提高，它也会通过光合系统影响其他生理过程。但是，过量的硒也会降低光系统的能量吸收效率、抑制光合电子的传递，从而降低光量子产量，抑制光合作用（Labanowska et al.，2012）。叶绿体结构和功能的完整对植物有着十分重要的意义，适量硒可以增加叶绿素含量、保护叶绿体，从而抵抗重金属对植物的危害，同样是硒缓解重金属胁迫的重要机制之一。

三、硒与植物螯合肽

与植物对重金属耐受性有关的植物螯合肽（phytochelatin，PC），其形成也受硒的影响。重金属进入细胞质后，可激活 PC 合成酶（γ-谷胺酰半胱氨酸二肽酶）（Grill et al.，1989），然后在细胞质内以谷胱甘肽（glutathione，GSH）为底物合成 PC。PC 是一种由谷氨酸、半胱氨酸和甘氨酸组成的含巯基螯合多肽，含有大量巯基，对重金属的亲和力大，能够与多种重金属离子进行配合，使重金属离子失去活性，从而降低重金属对植物的毒性（Grill et al.，1985）。目前在很多植物中都检测到了重金属-PC 复合物。Speiser 等（1992）通过凝胶色谱发现，芥菜（*Brassica juncea*）地上部中的镉具有高分子质量和低分子质量两个峰，分别对应 Cd-GSH 和 Cd-PC3 的结合态。Salt 等（1995）通过扩展 X 射线吸收精细结构，检测到印度芥菜根部镉具有木质部汁液所不具有的 Cd-PC 四配位物质,并且根部的大部分镉均以此形态存在。重金属-PC 螯合物在 ATP 的作用下穿过液泡膜转运到液泡中储存(Pal and Rai,2010),例如,进入细胞中的大部分镉和汞与 PC 结合成复合物,然后进入液泡中通过隔离来降低其毒性(Park et al.,2012)。

GSH 是 PC 合成的前体,已有研究发现添加硒能够增加 PC 合成前体 GSH 的形成(Malik et al.,2012),从而缓解重金属的胁迫作用。小麦喷施硒酸钠后,叶片中谷胱甘肽过氧化物酶(glutathione peroxidase,GSH-Px)含量增加(薛文韬等,2010)。GSH-Px 是生物体内广泛存在的一种重要的过氧化物分解酶,能够将有害的过氧化物还原成无害的羟基化合物,清除自由基而减少氧化损伤,从而保护细胞膜的结构和功能不受过氧化物的干扰及损害,保护细胞膜的结构及功能的完整性(Yadav,2010)。而硒是 GSH-Px 的必需组分,硒代半胱氨酸是该酶活性中心的必需因子。硒可以增加 PC 的合成前体 GSH,并激活 PC 合成酶,使植物产生更多的 PC,配合更多的重金属离子,形成更多的重金属-PC 螯合物,从而使重金属离子失去活性,缓解植物重金属的胁迫。

四、硒与重金属复合物的形成

硒缓解重金属胁迫的另一种有效机制则是与重金属形成硒-重金属复合物,从而减少植物对重金属的吸收,以及降低重金属的生理生化毒性。重金属离子可与蛋白质及酶的巯基结合,而硒是比硫更软的碱,理论上,硒能优先和重金属离子结合形成硒-重金属复合物,使重金属离子变为稳定无毒的硒-重金属大分子配合物。对大豆(*Glycine max*)根系进行的分子排阻层析发现,大部分游离态的汞都与硒结合成大分子物质(>600 kDa)(Yathavakilla and Caruso,2007)。当根际存在有机酸时,硒与汞会形成难溶于水的硒化汞复合物,使汞的有效性下降,抑制了大葱对汞的吸收(McNear et al.,2012)。同时供给大蒜汞和硒后进行培养,也在其根系和鳞茎检测到有部分 Hg-Se 的存在(Zhao et al.,2013a)。同样,Zhang 等(2012)的研究也发现,生长在汞矿区的水稻根中硒、汞的摩尔比约为 1,但水稻籽粒中二者的含量却呈负相关,表明汞的向上运输受到了抑制,这极有可能是水稻根系及其根际周围土壤中的硒和汞形成了复合物。Mounicou 等(2006)在印度芥菜(*Brassica juncea*)根部检测到了以硒-汞高分子聚合物(≥70 kDa)形式存在的物质。

硒能够缓解重金属的胁迫,这是植物体内硒转化影响植物生理生化作用的综合效果。硒可以增强植物的抗氧化系统,修复细胞,影响 PC 合成酶和谷胱甘肽,进而影响植物螯合肽的合成,这些都与硒缓解植物对重金属的胁迫有关。

参 考 文 献

陈火云, 谢义梅, 周灵, 等. 2019. 施硒方式对油菜生长和籽粒硒、镉、铅含量的影响. 河南农业科学, 48(3): 49-54.

陈铭, 刘更另. 1996. 高等植物的硒营养及在食物链中的作用. 土壤通报, 27(4): 185-188.

陈平, 吴秀峰, 张伟锋, 等. 2006. 硒对镉胁迫下水稻幼苗叶片元素含量的影响. 中国生态农业学报, 14(3): 114-117.

陈平, 余士元, 陈惠阳, 等. 2002. 硒对镉胁迫下水稻幼苗生长及生理特性的影响. 广西植物, 22(3): 277-282.

方勇, 陈曦, 陈悦, 等. 2013. 外源硒对水稻籽粒营养品质和重金属含量的影响. 江苏农业学报, 29(4): 760-765.

高阿祥, 周鑫斌, 张城铭. 2017. 硒(Ⅳ)预处理下根表铁膜对水稻幼苗吸收和转运汞的影响. 土壤学报, 54(4): 989-998.

郭锋, 樊文华, 冯两蕊, 等. 2014. 硒对镉胁迫下菠菜生理特性、元素含量及镉吸收转运的影响. 环境科学学报, 34(2): 524-531.

呼艳姣, 陈美凤, 强瑀, 等. 2022. 镉胁迫下锌硒交互作用对水稻镉毒害的缓解机制. 生物技术通报, 38(4): 143-152.

胡莹, 黄益宗, 刘云霞. 2013. 砷-硒交互作用对水稻吸收转运砷和硒的影响. 环境化学, 32(6): 952-958.

黄青青, 刘艺芸, 徐应明, 等. 2018. 叶面硒肥与海泡石钝化对水稻镉硒累积的影响. 环境科学与技术, 41(4): 116-121, 159.

李虹颖, 唐杉, 王允青, 等. 2016. 硒对水稻镉含量及其在亚细胞中的分布的影响. 生态环境学报, 25(2): 320-326.

李荣林, 王胜兵, 李优琴. 2008. 锌硒对油菜吸收和累积镉的影响. 江苏农业学报, 24(3): 274-278.

刘达, 涂路遥, 赵小虎, 等. 2016. 镉污染土壤施硒对植物生长及根际镉化学行为的影响. 环境科学学报, 36(3): 999-1005.

刘晓娟, 程滨, 赵瑞芬, 等. 2019. 硒对铬胁迫下茄子生理特性及铬吸收的影响. 水土保持学报, 33(4): 360-366.

刘杨, 齐明星, 王敏, 等. 2021. 不同外源硒对镉污染土壤中小白菜生长及镉吸收的影响. 环境科学, 42(4): 2024-2030.

龙友华, 张承, 龚芬, 等. 2016. 叶面施硒对猕猴桃含硒量、镉铅积累及品质的影响. 食品科学, 37(11): 74-78.

吕选忠, 宫象雷, 唐勇. 2006. 叶面喷施锌或硒对生菜吸收镉的拮抗作用研究. 土壤学报, 43(5): 868-870.

庞晓辰, 王辉, 吴泽嬴, 等. 2014. 硒对水稻镉毒性的影响及其机制的研究. 农业环境科学学报, 33(9): 1679-1685.

齐田田. 2017. 硒对生菜吸收转运镉的影响. 北京: 中国农业大学硕士学位论文.

谭骏, 潘丽萍, 黄雁飞, 等. 2020. 叶面阻隔联合土壤钝化对水稻镉吸收转运的影响. 农业资源与环境学报, 184(6): 195-201.

铁梅, 刘阳, 李华为, 等. 2014. 硒镉处理对萝卜硒镉吸收的影响及其交互作用. 生态学杂志, 33(6): 1-7.

万亚男, 刘哲, Camara A Y, 等. 2018. 水分管理和外源硒对水稻吸收累积铅的影响. 环境科学, 39(10): 4759-4765.

万亚男, 余垚, 齐田田, 等. 2017. 硒对植物吸收转运镉影响机制的研究进展. 生物技术进展, 7(5): 473-479.

万亚男. 2018. 硒对水稻吸收、转运及累积镉的影响机制. 北京: 中国农业大学博士学位论文.

王康. 2021. 不同形态硒对水稻吸收、积累砷的影响机制. 北京: 中国农业大学博士学位论文.

吴永尧, 卢向阳, 彭振坤, 等. 2000. 硒在水稻中的生理生化作用探讨. 中国农业科学, 33(1): 100-103.

徐境懋, 顾明华, 韦燕燕, 等. 2021. 纳米硒和亚硒酸盐对镉污染土壤中水稻镉积累的影响. 南方农业学报, 52(10): 2727-2734.

许自成, 邵惠芳, 孙曙光, 等. 2011. 土壤施硒对烤烟生理指标的影响. 生态学报, 31(23): 7179-7187.

薛瑞玲, 梁东丽, 吴雄平, 等. 2010. 亚硒酸钠和硒酸钠对小白菜生长生理特性的影响. 西北植物学报, 30(5): 974-980.

薛文韬, 严俊, 杨荣志, 等. 2010. 硒酸钠对小麦谷胱甘肽过氧化物酶活性的影响. 贵州农业科学, 38(12): 89-91.

余垚. 2019. 亚硒酸盐和硒酸盐影响青菜中镉累积的机制研究. 北京: 中国农业大学博士学位论文.

张驰, 吴永尧, 彭振坤, 等. 2002. 硒对油菜苗期叶片色素的影响研究. 湖北民族学院学报(自然科学版), 20(3): 63-65.

张翠翠, 常介田, 赵鹏. 2013. 叶面施硒对西瓜镉和铅积累的影响. 华北农学报, 28(3): 159-163.

张海英, 韩涛, 田磊, 等. 2011. 草莓叶面施硒对其重金属镉和铅累积的影响. 园艺学报, 38(3): 409.

张志元, 张翼, 郭清泉, 等. 2011. 含硒植物营养剂对桃和梨吸收铅、镉、汞的拮抗作用. 作物研究, 25(4): 368-369.

赵家印, 王永杰, 钟寰. 2019. 三种修复剂对稻米甲基汞富集的影响研究. 农业环境科学学报, 38(2): 284-289.

郑春荣. 2002. 土壤中污染元素的交互作用对其行为的影响. 见: 陈怀满等. 土壤中化学物质的行为与环境质量. 北京: 科学出版社: 422-442.

Afton S E, Caruso J A. 2009. The effect of Se antagonism on the metabolicfate of Hg in *Allium fistulosum*. Journal of Analytical Atomic Spectrometry, 24: 759-766.

Batista B L, Souza J M O, De Souza S S, et al. 2011. Speciation of arsenic in rice and estimation of daily intake of different arsenic species by Brazilians through rice consumption. Journal of Hazardous Materials, 191(1-3): 342-348.

Björnberg A, Håkanson L, Lundbergh K. 1988. A theory on the mechanisms regulating the bioavailability of mercury in natural-waters. Environmental Pollution, 49(1): 53-61.

Cai M M, Hu C X, Wang X, et al. 2019. Selenium induces changes of rhizosphere bacterial characteristics and enzyme activities affecting chromium/selenium uptake by pak choi (*Brassica campestris* L. ssp. *Chinensis makino*) in chromium contaminated soil. Environmental Pollution, 249: 716-727.

Cai M M, Zhao X H, Wang X, et al. 2021. Se changed the component of organic chemicals and Cr bioavailability in pak choi rhizosphere soil. Environmental Science and Pollution Research, 28: 67331-67342.

Camara A Y, Wan Y N, Yu Y, et al. 2018. Effect of selenium on uptake and translocation of arsenic in rice seedlings (*Oryza sativa* L.). Ecotoxicology and Environmental Safety, 148: 869-875.

Chang H, Zhou X B, Wang W H, et al. 2013. Effects of selenium application in soil on formation of iron plaque outside roots and cadmium uptake by rice plants. Advanced Materials Research, 750: 1573-1576.

Chi S L, Xu W H, Liu J, et al. 2017. Effect of exogenous selenium on activities of antioxidant enzymes, cadmium accumulation and chemical forms of cadmium in tomatoes. International Journal of Agriculture and Biology, 19(6): 1615-1622.

Cui J H, Liu T X, Li Y D, et al. 2018. Selenium reduces cadmium uptake into rice suspension cells by regulating the expression of lignin synthesis and cadmium-related genes. Science of the Total Environment, 644: 602-610.

Ding Y Z, Feng R W, Wang R G, et al. 2014. A dual effect of Se on Cd toxicity: evidence from plant growth, root morphology and responses of the antioxidative systems of paddy rice. Plant Soil, 375: 289-301.

Ebbs S, Weinstein L. 2001. Alteration of selenium transport and volatilization in barley (*Hordeum vulgare*) by arsenic. Journal of Plant Physiology, 158: 1231-1233.

Fargašová A, Pastierová J, Svetková K. 2006. Effect of Se-metal pair combinations (Cd, Zn, Cu, Pb) on photosynthetic pigments production and metal accumulation in *Sinapis alba* L. seedlings. Plant Soil and Environment, 52 (1): 8-15.

Feng R W, Wang L Z, Yang J G, et al. 2021. Underlying mechanisms responsible for restriction of uptake and translocation of heavy metals (metalloids) by selenium via root application in plants. Journal of Hazardous Materials, 402: 123570.

Feng R W, Wei C Y, Tu S X, et al. 2009. Interactive effects of selenium and arsenic on their uptake by *Pteris vittata* L. under hydroponic conditions. Environmental and Experimental Botany, 65(2-3): 363-368.

Feng R W, Wei C Y, Tu S X, et al. 2013. A dual role of Se on Cd toxicity: Evidences from the uptake of Cd and some essential elements and the growth responses in paddy rice. Biological Trace Element Research, 151: 113-121.

Filek M, Keskinen R, Hartikainen H, et al. 2008. The protective role of selenium in rape seedlings subjected to cadmium stress. Journal of Plant Physiology, 165(8): 833-844.

Filek M, Kościelniak J, Łabanowska M, et al. 2010. Selenium-induced protection of photosynthesis activity in rape (*Brassica napus*) seedlings subjected to cadmium stress. Fluorescence and EPR measurements. Photosynthesis Research, 105(1): 27-37.

Gao M, Zhou J, Liu H L, et al. 2018. Foliar spraying with silicon and selenium reduces cadmium uptake and mitigates cadmiumtoxicityinrice. Science of The Total Environment, 631-632: 1100-1108.

Grill E, Loffler S, Winnacker E L, et al. 1989. Phytochelatins, the heavy-metal-binding peptides of plants, are synthesized from glutathione by a specific gamma-glutamylcysteine dipeptidyl transpeptidase (phytochelatin synthase). Proceedings of the National Academy of Sciences of the United States of America, 86(18): 6838-6842.

Grill E, Winnacker E L, Zenk M H. 1985. Phytochelatins: the principal heavy metal complexing peptides of higher plants. Science, 230: 674-676.

Han D, Xiong S L, Tu S X, et al. 2015. Interactive effects of selenium and arsenic on growth, antioxidant system, arsenic and selenium species of *Nicotiana tabacum* L. Environmental and Experimental Botany, 117: 12-19.

Handa N, Kohli S K, Sharma A, et al. 2018. Selenium ameliorates chromium toxicity through modifications in pigment system, antioxidative capacity, osmotic system, and metal chelators in *Brassica juncea* seedlings. South African Journal of Botany, 119: 1-10.

Handa N, Kohli S K, Thukral A K, et al. 2017. Role of Se(Ⅳ) incounteracting oxidative damage in *Brassica juncea* L. under Cr(Ⅳ) stress. Acta Physiologiae Plantarum, 39: 51.

He P P, Lv X Z, Wang G Y. 2004. Effects of Se and Zn supplementation on the antagonism against Pb and Cd in vegetables. Environment International, 30: 167-172.

Hu Y, Norton G J, Duan G L, et al. 2014. Effect of selenium fertilization on the accumulation of cadmium and lead in rice plants. Plant Soil, 384: 131-140.

Huang Q Q, Xu Y M, Liu Y Y, et al. 2018. Selenium application alters soil cadmium bioavailability and reduces its accumulation in rice grown in Cd-contaminated soil. Environmental Science and Pollution Research, 25: 31175-31182.

Hussain B, Lin Q, Hamid Y, et al. 2020. Foliage application of selenium and silicon nanoparticles alleviates Cd and Pb toxicity in rice (*Oryza sativa* L.). Science of the Total Environment, 712: 136497.

Ismael M A, Elyamine A M, Moussa M G, et al. 2019. Cadmium in plants: uptake, toxicity, and its interactions with selenium fertilizers. Metallomics, 11: 255-277.

Jayaweera G R, Biggar J W. 1996. Role of redox potential in chemical transformations of selenium in soils. Soil Science Society of America Journal, 60: 1056-1063.

Jiang H Y, Lin W Q, Jiao H P, et al. 2021. Uptake, transport and metabolism of selenium and its protective effects against toxic metals in plants: a review. Metallomics, 13(7): mfab040.

Khattak R A, Page A L, Parker D R, et al. 1991. Accumulation and interactions of arsenic, selenium, molybdenum and phosphorus in Alfalfa. Journal of Environmental Quality, 20: 165-168.

Kramárová Z, Fargašová A, Molnárová M, et al. 2012. Arsenic and selenium interactive effect on alga *Desmodesmus quadricauda*. Ecotoxicology and Environmental Safety, 86: 1-6.

Kumar A, Singh R P, Singh P K, et al. 2014. Selenium ameliorates arsenic induced oxidative stress through modulation of antioxidant enzymes and thiols in rice (*Oryza sativa* L.). Ecotoxicology, 23(7): 1153-1163.

Kumar M, Bijo A J, Baghel R S, et al. 2012. Selenium and spermine alleviate cadmium induced toxicity in the red seaweed *Gracilaria dura* by regulating antioxidants and DNA methylation. Plant Physiology and Biochemistry, 51: 129-138.

Labanowska M, Filek M, Koscielniak J, et al. 2012. The effects of short-term selenium stress on Polish and Finnish wheat seedlings-EPR, enzymatic and fluorescence studies. Journal of Plant Physiology, 169(3): 275-284.

Li H F, Lombi E, Stroud J L, et al. 2010. Selenium speciation in soil and rice: influence of water management and Se fertilization. Journal of Agricultural and Food Chemistry, 58: 11837-11843.

Li H F, McGrath S P, Zhao F J. 2008. Selenium uptake, translocation and speciation in wheat supplied with selenate or selenite. New Phytologist, 178: 92-102.

Li M Q, Hasan M K, Li C X, et al. 2016. Melatonin mediates selenium-induced tolerance to cadmium stress in tomato plants. Journal of Pineal Research, 61: 291-302.

Li Y F, Zhao J T, Li Y Y, et al. 2015. The concentration of selenium matters: a field study on mercury accumulation in rice by selenite treatment in qingzhen, Guizhou, China. Plant and Soil, 391(1/2): 195-205.

Lin L, Zhou W H, Dai H X, et al. 2012. Selenium reduces cadmium uptake and mitigates cadmium toxicity in rice. Journal of Hazardous Materials, 235-236: 343-351.

Liu W, Shang S, Feng X, et al. 2015. Modulation of exogenous selenium in cadmium-induced changes in antioxidative metabolism, cadmium uptake, and photosynthetic performance in the 2 tobacco genotypes differing in cadmium tolerance. Environmental Toxicology and Chemistry, 34(1): 92-99.

Lv H Q, Chen W X, Zhu Y M, et al. 2020. Efficiency and risks of selenite combined with different water conditions in reducing uptake of arsenic and cadmium in paddy rice. Environmental Pollution, 262: 114283.

Ma J F, Tamai K, Yamaji N, et al. 2006. A silicon transporter in rice. Nature, 440(7084): 688-691.

Ma J F, Yamaji N, Mitani N, et al. 2008. Transporters of arsenite in rice and their role in arsenic accumulation in rice grain. Proceedings of the National Academy of Sciences of the United States of America, 105(29): 9931-9935.

Malik A J, Goel S, Kaur N, et al. 2012. Selenium antagonises the toxic effects of arsenic on mungbean (*Phaseolus aureus* Roxb.) plants by restricting its uptake and enhancing the antioxidative and detoxification mechanisms. Environmental and Experimental Botany, 77: 242-248.

McNear D H, Afton S E, Caruso J A. 2012. Exploring the structural basis for selenium/mercury antagonism in *Allium fistulosum*. Metallomics, 4(3): 267-276.

Meharg A A, Hartley-Whitaker J. 2002. Arsenic uptake and metabolism in arsenic resistant and nonresistant plant species. New Phytologist, 154(1): 29-43.

Meharg A A, Jardine L. 2003. Arsenite transport into paddy rice (*Oryza sativa*) roots. New Phytologist, 157(1): 39-44.

Meharg A A, Macnair M R. 1992. Suppression of the high affinity phosphate uptake system: A mechanism of arsenate tolerance in *Holcus lanatus* L. Journal of Experimental Botany, 43(249): 519-524.

Meharg A A, Naylor J, Macnair M R. 1994. Phosphorus nutrition of arsenate-tolerant and nontolerant phenotypes of velvetgrass. Journal of Environmental Quality, 23(2): 234-238.

Meng B, Feng X B, Qiu G L, et al. 2014. Localization and speciation of mercury in brown rice with implications for Pan-Asian public health. Environmental Science and Technology, 48: 7974-7981.

Mounicou S, Shah M, Meija J, et al. 2006. Localization and speciation of selenium and mercury in *Brassicajuncea* - implications for Se-Hg antagonism. Journal of Analytical Atomic Spectrometry, 21(4): 404-412.

Mozafariyan M, Shekari L, Hawrylak-Nowak B, et al. 2014. Protective role of selenium on pepper exposed to cadmium stress during reproductive stage. Biological Trace Element Research, 160(1): 97-107.

Mroczek-Zdyrska M, Wójcik M. 2012. The influence of selenium on root growth and oxidative stress induced by lead in *Vicia faba* L. minor plants. Biological Trace Element Research, 147(1-3): 320-328.

Muñoz A H S, Wrobel K, Corona J F G, et al. 2007. The protective effect of selenium inorganic forms against cadmium and silver toxicity in mycelia of *Pleurotus ostreatus*. Mycological Research, 111: 626-632.

Norton G J, Dasgupta T, Islam M R, et al. 2010a. Arsenic influence on genetic variation in grain trace element nutrient content in Bengal Delta grown rice. Environmental Science and Technology, 44(21): 8284-8288.

Norton G J, Islam M R, Duan G L, et al. 2010b. Arsenic shoot-grain relationships in field grown rice cultivars. Environmental Science and Technology, 44(4): 1471-1477.

Pal R, Rai J P N. 2010. Phytochelatins: peptides involved in heavy metal detoxification. Applied Biochemistry and Biotechnology, 160: 945-963.

Park J Y, Song W Y, Donghwi K O, et al. 2012. The phytochelatin transporters AtABCC1 and AtABCC2 mediate tolerance to cadmium and mercury. The Plant Journal, 69: 278-288.

Pedrero Z, Madrid Y, Hartikainen H, et al. 2008. Protective effect of selenium in Broccoli (*Brassica oleracea*) plants subjected to cadmium exposure. Journal of Agricultural and Food Chemistry, 56(1): 266-271.

Plant J A, Kinniburgh D G, Smedley P L, et al. 2006. Arsenic and selenium. Treatise on Geochemistry, 9: 17-66.

Qing X J, Zhao X H, Hu C X, et al. 2015. Selenium alleviates chromium toxicity by preventing oxidative stress in cabbage (*Brassica campestris* L. ssp. *Pekinensis*) leaves. Ecotoxicology and Environmental Safety, 114: 179-189.

Ramos S J, Faquin V, Guiherme, et al. 2010. Selenium biofortification and antioxidant activity in lettuce plants fed with selenate and selenite. Plant Soil and Environment, 56: 584-588.

Saidi I, Chtourou Y, Djebali W. 2014. Selenium alleviates cadmium toxicity by preventing oxidative stress in sunflower (*Helianthus annuus*) seedlings. Journal of Plant Physiology, 171: 85-91.

Salt D E, Prince R C, Pickering I J, et al. 1995. Mechanisms of cadmium mobility and accumulation in Indian mustard. Plant Physiology, 109(4): 1427-1433.

Shanker K, Mishra S, Srivastava S, et al. 1995. Effect of selenite and selenate on plant uptake of cadmium by kidney bean (*Phaseolus mungo*) with reference to Cd-Se interaction. Chemical Speciation and Bioavailability, 7: 97-100.

Shanker K, Mishra S, Srivastava S, et al. 1996a. Effect of selenite and selenate on plant uptake and translocation of mercury by tomato (*Lycopersicum esculentum*). Plant Soil, 183: 233-238.

Shanker K, Mishra S, Srivastava S, et al. 1996b. Study of mercury-selenium (Hg-Se) interactions and their impact on Hg uptake by the radish (*Raphanus sativus*) plant. Food and Chemical Toxicology, 34(9): 883-886.

Singh R, Upadhyay A K, Singh D P. 2018. Regulation of oxidative stress and mineral nutrient status by selenium in arsenic treated crop plant *Oryza sativa*. Ecotoxicology and Environmental Safety, 148: 105-113.

Sors T G, Ellis D R, Salt D E. 2005. Selenium uptake, translocation, assimilation and metabolic fate in plants. Photosynthesis Research, 86(3): 373-389.

Speiser D M, Abrahamson S L, Banuelos G, et al. 1992. *Brassica-juncea* produces a phytochelatin-cadmium-sulfide complex. Plant Physiology, 99(3): 817-821.

Srivastava M, Ma L Q, Rathinasabapathi B, et al. 2009. Effects of selenium on arsenic uptake in arsenic hyperaccumulator *Pteris vittata* L. Bioresource Technology, 100(3): 1115-1121.

Srivastava S, Shanker K, Srivastava S, et al. 1998. Effect of selenium supplementation on the uptake and translocation of chromium by spinach (*Spinacea oleracea*). Bulletin of Environmental Contamination & Toxicology, 60(5): 750-758.

Sun H Y, Dai H X, Wang X Y, et al. 2016a. Physiological and proteomic analysis of selenium-mediated tolerance to Cd stress in cucumber (*Cucumis sativus* L.). Ecotoxicology and Environmental Safety, 133: 114-126.

Sun H Y, Wang X Y, Wang Y N, et al. 2016b. Alleviation of cadmium toxicity in cucumber (*Cucumis sativus*) seedlings by the application of selenium. Spanish Journal of Agricultural Research, 14(4): 1-12.

Tang H, Liu Y, Gong X, et al. 2015. Effects of selenium and silicon on enhancing antioxidative capacity in ramie (*Boehmeria nivea* L. Gaud.) under cadmium stress. Environmental Science and Pollution Research, 22(13): 9999-10008.

Tran T A T, Dinh Q T, Cui Z W, et al. 2018. Comparing the influence of selenite (Se^{4+}) and selenate (Se^{6+}) on the inhibition of the mercury (Hg) phytotoxicity to pak choi. Ecotoxicology and Environmental Safety, 147: 897-904.

Wan Y N, Camara A Y, Huang Q Q, et al. 2018a. Arsenic uptake and accumulation in rice (*Oryza sativa* L.) with selenite fertilization and water management. Ecotoxicology and Environmental Safety, 156: 67-74.

Wan Y N, Camara A Y, Yu Y, et al. 2018b. Cadmium dynamics in soil pore water and uptake by rice: Influence of soil-applied selenite with different water managements. Environmental Pollution, 240: 523-533.

Wan Y N, Wang K, Liu Z, et al. 2019. Effect of selenium on the subcellular distribution of cadmium and oxidative stress induced by cadmium in rice (*Oryza sativa* L.). Environmental Science and Pollution Research, 26: 16220-16228.

Wan Y N, Yu Y, Wang Q, et al. 2016. Cadmium uptake dynamics and translocation in rice seedling: Influence of different forms of selenium. Ecotoxicology and Environmental Safety, 133: 127-134.

Wang K, Wang Y Q, Wan Y N. 2021. The fate of arsenic in rice plants (*Oryza sativa* L.): Influence of different forms of selenium. Chemosphere, 264: 128417.

Wang Y J, Dang F, Evans R D, et al. 2016a. Mechanistic understanding of MeHg-Se antagonism in soil-rice systems: The key role of antagonism in soil. Scientific Reports, 6: 19477.

Wang Y J, Dang F, Zhao J T, et al. 2016b. Selenium inhibits sulfate-mediated methylmercury production in rice paddy soil. Environmental Pollution, 213: 232-239.

Williams P N, Islam S, Islam R, et al. 2009. Arsenic limits trace mineral nutrition (selenium, zinc, and nickel) in Bangladesh rice grain. Environmental Science and Technology, 43(21): 8430-8436.

Wu C, Dun Y, Zhang Z J, et al. 2020. Foliar application of selenium and zinc to alleviate wheat (*Triticum aestivum* L.) cadmium toxicity and uptake from cadmium-contaminated soil. Ecotoxicology and Environmental Safety, 190: 110091.

Wu Z C, Ren H Y, McGrath S P, et al. 2011. Investigating the contribution of the phosphate transport pathway to arsenic accumulation in rice. Plant Physiology, 157(1): 498-508.

Xie Y, Su L, He Z, et al. 2021. Selenium inhibits cadmium absorption and improves yield and quality of cherry tomato (*Lycopersicon esculentum*) under cadmium stress. Journal of Soil Science and Plant Nutrition, 1125-1133.

Xu X H, Yan M, Liang L C, et al. 2019. Impacts of selenium supplementation on soil mercury speciation, andinorganic mercury and methylmercury uptake in rice (*Oryza sativa* L.). Environmental Pollution, 249: 647-654.

Xu Z R, Cai M L, Chen S H, et al. 2021. High-affinity sulfate transporter Sultr1;2 as a major transporter for Cr(Ⅳ) uptake in plants. Environmental Science and Technology, 55(3): 1576-1584.

Xue T L, Hartikainen H, Piironen V. 2001. Antioxidative and growth-promoting effect of selenium on senescing lettuce. Plant and Soil, 237: 55-61.

Yadav S K. 2010. Heavy metals toxicity in plants: An overview on the role of glutathione and phytochelatins in heavy metal stress tolerance of plants. South African Journal of Botany, 76: 167-179.

Yathavakilla S K V, Caruso J A. 2007. A study of Se-Hg antagonism in *Glycine max* (soybean) roots by size exclusion and reversed phase HPLC-ICP MS. Analytical and Bioanalytical Chemistry, 389: 715-723.

Yu Y, Fu P N, Huang Q Q, et al. 2019. Accumulation, subcellular distribution, and oxidative stress of cadmium in *Brassica chinensis* supplied with selenite and selenate at different growth stages. Chemosphere, 216: 331-340.

Zhang H, Feng X B, Jiang C X, et al. 2014a. Understanding the paradox of selenium contamination in mercury mining areas: high soil content and low accumulation in rice. Environ Pollution, 188: 27-36.

Zhang H, Feng X B, Zhu J M, et al. 2012. Selenium in soil inhibits mercury uptake and translocation in rice

(*Oryza sativa* L.). Environmental Science and Technology, 46(18): 10040-10046.

Zhang L H, Hu B, Li W, et al. 2014b. OsPT2, a phosphate transporter, is involved in the active uptake of selenite in rice. New Phytologist, 201: 1183-1191.

Zhao J T, Gao Y X, Li Y F, et al. 2013a. Selenium inhibits the phytotoxicity of mercury in garlic (*Allium sativum*). Environmental Research, 125: 75-81.

Zhao J T, Hu Y, Gao Y X, et al. 2013b. Mercury modulates selenium activity via altering its accumulation and speciation in garlic (*Allium sativum*). Metallomics, 5: 896-903.

Zhao J T, Li Y F, Li Y Y, et al. 2014. Selenium modulates mercury uptake and distribution in rice (*Oryza sativa* L.) in correlation with mercury species and exposure level. Metallomics, 6: 1951-1957.

Zhao X Q, Mitani N, Yamaji N, et al. 2010. Involvement of silicon influx transporter OsNIP2;1 in selenite uptake in rice. Plant Physiology, 153(4): 1871-1877.

Zhao Y Y, Hu C X, Wang X, et al. 2019. Selenium alleviated chromium stress in Chinese cabbage (*Brassica campestris* L. ssp. *Pekinensis*) by regulating root morphology and metal element uptake. Ecotoxicology and Environmental Safety, 173: 314-321.

Zhou J, Chen Z, Du B Y, et al. 2021. Soil and foliar applications of silicon and selenium effects on cadmium accumulation and plant growth by modulation of antioxidant system and Cd translocation: Comparison of soft vs. durum wheat varieties. Journal of Hazardous Materials, 402: 1123546.

Zhou X B, Gao A X, Lai F, et al. 2017. The role of selenium in soil: effect on the uptake and translocation of arsenic in rice (*Oryza Sativa* L.). International Journal of Agriculture and Biology, 19: 1227-1234.

第八章 农业环境中硒研究的前沿和展望

我国作为缺硒大国，植物中硒的吸收和积累是提高人体中硒摄入量的关键。而土壤中硒的生物有效性决定了植物中硒的含量。我国针对富硒农业措施开展了大量卓有成效的研究，如硒富集植物或真菌的培养、硒叶面肥的使用等，开发了许多新技术和应用途径。但目前仍有很多问题亟待解答，特别是机理方面的研究有待加强。

第一节 土壤化学过程

硒在农田土壤中的化学行为和迁移过程相对比较清楚，但其生物有效性仍是一个需要进一步加强的研究方向。我国南方一些地区，其土壤中硒浓度已达到富硒水平，但作物中硒的含量仍处于缺乏状态。虽然作物品种间存在一定差异，会影响硒在植物中的积累，但土壤中硒的生物有效性偏低仍然是影响植物吸收和积累硒的关键因素。如何在不影响作物产量和品质的前提下，进一步活化土壤中的硒，提高其生物有效性，进而提高作物中的硒含量，充分利用富硒土壤，成为目前我国农业生产中急需解决的主要问题。

土壤氧化还原状态对硒的形态有较大影响，从而影响硒的生物有效性，如何原位探索土壤氧化-还原状态的改变对硒的吸附-解吸和形态转化的影响？土壤有机质含量显著影响土壤中硒的迁移和生物有效性，而土壤有机质成分复杂，仅可溶性有机质中的化学物质已多达数千至上万种，不可溶解的有机质化学物质的多样性可能更高。这些复杂的土壤有机质如何通过与硒结合影响硒的生物有效性，哪些类型的有机质可提高硒的生物有效性，哪些有机质会降低硒的生物有效性，目前仍然未知。关于土壤中黏土的矿物类型、pH和氧化还原状态如何影响硒的生物有效性，多因素状态下硒的生物有效性变化规律及主要驱动机制，目前仍缺乏深入研究和探讨。

第二节 土壤生物学过程

土壤中蕴藏着丰富的微生物，其生物多样性非常高，每克表层土壤包含上亿个细菌和古菌、数万亿个病毒。土壤中矿物质不仅是微生物菌落的物理基质，还可提供关键元素，被称为地球元素循环的引擎。土壤中硒的吸附-解吸、溶解-固定、形态转化、迁移挥发等过程均有各种微生物参与其中。植物根际微生物数量远远超过非根际微生物，是非根际土壤微生物数量的5—20倍。这些土壤微生物对硒生物地球化学循环的贡献巨大，但在不同生境下，哪些微生物的作用为主，其贡献率有多少，不同种类微生物之间如何相互配合实现对硒的代谢，目前仍不清楚。关于微生物对硒的转化研究，以前主要集中在单一菌株对硒的转化过程，包括参与的基因、蛋白等；而关于微生物群落对硒生物地球化学循环的驱动过程，则研究较少。相关代谢基因的丰度变化规律，不同类型微生物（病毒、细菌、古菌、真菌等）对硒的氧化、还原、甲基化等方面的贡献及其机制，也仍缺乏全面、系统、深入的研究。根际微生物对硒的转化与植物对硒的吸收和积累机制，同样缺乏深入探讨。

第三节 植物对硒的吸收、转运及积累机制

植物吸收和积累硒是硒进入食物链的主要途径，在硒缺乏地区，如何调控土壤中硒向植物可食部分转运是功能农业研究的重点之一。硒高富集地区，如湖北恩施，调控作物中硒浓度不高于可食用范围，对食用人群的健康不产生潜在威胁，同样值得研究。由于人们已逐步认识到硒对健康的重要性，同时认识到植物是人体中硒的主要来源，近年来植物对硒的吸收以及硒在植物体内的转运、转化机制也得到了广泛的研究，明确了多种植物体内硒相关的代谢通路、关键蛋白（酶）及基因，提出了植物硒代谢的主要通路。但超积累植物中独特的硒代谢过程及其调控基因，仍缺乏足够的认识。另外，许多高等食用真菌具有较强的硒富集能力，作为一种理想的富硒载体，在富硒产品的开发上具有广阔的发展前景。但目前的相关研究主要集中在高等食用菌中硒含量、种类测定，不同食用菌的富硒培养途径，以及真菌中硒的吸收、代谢、转化机制尚不十分清楚。硒可以阻控植物对土壤中多种重（类）金属的吸收和积累，从而降低植物可食部分重金属的浓度，减少重金属对食用人群健康的潜在威胁。但植物体内硒与不同重金属的拮抗机制，目前还没有全面的阐述，需要深入探索。

第四节 植物内生菌、叶际微生物对硒积累的作用及机制

由于土壤硒浓度较低，加之植物根系吸收效率不高，造成作物中硒含量较低。为了提高作物的硒含量，实际生产中常常通过外源喷施含硒叶面肥来实现植物富硒。叶面喷施硒肥可快速有效提高作物可食部位硒含量。与土壤施加硒肥相比，叶面喷施硒肥可有效减少土壤吸附、化学或生物转化造成的硒损失，并且减少了硒在根系和其他非可食部位的截留，提高了可食部位硒富集效率。

有机硒和无机硒均可被叶片吸收，但植物叶面对硒的吸收机制缺少系统深入的探索。植物叶片表面通常覆盖一层不亲水的蜡质层。研究显示，蜡质覆盖的叶片表皮是叶面硒肥主要的吸收途径。叶面允许小分子的无机硒通过，直径较大的有机硒与纳米硒等分子透过效率较差。叶面硒肥的吸收、植物体内的转运和代谢机制，目前尚不清楚，包括硒从叶片向籽粒的长距离运输、叶片细胞韧皮部硒的装载和卸载、硒在木质部-韧皮部的转换和卸载等过程，均可能有诸多转运蛋白和通道蛋白参与，但目前仍未知。

另外，植物叶表面含有多种微生物，即叶际微生物，这些微生物可通过多种方式影响植物的健康和适应性。已有研究表明，叶际微生物能帮助植物吸收营养并调节植物代谢平衡，但是目前的研究都不够深入。植物叶际微生物的研究才刚刚起步，关于叶际微生物的功能及其与宿主、环境的相互关系仍需进一步了解。在叶面硒肥施用过程中，高浓度硒肥对叶际微生物的影响，以及叶际微生物对植物吸收硒的作用和贡献同样是研究空白，未来仍需探索叶面肥、叶际微生物及植物之间的复杂关系。

植物内生菌（内生真菌、内生细菌和内生放线菌）与宿主形成互惠共生体，进化过程中不仅获得了与宿主产生相同或相似产物的能力，还可产生宿主所不能产生的物质，对植物中元素的吸收、积累和代谢有重要影响。当前的研究主要集中在内生菌的分离培养，以及纯培养条件下内生菌对硒的积累、转化等方面，而不同植物内生菌具有宿主特异性，其在植物体内参与硒的代谢和转化过程及机制，目前仍为研究空白。